Primates in Fragments

Ecology and Conservation

Primates in Fragments
Ecology in Conservation

Edited by

Laura K. Marsh

Los Alamos National Laboratory
Los Alamos, New Mexico

Kluwer Academic / Plenum Publishers
New York, Boston, Dordrecht, London, Moscow

ISBN 0-306-47696-7

©2003 Kluwer Academic/Plenum Publishers, New York
233 Spring Street, New York, New York 10013

http://www.wkap.com

10 9 8 7 6 5 4 3 2 1

A C.I.P. record for this book is available from the Library of Congress

Printed in the United States of America

This book is dedicated to my mom, Janet G. Marsh,
for going with me to Kenya when I was 14 and
for sending me by myself when I was 18 to meet strangers
in Panamá to do primate research.
Her unending support has been truly appreciated.

PREFACE

This volume was created initially from a symposium of the same name presented at the International Primatological Society's XVIII Congress in Adelaide, South Australia, 6–12 January 2000. Many of the authors who have contributed to this text could not attend the symposium, so this has become another vehicle for the rapidly growing discipline of Fragmentation Science among primatologists. Fragmentation has quickly become a field separate from general ecology, which underscores the severity of the situation since we as a planet are rapidly losing habitat of all types to human disturbance.

Getting ecologists, particularly primatologists, to admit that they study in fragments is not easy. In the field of primatology, one studies many things, but rarely do those things (genetics, behavior, population dynamics) get called out as studies in fragmentation. For some reason "fragmentation primatologists" fear that our work is somehow "not as good" as those who study in continuous habitat. We worry that perhaps our subjects are not demonstrating as robust behaviors as they "should" given fragmented or disturbed habitat conditions. I had a colleague openly state that she did not work in fragmented forests, that she merely studied behavior when it was clear that her study sites, every one of them, was isolated habitat. Our desire to be just another link in the data chain for wild primates is so strong that it makes us deny what kinds of habitats we are working in. However, the time has come to embrace Fragmentation Science as a field and to pursue it vigorously within primatology. We as scientists have chosen to study in undesirable habitats as our field of concern. This is unfortunate because habitats are degraded, and unfortunate because we all enjoy pristine habitats. But frankly, it is a good thing we chose the "less desirable" path since the future of primate conservation may depend on it.

One of the primary goals of this book is to be a reference not only in the primate literature, but in Fragmentation Science. The contributors to this volume realize the significance of working within fragments as a whole no matter how disturbed and that these systems respond to and depend upon the matrix they are embedded within. Given that many of the researchers in this volume are also fluent in botany and general ecology, we endeavor to make the text accessible to students and fellow fragment biologists who are interested in this field.

This volume seeks to address several key questions regarding primates in fragments. We attempt to clarify some of the issues, but perhaps in trying we bring to light the complexity of the situation with primates in disturbed habitats. There are few clear patterns that have emerged demonstrating how primates respond to fragmentation. This volume introduces possible trends for some species, but predictive powers for species other than those discussed in the literature are less reliable. There was no attempt to standardize nomenclature across the volume, but within each chapter nomenclature is consistent.

The book is divided into sections based on broad categories of research on primates in fragments. In the Genetics and Population Dynamics section, the authors cover topics in viability, metapopulation, and species that remain in remnant forests. In the Behavioral Ecology section, authors take a closer look at feeding, ranging, and other behaviors that allow primates to remain in or disperse between fragments. In Conservation and Management, authors bring knowledge of species who remain in fragments together with plans to implement strategies for their long-term viability. And, finally, in the Integration and Future Directions section, authors synthesize the information in this volume and make recommendations for future and continued work in this field. We hope this volume will provoke more research and inquiry to this quickly developing discipline in primatology.

One of the greatest challenges we have facing conservation today is apathy. We encounter this in the consumers of industrialized countries as well as villagers in source nations. Engaging all people, but especially those in tropical nations, to become interested and dedicated as scholars, researchers, teachers, and supporters of conservation efforts within their native countries is vital[1]. This volume represents work conducted in 10 countries. It is remarkable that out of 51 authors, 37 (75%) are native to the countries they work in. It is my personal hope that this is a growing global trend. If it is, then perhaps primates will have a future in the wild.

Given the recent increase in books published on various aspects of Fragmentation Science, I have provided a list as a supplement to this volume. We no longer have the luxury of studying a single species without having a complete understanding of its habitat and its place within it. These references have become the staples within the discipline.

I would like to express my thanks to all of the authors contributing to this volume. I think we have accomplished a tremendous effort. In particular, without the participants of the International Primatological Society Symposium in Adelaide this volume would never have come into being. I would especially like to thank all of the authors for contributing references and text to the introduction and summaries before each chapter. I hope I represented all of our similar thoughts. Funding for my travel to Australia would not have been possible without the support and encouragement from Fairley Barnes. Without her persistence, I could never have done this project. I thank Andrea Macaluso for asking me to do this book in the first place. Huge thanks must be given to Los Alamos National Laboratory, particularly Tim Haarmann, Ted Doerr, and Diana Webb for their internal support. I cannot thank Teresa Hiteman and Hector Hinojosa enough. They will continue to be richly rewarded for all of their amazing work and dedication to editing and preparing the entire volume (camera ready copy!) before submission to Kluwer. I would also like to thank Winters Red Star, Julie Hill, Rhonda Robinson, and Marjorie Wright for help with maps, figures, and slides for talks having to do with this volume. Many thanks to Bill Laurance and Colin Chapman for great advice and constant inspiration. Special love and thanks to Sal for being a feathery support. Since this is my first edited

---

[1] There are obvious regions missing from this volume that suffer from severe fragmentation, such as Madagascar and Southeast Asia. It is my hope that in future volumes these countries will be included.

book, I would also like to thank Thelma Rowell, Marina Cords, Tsingalia Mugatsia, and Dennis Rasmussen for their support and encouragement during my formative years as a young primatologist. Look guys, I finally did something! And finally, I thank Jeanne Fair for enduring the painful process of the first book and for all of the helpful comments throughout.

Santa Fe, New Mexico   LKM
May 2002

FOREWORD: PRIMATES AS ICONS FOR CONSERVATION

William F. Laurance*

I'm not a primatologist, but in my career I've been dazzled by many primates in the wild, from orangutans and gibbons in Borneo to white-faced sakis and spider monkeys in Central America and the Amazon. I've been equally intrigued by their ecological analogs—the possums, cuscuses, and tree-kangaroos—that I've studied in the rain forests of Australia and New Guinea.

I'm not alone, of course. The apes, monkeys, and lemurs—our closest living relatives in the world—hold a special magic for most people, perhaps rivaled only by whales and dolphins in their ability to capture the popular imagination. In the 1980s, the World Wildlife Fund discovered that the public was far more strongly galvanized by exhortations to "Save the Primates!" than "Save the Rain Forests!" despite the fact that these goals overlap considerably.

Primates are icons for conservation because we identify with them, empathize with them, and feel for their plight. And so many primates are direly threatened. Of 600 primate species in the world, fully a fifth are seriously endangered, and half of those are at imminent risk of extinction.

Primates are imperiled by the same forces that threaten much of the world's biodiversity—habitat destruction and fragmentation, logging and fires, and over hunting. They are icons for conservation because they reside in many of the world's most mega-diverse regions. Of the 25 most gravely endangered primate species in the world, 24 are found exclusively in recognized "biodiversity hotspots"—regions such as Madagascar, Indochina, Sundaland (Borneo, Java, Sumatra), Brazil's Atlantic forest, and the Guinean forests of West Africa that sustain a disproportionate fraction of the world's biological diversity and have been ravaged by human activities. To conserve viable populations of wild primates is to protect many of the world's most critically endangered ecosystems.

Primates are icons for conservation because they play vital roles in ecosystem structure and functioning. They disperse fruits and seeds; play integral roles in food webs as consumers of insects, fruits, and foliage and as prey for mammalian carnivores,

---

* Smithsonian Tropical Research Institute, Apartado 2072, Balboa, Republic of Panamá, and Biological Dynamics of Forest Fragments Project, National Institute for Amazonian Research (INPA), Manaus, AM 69011-970, Brazil. Correspondence to W. Laurance (email: laurancew@tivoli.si.edu).

snakes, and raptors; and participate in a diverse array of coevolved relationships with other species (in the Amazon, for example, certain dung beetles rely exclusively on the droppings of howler monkeys, and, in burying the seed-rich dung to feed their larvae, greatly increase the germination and survival of some plant species). A forest that has lost its primates is an unhealthy, dysfunctional forest.

For these and many other reasons, primates are special, and this book is a timely and impressively comprehensive effort to describe the myriad impacts of habitat fragmentation on their survival, population dynamics, genetic structure, ecological interactions, biogeography, and behavior. It is an unquestionably important contribution, following upon earlier books such as *Tropical Forest Remnants* (Laurance and Bierregaard, 1997) and *Lessons from Amazonia: Ecology and Conservation of a Fragmented Forest* (Bierregaard et al., 2001) that provide more general insights into the ecological consequences of habitat fragmentation.

Primates are icons for conservation because, throughout much of the world, their formerly intact habitats are in desperate retreat. These captivating and intelligent creatures are figureheads for conservation—as they should be—and our success or failure in conserving their rapidly dwindling populations may presage the fate of much of the natural world.

REFERENCES

Bierregaard, Jr., R. O., Gascon, C., Lovejoy, T. E., and Mesquita, R., 2001, *Lessons from Amazonia: The Ecology of Conservation of a Fragmented Forest*, Yale University Press, New Haven.
Laurance, W. F., and Bierregaard, Jr., R. O., 1997, *Tropical Forest Remnants: Ecology, Management and Conservation of Fragmented Communities*, University of Chicago Press, Chicago.

CONTENTS

Baoguo Li, Zhiyun Jia, Ruliang Pan, and Baoping Ren

Ariel R. Rodríguez-Vargas

PRIMATE SURVIVAL IN COMMUNITY-OWNED FOREST FRAGMENTS:
ARE METAPOPULATION MODELS USEFUL AMIDST INTENSIVE
USE? 63

Colin A. Chapman, Michael J. Lawes, Lisa Naughton-Treves, and Thomas Gillespie

RELATIONSHIPS BETWEEN FOREST FRAGMENTS AND HOWLER
MONKEYS (*ALOUATTA PALLIATA MEXICANA*) IN SOUTHERN
VERACRUZ, MEXICO 79

Erika M. Rodriguez-Toledo, Salvador Mandujano, and Francisco Garcia-Orduña

PRIMATES OF THE BRAZILIAN ATLANTIC FOREST: THE INFLUENCE OF
FOREST FRAGMENTATION ON SURVIVAL

Adriano G. Chiarello

DYNAMICS OF PRIMATE COMMUNITIES ALONG THE SANTARÉM-
CUIABÁ HIGHWAY IN SOUTH-CENTRAL BRAZILIAN
AMAZONIA

Stephen F. Ferrari, Simone Iwanaga, André L. Ravetta, Francisco C. Freitas, Belmira A.
R. Sousa, Luciane L. Souza, Claudia G. Costa, and Paulo E. G. Coutinho

PRIMATES AND FRAGMENTATION OF THE AMAZON FOREST 145

Kellen A. Gilbert

SECTION II: BEHAVIORAL ECOLOGY

Laura K. Marsh

Marilyn A. Norconk and Brian W. Grafton

Ernesto Rodríguez-Luna, Laura E. Domínguez-Domínguez, Jorge E. Morales-Mávil, and
Manuel Martínez-Morales

HOW DO HOWLER MONKEYS COPE WITH HABITAT
    FRAGMENTATION? 283

Júlio César Bicca-Marques

SECTION III: CONSERVATION AND MANAGEMENT

Laura K. Marsh

EFFECTS OF HABITAT FRAGMENTATION ON THE CROSS RIVER GORILLA (*GORILLA GORILLA DIEHLI*): RECOMMENDATIONS FOR CONSERVATION 343

Edem A. Eniang

THE NATURE OF FRAGMENTATION

Laura K. Marsh[*]

1. INTRODUCTION

The numbers are staggering and grotesque. They are the opening factoids for every paper where research in tropical forest is conducted. Tropical forests are disappearing faster than any other biome (Myers, 1991). Tropical forests once covered up to 15% of the earth's surface and currently cover only 6% to 7%, but contain more than 50% and possibly as much as 90% of all species of plants and animals (WRI, 1990). Rain forests are being systematically reduced by 150,000 km^2 (10 to 15 million ha) per year (Whitmore, 1997; Achard et al., 2002), which is more than a full percentage point (1.2%, Laurance, 1997). Assuming this rate maintains, the last rain forest tree will fall in 2027. Many of us will witness this within our lifetimes. Additionally (to make matters worse), only 3% to 7% of the world's land area is officially protected as national parks or forest reserves (Chapman and Peres, 2001). The travesty of rampant deforestation will have profound effects, not simply on forests and their inhabitants, but for all humanity. Thus, the resulting fragmented patchwork of habitat and the species remaining within become central to the challenge of conservation.

Fragmentation of tropical forests affects the viability of primate populations worldwide. Of all primate species, 90% are found in tropical regions and most depend upon forests (Mittermeier and Cheney, 1987). Almost half of the 250 primate species are considered to be of conservation concern by the Primate Specialist Group of the World Conservation Union, and one in five is either endangered or critically endangered, meaning that without better protection they will be extinct in the next 20 years (IUCN, 1996; Stevenson et al., 1992; Rylands et al., 1995; Mittermeier, 1996). Habitat destruction is a significant threat to the survival of nonhuman primate populations in most parts of the world (Mittermeier, 1991). A recent assessment of habitat loss by the Food and Agriculture Organization estimated extant forest loss between 1980 and 1995 to be 9.7% for Latin America and the Caribbean, 10.5% for Africa, and 6.4% for Asia and Oceania (Chapman and Peres, 2001).

---

[*]Los Alamos National Laboratory, Ecology Group (RRES-ECO), Mail Stop M887, Los Alamos, New Mexico 87545, USA, email: lkmarsh@lanl.gov.

Primates in Fragments: Ecology and Conservation
Edited by L. K. Marsh, Kluwer Academic/Plenum Publishers, 2003

The nature of fragmentation is crucial to the survival of primates in a disturbed habitat. How a fragment is created is significant in understanding any remnant population's viability, density, behavior, persistence, and, ultimately, their conservation and management. This volume presents a first collective look at one of the world's most ecologically important orders that has suffered and sustained intense habitat and population loss in recent decades. The force of development is driving faster and more ominously into otherwise untouched habitat all over the world. As this force begins to dominate the landscape over intact habitat, primates and other species must face life threatening changes. Thus, it is critical to evaluate the ecological role of primates in a fragmented forest for many reasons. Primates are 1) often important as seed dispersers; 2) a high proportion of the animal biomass in a forest remnant; 3) integral to the energy flow in the system; 4) conservation ambassadors for tropical habitats, and 5) important educational icons for local people (Laurance, this volume). We simply do not know enough about tropical ecosystems to fully predict all of the system-wide consequences that may occur if primates are eliminated.

2. FRAGMENTATION SCIENCE

Fragmentation scientists are rapidly learning the dynamics of rain forests. The vast adaptations and interdependencies that may allude us within intact forests are being systematically deconstructed within fragments. Where all of the parts once worked in perfect harmony, we see the unraveling of an ecosystem, and the disintegration of processes. It is almost as though we are witnessing forest development in reverse. Because of this, those who work in fragmented ecosystems tend to have knowledge of the forces working for and against the primates they study.

The science of fragmented forests has been discussed in the literature and, until recently, most of the studies used theories of island biogeography to predict species numbers (MacArthur and Wilson, 1967; Wilcox, 1980; Case and Cody, 1987; Shafer 1995); debated the optimum size of fragments for species conservation (e.g., SLOSS; Single Large Or Several Small: Simberloff and Able 1976; Gilpin and Diamond 1980; Soulé and Simberloff, 1986; Shafer, 1995); or examined species survival within fragments (Soulé and Wilcox, 1980; Harris 1984; Bierregaard and Lovejoy 1986; Lovejoy and Bierregaard, 1990; Saunders et al., 1991; Redford, 1992; Bierregaard and Dale, 1996). Issues important to the study of fragmented forests include edge and isolation effects, invasions, secondary forest succession, disruption of biological processes, non-equilibrium dynamics, and human use of the remaining forest. Many of these ecological processes are currently being studied in depth (c.f., Suggested Reading, this volume).

Forests are rarely fragmented with an eye toward future habitat requirements of the remaining occupants. Similarly, most fragments are remnants of forest covering unwanted or "useless" land in terms of human exploitation, or are in a topographically difficult area to destroy (c.f., Eniang, Rodriguez-Toledo et al., this volume). Forest remnants can act as reserves for species that are represented in larger forest tracts. Since there is an increase in secondary plant species, there are often habitat zones within fragments not found, or rarely found, in continuous forest. Forest remnants provide habitat for many organisms with small home ranges, for organisms that range widely over the landscape, and for an array of species that tolerate human disturbance. Whatever the barrier, be it human or natural, fragmentation causes ecological disruption because of the combination of the surrounding matrix, edge effect, fragment size, shape, location in the landscape, and type of ownership (Saunders et al., 1993; Laurance, 1993). Primates themselves often determine the degree of isolation depending on whether they are specialists or generalists and how well they tolerate the fragmentation process.

2.1. Edge Effect

One of the distinguishing features in a fragmented forest is "edge effect" (Janzen, 1986; Lovejoy et al., 1986). It is defined in conservation biology as the creation of a distinct edge between previously undisturbed forest and a deforested clearing (e.g., for pasture) (Lovejoy et al., 1986). This definition differs from the ecological term of ecotone where there is a change of habitat at a natural edge or ecological boundary (e.g., where prairie meets forest; Odum, 1989). In fragmented forests, there is generally not an increase in species in this zone, although the term "edge effect" in ecology also describes this phenomenon.

"Edge" is important in fragments because plant species that grow along the boundaries are often very different from the plants at the core of the fragment (Bierregaard and Dale 1996). Often edges have more lianas and vines present because of intense light conditions (Hegarty and Caballé, 1991), and, over time, older forest edges may "seal" the core forest with a wall of dense vegetation aiding in the preservation of core species and microclimates (Kapos et al., 1997). Edges are important for many primate species, but for others they become complete barriers.

There are two main effects that determine the characteristics of a fragment: 1) alteration of the microclimate (solar radiation, water fluxes, wind and edge effects) and 2) isolation (time since isolation, distance from other remnants, connectivity, changes in the surrounding landscape) (Saunders et al., 1991). Edge effect can increase wind-throw and ambient temperature, as well as reduced humidity near fragment boundaries, which results in a sharp increase in tree mortality, tree damage, and the formation of canopy gaps (Schelhas and Greenberg, 1996). Interactions between the forest interior and the adjacent habitat along the edge include abiotic effects (changes in such environmental conditions as air temperature, humidity, light intensity), direct biological effects (changes in species distribution and abundance due to the modifications of environmental conditions), and indirect biological effects (higher-order changes resulting from alterations in species interactions) (Murcia, 1995; Turton and Freiburger, 1997; Sizer and Tanner, 1999). Additional changes in dynamics that differ from intact forest are influxes of invading or "homeless" individuals, invasions by exotics or disturbance specialists, and decrease in the regeneration of primary forest trees (c.f., Laurance and Bierregaard, 1997).

2.2. Matrix

The matrix or remaining landscape surrounding fragments, consists of various 'types,' such as livestock pasture, agriculture, downed vegetation from timber extraction, villages or other structures, small kitchen plots, mixtures of agro-forestry, secondary forest, and slash and burn (for various uses). No two fragments experience the same effects in part because each fragment is located in a unique section of the altered landscape with random localized habitats and topography remaining. Matrix habitat appears important in the evolution of fragmented landscape dynamics because it acts as a selective filter for movement of species between fragments and other landscape features, facilitating movements of some species while impeding others (Gascon et al., 2000). Additionally, disturbance-adapted species will be present in the matrix and may invade forest patches and edge habitat. The matrix type can influence the severity of the edge effects in forest patches (Gascon et al., 2000).

It is beneficial for a remnant species to be able to utilize the surrounding matrix including secondary growth, edge habitats, or cropland, and to be able to effectively disperse between forest fragments. Laurance (1991) found a "significant, negative relationship between matrix tolerance and extinction proneness for 16 species of rainforest mammals; species that used or exploited the matrix often remained stable or increased in fragments, whereas those that avoided the matrix declined or disappeared." Laurance (1994) also noted three key advantages of matrix-tolerant species: 1) they can disperse between fragments, or between continuous forest and fragments, and therefore increase genetic viability of the population; 2) they can recolonize fragments following local extinctions; and 3) since they tend to be generalists, they often exploit ecological changes in fragments, like edge effects. A primate that is able to negotiate the matrix may have greater potential for long-term viability in a fragmented landscape, but other factors, such as availability of appropriate foods may limit success for some species.

2.3. Landscape Patterns

Fragmentation can result in a patchwork with forests of various sizes, shapes, and degrees of isolation, requiring species to respond differently to each site. Harris and Silva-Lopez (1992) describe various types of fragmentation. It is worth repeating here since different primate species may respond according to the type of fragment they are in (Figure 1). "Regressive fragmentation" is fragmentation that occurs along only one edge of continuous forest and continually over time pushes back at the forest. Species may respond by shifting further into the remaining forest and away from the encroaching edge. "Enveloping fragmentation" occurs when forest is contracted from all sides. This is one of the most studied types of fragmentation, since it has rapid and dramatic consequences for species that reside within it. "Divisive fragmentation" results from a strip through intact forest, most typically a road. The effect on species greatly depends on the use of the road. If low use, species which can cross it may not be impacted too greatly, but if the use is great or development follows, then there are obvious additional consequences for species use of the area. "Intrusive fragmentation" occurs from within. This type of disturbance impacts the structural integrity of the internal forest as opposed to the surrounding matrix. Impacts most frequently seen in the tropics are kitchen gardens, cattle grazing, clearing through to rivers for water access, and logging. Finally, "encroaching fragmentation" is where forest is left in linear tracts, often as gallery forest along waterways. These can be fragments alone or can sometimes act as corridors between larger fragments or remaining forest. It is possible to have several fragment conditions impacting one area, particularly when there are multiple sources of deforestation.

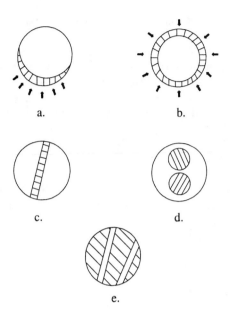

Figure 1. Fragmentation "types" where (a) regressive, (b) enveloping, (c) divisive, (d) intrusive, and (e) encroaching can be found as individual impacts or, if combined, multiple impacts (redrawn from Harris and Silva-Lopez, 1992). Illustration by Rhonda Robinson.

2.4. Cumulative Pressures

Even if species survive the initial fragmentation, cascading abiotic and biotic effects may eventually eliminate them. Bierregaard and Dale (1996) stated: "These data lead us to believe that the absolute maximum forest fragment size that could be considered viable for a substantial percentage of the species in the Amazon forest is 100 hectares." According to them, smaller fragmented forests have the potential for being "empty forests" where the trees are standing, but the species composition and complexity are missing often due to further disruption by the local human population (Redford, 1992; Robinson, 1996). This only takes into consideration those species that are permanent residents of a single fragment, not species that are able to disperse out of the fragment or can use the matrix (Section II, this volume). Contrary to the Biological Dynamics of Forest Fragments Project (BDFFP) (Lovejoy et al., 1986) size seemed to matter the least of many habitat parameters for some species studied in this volume (Bicca-Marques, Ferarri et al., Chapman et al., Rodriguez-Luna et al., Chiraello, Umapathy and Kumar) while for others it was critical (Chapman et al., Rodriguez-Vargas, Gilbert). See Marsh et al. this volume for a full discussion of fragment size.

Often fragments can recover in a functional sense for primates, even after the initial destruction as long as there are no other cumulative pressures from human disturbance. Common compounding impacts are hunting, harvesting of non-timber forest products, and capturing live primates for pets within fragmented forests or the surrounding matrix. One of the results of this "internal" fragmentation for primates is an increase in distancing between family groups, scattering of intact groups, removing larger fission-fusion subgroups, and, in some extreme cases, removing entire populations (Umapathy, Eniang, this volume).

When human populations are low and hunting is consistent with traditional food needs, then internal fragmentation is not out of balance with replacement rates (Robinson and Redford, 1991). There are unfortunately many places where this balance is not being maintained, and "bushmeat" is harvested in a non-sustainable manner. Bushmeat is wild caught game that is killed and consumed for protein. There is a rising global concern as secondary species, such as primates, become primary targets as more common game species are depleted (BCTF 2000). Access is increasing in areas where logging, oil excavation, and large-scale deforestation continue. For instance, in the Congo Basin (comprising 6 countries) up to 45% of all bushmeat taken is primates (Wilke and Carpenter 1999). Primates are one of the most affected by hunters and contribute with other species 1 million metric tons of wildlife that is killed annually for protein in West Central Africa alone (BCTF 2000). With the current rates of over-harvesting some primate species, internal fragmentation continues to be a serious issue for fragmentation primatologists.

3. PRIMATES IN FRAGMENTS

Tropical deforestation and the resultant fragmentation may affect primate populations in at least two ways. First, the process of fragmentation randomly distributes primates among forest fragments, some with and some without primates. Second, due to inadequate remnant size or other factors, primates may become locally extinct following fragmentation. The lack of a continuously available food supply as a result of forest reduction is a potentially limiting factor to primate populations that have specialized diets (van Roosmalen and Klein, 1988). The availability of key food resources may aid in the prediction of primate survival in a given forest remnant (Terborgh, 1986; Terborgh and Winter, 1980), even though the ecological concept of 'keystone' species is under debate (Mills et al., 1993; Marsh et al., this volume).

Fragmentation of contiguous habitat can 1) exclude a primate species from residence in a given forest fragment, effectively causing localized extinctions (Lovejoy et al., 1986), 2) alter the group sizes or population densities of species still able to inhabit the fragmented landscape (Milton, 1982; Estrada and Coates-Estrada, 1988, 1996; Terborgh et al., 1997; Tutin and White, 1999; Tutin, 1999), 3) alter the dietary strategies of species able to reside in fragments (Johns and Skorupa, 1987; Tutin, 1999), and 4) affect gene flow among resident populations (Estrada and Coates-Estrada, 1996; Pope, 1996; Cosson et al., 1999; Gravitol et al., 2001).

At present, there are few generalizations that work across primate taxa to predetermine success or failure in a fragment (c.f., Marsh et al., this volume). One of the things fragmentation scientists are quickly learning is how non-patterned many species responses are to fragments. For primates, finding a common thread of responses for species is proving difficult. Individuals of the same group may have different responses to fragment pressures, and inter- and intra-specific differences vary depending on the location and nature of the fragmentation. The study of fragmentation has been made difficult by the fact that when researchers talk about fragments, they often have different definitions for areas that vary dramatically in size. In some sites researchers are concerned with fragments of a few hectares, while others have orders of magnitude larger sites. In both cases, these are often the only forests left outside of purely protected areas. Authors in this volume do seem to agree that fragments can be defined by size, such that small is 1 to 10 ha, medium is 10 to 100 ha, large is 100 to 1,000 ha, and extra large is 1,000 to 10,000 ha, although this scale generally varies with the maximum size of the largest fragment in the study area.

However inconsistent, we have seen trends in species that survive within fragments. There are four main characteristics of primates that may influence their ability to live within forest fragments: home range size, the degree of frugivory in the diet, behavioral and dietary plasticity, and the ability to utilize the matrix (Tutin and White, 1999; Onderdonk and Chapman, 2000; Estrada and Coates-Estrada, 1996; Lovejoy et al., 1986; Laurance, 1991; Bicca-Marques, this volume, Silver and Marsh, this volume). The interaction between fragment size, home range size, and diet type is complex; the limited area resulting from fragmentation reduces the diversity of plant species and the number of food plants available to consumers (Tutin and White, 1999). Fruit as a resource is highly heterogeneous in terms of its spatial and temporal distribution, and larger frugivorous primates usually require large tracts of forest to provide enough resources to support viable populations (Johns and Skorupa, 1987; Turner, 1996; Onderdonk and Chapman, 2000).

In general, predictions of those species who might be the most vulnerable have been said to demonstrate one or more of the following characteristics: avoidance or infrequent use of the matrix, intolerant of changes within fragments (micro- and macro-climates), have large home range requirements, are large bodied, are vulnerable to hunting pressures or other forms of exploitation (e.g., live capture), have low population densities, have unstable or highly variable populations, are dependent upon certain species because of strong or obligate ecological relationships, have limited dispersal abilities, and/or have low fecundidty (Gilpin and Soulé, 1986; Terborgh and Winter, 1980; Laurance, 1994; Bierregaard et al., 1992; Laurance and Bierregaard, 1997; Schelhas and Greenberg 1996). Many of these are proving correct, but again it is still too early to draw conclusions based on the limited data available. Probably one of the keys for primate success in forest fragments is their ability to use the matrix. This seems to hold true for several species of primates discussed in this volume including *Alouatta* sp., *Pan troglodytes*, *Ateles geoffroyi*, *Gorilla gorilla dieli*, *Cercopithecus ascanius*, and *Trachypithecus johnii* (Bicca-Marques, Reynolds et al., Umapathy and Kumar, Chiarello, Chapman et al., Reynolds et al., McCann et al., Gilbert, Eniang, and Ramos-Fernandez and Ayala-Orozco, this volume).

4. SUMMARY

The nature of fragmentation can greatly affect primates in remnant forests. All aspects of primate livelihood can be affected by fragmentation and it is imperative to understand the basics of fragmentation to better understand a species' response to it. In this chapter, I introduce concepts of edge, matrix, landscape patterns, and cumulative pressures that contribute to internal fragmentation. A general discussion of primates in fragments introduces the concepts of what species might thrive in a fragment and which species may struggle. The need for generalizations is great, but we are only beginning to understand relationships between primates and the damaged forests they live in.

5. REFERENCES

Achard, F., Eva, H. D., Stibig, H. J., Mayaux, P., Gallego, J., Richards, T., and Malingreau, J. P., 2002, Determination of deforestation rates of the world's humid tropical forests, *Science* 297:999–1,002.

BCTF, 2000, Bushmeat: A wildlife crisis in west and central Africa and around the world, briefing document, www.bushmeat.org.

Bierregaard, R. O., Jr., and Dale, V. H., 1996, Islands in an ever-changing sea: The ecological and socioeconomic dynamics of Amazonian rainforest fragments, in: *Forest Patches in Tropical Landscapes*, J. Schelhas and R. Greenberg, eds., Island Press, CA, pp. 187–204.

Bierregaard, R. O., Jr., Lovejoy, T. E., Kapos, V., Augusto dos Santos, A., and Hutchings, R. W., 1992, The biological dynamics of tropical rainforest fragments, *BioScience* 42(11):859–866.

Bierregaard, R. O., Jr., and Lovejoy, T. E., 1989, Effects of forest fragmentation on Amazonian understory bird communities, *Acta Amazonica* 19:215–241.

Case, T. J., and Cody, M. L., 1987, Testing theories of island biogeography, *Amer. Scientist* 75:40–311.

Chapman, C. A., and Peres, C. A., 2001, Primate conservation in the new millennium: The role of scientists, *Evol. Anthrop.* 10:16–33.

Cosson, J. F., Ringuet, S., Claessens, O., de Massary, J. C., Dalecky, A., Villiers, J. F., Granjon, L., and Pons, J. M., 1999, Ecological changes in recent land-bridge islands in French Guiana, with emphasis on vertebrate communities, *Biol. Conser.* 91:213–222.

Estrada, A., and Coates-Estrada, R., 1988, Tropical rain forest conversion and perspectives in the conservation of wild primates (*Alouatta* and *Ateles*) in Mexico, *Amer. J. Primat.* 14:315–327.

Estrada, A., and Coates-Estrada, R., 1996, Tropical rain forest fragmentation and wild populations of primates at Los Tuxtlas, Mexico, *Int. J. Primat.* 17:759–783.

Gascon, C., Williamson, G. B., and Fonseca, G. A. B., 2000, Receding edges and vanishing reserves, *Science* 288:1,356–1,358.

Gilpin, M. E., and Soulé, M. E., 1986, Minimum viable populations: Process of species extinction, in: *Conservation Biology: The Science of Scarcity and Diversity,* M. E. Soulé, ed., Sinauer, Sunderland, MA, pp. 19–34.

Gilpin, M. E., and Diamond, J. M., 1980, Subdivision of nature reserves and the management of species diversity, *Nature* 285:567–568.

Gravitol, A. D., Ballou, J. D., and Fleischer, R. C., 2001, Microsatellite variation within and among recently fragmented populations of the golden lion tamarin (*Leontopithecus rosalia*), *Conser. Gen.* 2:1–9.

Harris, L. D., and Silva-Lopez, G., 1992, Forest fragmentation and the conservation of biological diversity, in: *Conservation Biology*, P. L. Fiedler and S. K. Jain, eds., Chapman and Hall, New York, pp. 197–238.

Harris, L. D., 1984, *The Fragmented Forest: Island Biogeography Theory and the Preservation of Biotic Diversity*, University of Chicago Press, Chicago.

Hegarty, E. E., and Caballé, G., 1991, Distribution and abundance of vines in forest communities, in: *The Biology of Vines*, F. E. Putz and H. A. Mooney, eds., Cambridge University Press, Cambridge, MA.

Horwich, R. H., and Lyon, J., 1993, Modification of tropical forest fragments and black howler monkey conservation in Belize, in: *Symposium for the Forest Remnants in the Tropical Landscape: Benefits and Policy Implications*, J. K. Doyle and J. Schelhas, eds., Smithsonian Migratory Bird Center, 10-11 September 1992, Washington D.C.

IUCN, 1996, Primate specialist list of endangered species, Gland Switzerland.

Janzen, D. H., 1986, The eternal external effect, in: *Conservation Biology: The Science of Scarcity and Diversity*, M. E. Soulé, ed., Sinauer Assoc., Sunderland, MA, pp. 286–303.

Johns, A. D., and Skorupa, J. P., 1987, Responses of rain-forest primates to habitat disturbance: A review, *Internat. J. Primatol.* 8:157–191.

Kapos, V., Wandelli, E., Camargo, J. L., and Ganade, G., 1997, Edge-related changes in environment and plant responses due to forest fragmentation, in: *Tropical Forest Remnants: Ecology, Management, and Conservation of Fragmented Communities*, W. F. Laurance and R. O. Bierregarrd, eds., University of Chicago Press, Chicago, pp. 33–44.

Laurance, W. F., 1993, The pre-European and present distributions of *Antechinus godmani* (Marsupialia: Dasyuridae), a restricted rainforest endemic, *Australian Mammal* 16:23–27.

Laurance, W. F., 1997, Introduction, in: *Tropical Forest Remnants: Ecology, Management and Conservation of Fragmented Communities*, W. F. Laurance and R. O. Bierregaard, Jr., eds., University of Chicago Press, Chicago, pp. 1–3.

Laurance, W. F., and Bierregaard, R. O., eds., 1997, *Tropical Forest Remnants: Ecology, Management, and Conservation of Fragmented Communities,* University of Chicago Press, Chicago, IL, 616 pgs.

Lovejoy, T. E., and Bierregaard, R. O., Jr., 1990, Central Amazonian forests and the Minimum Critical Size of Ecosystems Project, in: *Four Neotropical Rainforests*, A. H. Gentry, ed., Yale University Press, New Haven, pp. 60–74.

Lovejoy, T. E., Bierregaard, R. O., Jr., Rylands, A. B., Malcolm, J. R., Quintela, C. E., Harper, L. H., Brown, K. S., Powell, Jr., A. H., Powell, G. V. N., Schubart, H. O. R., and Hays, M. B., 1986, Edge and other effects on isolation on Amazon forest fragments, in: *Conservation Biology: The Science of Scarcity and Diversity*, M. E. Soulé, ed., Sinauer Assoc., Sunderland, MA.

MacArthur, R. H., and Wilson, E. O., 1967, *The Theory of Island Biogeography*, Princeton University Press, New Jersey.

Mills, L. S., Soulé, M. E., and Doak, D. F., 1993, The keystone-species concept in ecology and conservation, *BioScience* **43(4)**:219–224.

Milton, K., 1982, Dietary quality and demographic regulation in a howler monkey population, in: *Ecology of a tropical forest: seasonal rhythms and long-term changes*, E.G. Leigh, Jr., A.S. Rand, and D.M. Windsor, eds., Smithsonian Institution Press, Washington, D.C., pp. 273–289.

Mittermeier, R. A., 1996, Introduction, in: *The Pictorial Guide to the Living Primates*, N. Rowe, Pogonias Press, East Hampton, New York, p. 1.

Mittermeier, R. A., 1991, Hunting and its effect on wild primate populations in Suriname, in: *Neotropical Wildlife Use and Conservation*, J. G. Robinson and K. H. Redford, eds., University of Chicago Press, Chicago, IL, pp. 93–110.

Murcia, C., 1995, Edge effects in fragmented forests: implications for conservation, *TREE* **10**:58–62.

Myers, N., 1991, Tropical forests: present status and future outlook, *Climatic Change* **19**:3–32.

Odum, E. P., 1989, *Ecology and Our Endangered Life Support Systems*, Sinaur Assoc. Inc., MA, p. 49.

Onderdonk , D. A., and Chapman, C. A., 2000, Coping with forest fragmentation: The primates of Kibale National Park, Uganda, *Int. J. of Primat.* **21**:587–611.

Pope, T. R., 1996, Socioecology, population fragmentation, and patterns of genetic loss in endangered primates, in: *Conservation Genetics: Case Histories from Nature*, J. Avise and J. Hamrick, eds., Kluwer Academic Publishers, Norwell, MA., pp. 119–159.

Redford, K. H., 1992, The empty forest, *BioScience* **42(6)**:412–422.

Robinson, R. G., and Redford, K. H., 1991, Sustainable harvest of Neotropical forest animals, in: *Neotropical Wildlife Use and Conservation*, Robinson, R. G., Redford, K. H., eds., University of Chicago Press, Chicago, pp. 415–429.

Rylands, A. B., Mittermeier, R. A. and Rodriquez-Luna, E., 1995, A species list for the New World primates (*playrrhini*), Distribution by country, endemism, and conservation status according to the Mace-Lande system, *Neotrop. Primates* **3**:113–160.

Saunders, D. A., Hobbs, R. J., and Margules, C. R., 1991, Biological consequences of ecosystem fragmentation: A review, *Cons. Biol.* **5**:18–32.

Saunders, D. A., Hobbs, R., and Ehrlich, P. R., eds., 1993, *Nature Conservation 3: Reconstruction of Fragmented Ecosystems*, Surrey Beatty and Sons, Australia.

Schelhas, J., and Greenberg, R., 1996, The value of forest patches, in: *Forest Patches in Tropical Landscapes*, J. Schelhas and R. Greenberg, eds., Island Press, CA, pp. xv–xxxvi.

Shafer, C. L., 1995, Values and shortcomings of small reserves, *BioScience* **45(2)**:80–88.

Simberloff, D. S., and Abele, L. G., 1976, Island biogeography theory and conservation practice, *Science* **191**:285–286.

Sizer, N., and Tanner, E. V. J., 1999, Responses of woody plant seedlings to edge formation in a lowland tropical rainforest, Amazonia, *Biol. Conser.* **91**:135–142.

Soulé, M. E., and Simberloff, D., 1986, What do genetics tell us about the design of nature reserves? *Biol. Cons.* **35**:19–40.

Soulé, M. E., and Wilcox, B., eds., 1980, *Conservation Biology: An Evolutionary-Ecological Approach*, Sinauer Assoc., Sunderland, MA.

Stevenson, M., Baker, A., and Foose, T. J., 1992, Conservation assessment and management plan for primates, *IUCN/SSC Captive Breeding Specialist Group and IUCN/SSC Primate Specialist Group*, Apple Valley, MI.

Terborgh, J., 1986, Keystone plant resources in the tropical forest, in: *Conservation Biology: The Science of Scarcity and Diversity*, M. E. Soulé, ed., Sinauer Associates, Sunderland, MA, pp. 330–344.

Terborgh, J., and Winter, B., 1980, Some causes of extinction, in: *Conservation Biology: An Evolutionary-Ecological Perspective*, M. E. Soulé and B. A. Wilcox, eds., Sinauer Associates, Sunderland, MA.

Terborgh, J., Lopez, L., Tello, J., Yu, D., and Bruni, A. R., 1997, Transitory states in relaxing ecosystems of land bridge islands, in: *Tropical forest remnants: ecology, management, and conservation of fragmented communities*, W. F. Laurance and R. O. Bierregarrd, eds., University of Chicago Press, Chicago, IL, pp. 256–274.

Turner, I. M. and Corlett, R., 1996, The conservation value of small, isolated fragments of lowland tropical rain forest, *TREE* **11**:330–333.

Turton, S. M., and Freiburger, H. J., 1997, Edge and aspect effects on the microclimate of a small tropical forest remnant on the Atherton Tableland, Northern Australia, in: *Tropical forest remnants: ecology, management, and conservation of fragmented communities*, W. F. Laurance and R. O. Bierregarrd, eds., University of Chicago Press, Chicago, IL, pp. 45–54.

Tutin, C. E. G., 1999, Fragmented living: behavioural ecology of primates in a forest fragment in the Lopé Reserve, Gabon, *Primates* **40**:249–265.

Tutin, C. E. G., and White, L., 1999, The recent evolutionary past of primate communities: likely environmental impacts during the past three millenia, in: *Primate Communities*, J. G. Fleagle, C. Janson, and K. E. Reed, eds., Cambridge University Press, Cambridge, MA, pp. 220–236.

van Roosmalen, M. G. M., and Klein, L. L., 1987, The Spider Monkeys, Genus *Ateles*, in: *Ecology and Behavior of Neotropical Primates*, R. A. McHermeier and A. B. Rylands, eds., World Wildlife Federation, Washingtion, D.C.

Whitmore, T. C., 1997, Tropical forest disturbance, disappearance, and species loss, in: *Tropical Forest Remnants: Ecology, Management and Conservation of Fragmented Communities*, W. F. Laurance and R. O. Bierregaard, Jr. eds., University of Chicago Press, Chicago, IL, pp. 3–12.

Wilcox, B. A., 1980, Insular ecology and conservation, in: *Conservation Biology: An Evolutionary-Ecological Approach*, M. E. Soulé, and B. Wilcox, eds., Sinauer Assoc., Sunderland, MA, pp. 95–118.

Wilke, D. S., and Carpenter, J. F., 1999, Bushmeat hunting in the Congo Basin: An assessment of impacts and options for mitigation, *Biodiversity and Cons.* **8**:927–955.

WRI, 1990, *World Resources* 1990-1991, Oxford University Press, Oxford.

SECTION I: GENETICS AND POPULATION DYNAMICS

Laura K. Marsh

1. INTRODUCTION

The distribution, abundance, and demographic make up of primates living within fragments are an important component for understanding conservation needs of the species. Species response to habitat fragmentation and disturbance occurs in two phases (Wilcove et al., 1986). In the first phase, the species will try to adjust to modified habitats, while in the second phase, changes will occur in the demography of the population. The initial response to fragmentation includes changes in activity pattern, feeding, ranging, and other behaviors (Marsh, 1981; Johns, 1987; Johns and Skorupa, 1987; Menon, 1993), which in turn changes the demographics. Of primary concern is the genetic welfare of any species that remains in genetic or demographic isolation.

Inbreeding, lack of gene flow, and genetic drift are concerns for primatologists working in fragmented habitats (James, 1992; Pope, 1996; Gonçalves et al., this section). Distinct habitat fragmentation is an obvious barrier for some species, yet others are not as affected by it since they are able to utilize and travel through non-forest landscapes or the matrix (e.g., pasture, agriculture). Even in continuous habitat like African savannas, there may be hidden genetic barriers based on subtle habitat discontinuities (Templeton and Georgiadis, 1996). Species may be habitat specialists within their ecological niches at the landscape level. This separation is enough to create subspecies over time even though the populations may not be geographically isolated. Since not all populations can be watched closely, the potential for even random genetic events occurring (e.g., a migrant entering a new population from another area) can greatly impact the overall population (Templeton and Georgiadis, 1996).

Survival of species in forest fragments is related to landscape and habitat parameters. Among the parameters discussed are area, habitat variation, connectivity, and surrounding matrix. Knowing in advance what a species might persist in a fragment based on population size, landscape features, spatial distribution, gene flow, and matrix composition, will enhance the survival of the whole population or metapopulation for the region (Figure 1).

A metapopulation is a group of subpopulations within a larger area, where typically the migration of a subpopulation can occur. A subpopulation, or a local population, is

Primates in Fragments: Ecology and Conservation
Edited by L. K. Marsh, Kluwer Academic/Plenum Publishers, 2003

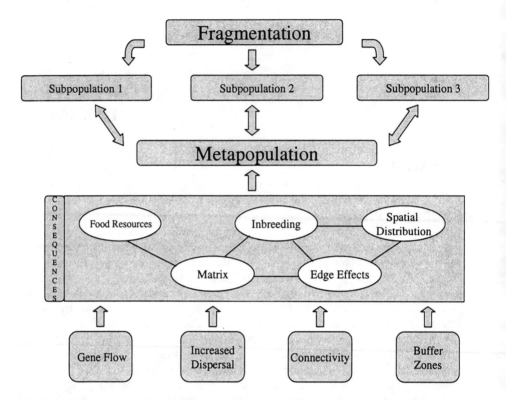

Figure 1. This is a conceptual diagram of fragmentation and primate metapopulations. Subpopulations are created by fragmentation, which collectively are the metapopulation. The metapopulation is affected by the consequences of fragmentation. This may potentially be to the detriment of the metapopulation as a whole. Some effects of fragmentation can be ameliorated and metapopulations enhanced by increasing gene flow, dispersal, connectivity, and buffer zones (c.f., Marsh, this volume). Illustration by Julie Hill.

described as a group of individuals that live in the same habitat. In other words, all individuals share a common environment (Hanski, 1998; Hanski and Simberloff, 1997). A metapopulation structure is defined as a chain of habitat fragments that are occupied by a metapopulation in which there exists a distribution of fragment areas and a migration rate between fragments (Hanski and Simberloff, 1997). A metapopulation might not be balanced, for instance where the rate of extinction exceeds the rate of colonization or vice versa. An extreme case occurs when the subpopulations are so far from each other that there is no migration between them and there is no possibility of recolonization (Hanski, 1996; Hanski and Gilpin, 1991). Additionally, the overall metapopulation may eventually become locally adapted to their particular resource base because of long-term fragmentation (Templeton and Georgiadis, 1996).

Even with these definitions, there is still vast interpretation of metapopulations and use of metapopulation theory (McCullough, 1996, Rodriguez-Vargas, this volume). Metapopulation, once it is consistently used and defined, when combined with landscape ecology, population dynamics, and behavioral ecology may prove to be one of the most viable foundations for primate conservation in the 21st century (Wiens, 1996; Sanderson and Harris, 2000). In this section, authors explore population dynamics and metapopulation theory for primates in fragments.

2. GENETICS AND VIABILITY

Gonçalves et al. in Chapter 2 discuss a case study of the effects of fragmentation on the genetic variability of the silvery marmoset (*Mico argentatus*). The authors conducted their study in the Flona Tapájos region in the extreme western part of Pará State, Brazil. Fieldwork was conducted at four sites representing a control (contiguous forest) and fragments: large (4,500 ha), medium (900 ha), and small (30 ha). Population density was much higher in the fragments as compared to contiguous forest, and there was a tendancy for increasing density with decreasing fragment size. Although the data are preliminary, the results show that the remnant populations are distict from one another genetically and that they appear to have suffered the effects of inbreeding and genetic drift. However, there is no clear indication that population size has a negative effect on genetic diversity, particularly since the smallest population in the 30 ha fragment returned the highest mean heterozygosity.

3. CHRONOLOGY OF FRAGMENTATION

Fragmentation for various reasons, geologic or anthropomorphic, have occurred in some regions over centuries rather than decades. Li et al. in Chapter 3 give a summary of the snub-nosed or golden monkey (*Rhinopithecus*) distribution in China from the Pliestocene to the present. The authors provide details on the fossil record and paleoecology of the original extent of golden monkeys. They bring us up through the centuries by providing distribution and habitat conditions for each of the *Rhinopithecus* species. They provide historical perspectives on the reasons for changes in distribution, particularly in terms of human impact over the last 400 years. And finally, they provide present day information on the survival needs of golden monkeys in their most fragmented habitats. The shift of this species from tropical/subtropical low-elevation mountains to tempertate high-elevation mountains has been a tranformation that has taken place through orogenesis, climate change, and an increase in human impacts for centuries.

4. METAPOPULATION THEORY

Rodriguez-Vargas discussed in Chapter 4 the usefulness of modeling a metapopulation for squirrel monkeys (*Saimiri oerstedii*) in the Chiriqui Province of Panamá. The author uses available subpopulation data and extrapolates for unknown populations by using ArcView. Analyses were done on fragments in the region and their potential suitability for squirrel monkeys. The author argues that by knowing the metapopulation organization in the region, conservation and management efforts will be better focused.

Chapman et al. discuss in Chapter 5 the effectiveness of metapopulation models, particularly incidence-function models, to the management of a community-owned forest fragment system in Kibale National Park, Uganda. The authors document the changes in forest structure in a series of fragments over a 5-year period, and in that time 19% of all fragments were cleared. The mathematical incidence-function model indicated that some species, such as *Colobus guereza*, could survive in the fragmented system even though its numbers declined along with habitat loss and its reproduction rate was low. *Procolobus badius* behaved as the model suggested, but more mobile species *Pan troglodytes* and *Cercopithecus ascanius* were outside of the model's predictive capabilities. The authors suggest that understanding a species' natural history characteristics is mandatory before committing to the outcomes of metapopulation models.

5. CASE STUDIES

Rodriguez-Toledo et al. in Chapter 6 show relationships between fragment size, shape, isolation, degree of vegetation disturbance, the presence of *Alouatta palliata mexicana*, and the numbers of individuals per group in the Los Tuxtlas Special Biosphere Reserve, Veracruz, Mexico. Of the 64 fragments studied, howlers were in only 19% of them. Howlers were found in 12% of the fragments under 10 ha, and in 80% of the fragments larger than 10 ha (up to 67 ha). Group presence was independent of fragment shape and straight-line distance, continuous forest, and nearest town. As distance between fragments decreased, there was an increase in primate presence. The authors suggest that for conservation purposes, fragments >10 ha should be considered.

Chiarello presented in Chapter 7 the influence of forest fragmentation on primate survival in the Atlantic Forest region of Brazil. Research was conducted in Espírito Santo and São Paulo States. The success of the primates varied between species and sites. The overall abundance showed no clear relationship with fragment area for any site. Frugivores appeared to be more affected by the absence of fruiting trees than fragment area. The author suggests that the abundance of second growth plant species, such as lianas or *Inga* sp. and the increase in insects, allow some primates like *Alouatta guariba*, *Callithrix geoffroyi*, *Callicebus personatus*, and *Cebus nigritus* to thrive.

Ferarri et al. in Chapter 8 discuss the dynamics of primate communities along the Santarém-Cuiaba Highway in south-central Amazonia, Brazil. The most frequent primates found in small fragments were *Alouatta belzebul*, *Cebus apella*, *Mico argentatus*, and *Callicebus moloch*. Over all study sites no clear relationship was found between fragment size and species number. The only universally consistent pattern of diversity was the presence of howlers and capuchins at all sites and the absence of spider monkeys in the smallest fragments. In every case, there was a combination of effects influencing primate presence, not just fragment area alone. The authors felt that rather than defining pattern and process, the results emphasized the complexity of the effects of anthropic habitat fragmentation.

Gilbert in Chapter 9 reviews the distribution of primates within the Biological Dynamics of Fragmented Forest Project experimental sites near Manaus, Brazil. When the sites were isolated in 1985, the immediate effect was loss of large frugivores, such as *Ateles paniscus*, *Chiropotes satanas*, and *Cebus apella* that seemed to require fragments with more area than 100 ha. *Alouatta seniculus* was able to survive and reproduce in even the smallest fragments (<10 ha). A low turnover in the number of species present suggests recolonization is not occurring rapidly in the isolated fragments. Groups of howlers (*Alouatta seniculus*), white-faced sakis (*Pithecia* spp.), and golden-handed tamarins (*Saguinus midas*) have lived in the 100-ha fragments for 20 years.

6. SUMMARY

There are few species that allow us to draw general conclusions about their presence and persistence in fragments. One common thread for the authors in this section is that the size of the fragment alone may not necessarily determine presence of primates. Species that traditionally use secondary habitats within contiguous forest (i.e., regrowth areas, streamsides, tree falls), such as marmosets and tamarins, were found to be successful in most fragments, including some of the smallest. Those with the ability to use the matrix or disperse through it were also more present in fragmented habitats, such as howlers, chimpanzees, and red-tailed monkeys. While it is clear primates are present in fragments throughout the tropics, only a few species are beginning to potentially show consistent patterns. However, *Cebus apella* was both present in smaller fragments (Ferrari, this section) and absent from small fragments (Gilbert, this section). More work needs to be done on the floristic comparisons across these sites for this species. Finally, the metapopulation may be the level of focus for some conservation efforts, but before landscape-level actions take place, there needs to be detailed enough information about the natural history of the species within the fragments in question to determine if the entire metapopulation should be managed.

7. REFERENCES

Hanski, I., 1998, Metapopulation dynamics, *Nature* **396**:41–49.
Hanski, I., and Gilpin, M. E., 1997, Metapopulation dynamics: Brief history and conceptual domain, *Biol. J. Linn. Soc.* **40**:3–16.
Hanski, I., and Simberloff, D., 1997, The metapopulation approach, its history, conceptual domain, and application to conservation, in: *Metapopulation Biology, Ecology, Genetics, and Evolution*, I. Hanski and M. E. Gilpin, eds., Academic Press, San Diego, CA, pp. 5–26.
James, R. A., 1992, Genetic variation in Belizean black howler monkeys (*Alouatta pigra*). Ph.D. dissertation, Rutgers University, New Brunswick, NJ.
Johns, A. D., 1987, The use of primary and selectively logged rainforest by Malaysian hornbills (*Bucerotidae*) and implications for their conservation, *Biol. Conserv.* **40**:179–190.
Johns, A. D., and Skorupa, J. P., 1987, Responses of rain-forest primates to habitat disturbance: A review, *Int. J. Primatol.* **8**:157–191.
Marsh, C. W., 1981, Diet choice among red colobus (*Colobus badius rufomitrans*) on the Tana River, Kenya, *Fol. Primatol.* **35**:147–178.
Menon, S., 1993, Ecology and conservation of the endangered lion-tailed macaque (*Macaca silenus*) in the landscape mosaic of the Western Ghats, Ph.D. dissertation, Ohio State University, Columbus.
McCullough, D. R., 1996, Introduction, in: *Metapopulations and Wildlife Conservation*, D. McCullough, ed., Island Press, Covelo, CA, USA, pp. 1–10.
Pope, T. R., 1996, Socioecology, population fragmentation, and patterns of genetic loss in endangered primates, in: *Conservation Genetics: Case Histories from Nature*, J. C. Avise and J. L. Hamrick, eds., Chapman and Hall, New York, pp. 119–159.
Sanderson, J., and Harris, L. D., 2000, *Landscape Ecology: A Top-Down Approach*, Lewis Publishers, New York, pp. 246.

Templeton, A. R., and Georgiadis, N. J., 1996, A landscape approach to conservation genetics: Conserving evolutionary processes in the African bovidae, in: *Conservation Genetics: Case Histories from Nature*, J. C. Avise and J. L. Hamrick, eds., Chapman and Hall, New York, pp. 398–430.

Wiens, J. A., 1996, Wildlife in patchy environments: Metapopulations, mosaics, and management, in *Metapopulations and Wildlife Conservation*, D. McCullough, ed., Island Press, Covelo, CA, USA, pp 53–84.

Wilcove, D. S., McLellan, C. H., and Dobson, A. P., 1986, Habitat fragmentation in the temperate zone, in: *Conservation Biology: The Science of Scarcity and Diversity*, M. E. Soulé, ed., Sinauer Associates, Sunderland, MA, pp. 237–256.

EFFECTS OF HABITAT FRAGMENTATION ON THE GENETIC VARIABILITY OF SILVERY MARMOSETS, *MICO ARGENTATUS*

Evonnildo C. Gonçalves, Stephen F. Ferrari, Artur Silva, Paulo E.G. Coutinho, Elytânia V. Menezes, and Maria Paula C. Schneider [*]

1. INTRODUCTION

Habitat fragmentation has evaluated effects on parameters such as species diversity, population density, and behavioral patterns, as shown in other chapters in this volume, but there are as yet few data available on its influence on genetic variation in free-ranging primate populations (but see Pope, 1996, 1998). Loss of genetic variability and inbreeding depression are generally presumed to be the primary results of the fragmentation of continuous populations, although outbreeding depression may also be relevant in many cases (Templeton, 1986; Dudash and Fenster, 2000).

Whatever its exact effects on genetic parameters, habitat fragmentation and, in particular, the isolation of relatively small subsets of the original population are likely to have highly deleterious implications for the long-term viability of remnant populations (Lacy, 1997). Dietz et al. (2000), for example, found evidence of significant inbreeding depression in remnant populations of golden lion tamarins, *Leontopithecus rosalia*, a species belonging to the same subfamily (Callitrichinae) as silvery marmosets, *Mico argentatus*. In this case, inbreeding reduced predicted long-term probability of survival considerably in small populations.

In the present study, preliminary data are presented on the genetic variability of silvery marmoset populations at four sites in central Amazonia, representative of different degrees of habitat fragmentation. A surprising amount of genetic variability was found, even in the smallest population, but analyses of heterozygosity indicate that all populations are subject to inbreeding or genetic drift.

[*] Evonnildo C. Gonçalves, Stephen F. Ferrari, Artur Silva, Paulo E. G. Coutinho, Elytânia V. Menezes, and Maria Paula C. Schneider, Department of Genetics, Universidade Federal do Pará, Caixa Postal 8607, 66.075-900 Belém – PA, Brasil. Correspondence to E. C. Gonçalves (email: ecostag@ufpa.br).

Primates in Fragments: Ecology and Conservation
Edited by L. K. Marsh, Kluwer Academic/Plenum Publishers, 2003

2. STUDY AREA

The study area is located just over 50 km to the south of the city of Santarém in the western extreme of the Brazilian state of Pará. The region's original forest cover is typical Amazonian terra firme, although human colonization along the Santarém-Cuiabá Highway has created a mosaic of forest fragments of different dimensions, interspersed with open pasture and plantations (Ferrari et al., this volume). Within the study area, the highway forms the eastern limit of the Tapajós National Forest (Flona Tapajós), a protected area of more than a half million hectares of continuous forest.

Fieldwork was carried out at four sites (Table 1), representing different degrees of habitat fragmentation. The Flona Tapajós was treated as the control site, being representative of the original forest cover and the marmoset population that inhabited the region. The remaining sites were chosen as representatives of fragments of the order of dozens, hundreds, and thousands of hectares, following an experimental design similar to that of the Biological Dynamics of Forest Fragments Project (BDFFP), in northeastern Amazonas state (e.g., Rylands and Keuroghlian, 1988; and Gilbert, this volume). Capture sites are between 17 and 50 km apart (Table 2).

Eight primate species occur in the region. However, the white-fronted spider monkey (*Ateles marginatus*) appears to be locally extinct from the study area, and the primate fauna is greatly reduced in all the fragments surveyed here (Table 1). Nevertheless, silvery marmosets were relatively common in the study area in comparison with sites further south.

Table 1. Characteristics of the study sites in western Pará. Fragments were categorized following Ferrari et al. (this volume).

Study site	Geographic coordinates	Fragment category and size (ha)	Diurnal primate species, in addition to *M. argentatus*
Flona Tapajós	03°02'46"S, 54°57'18"W	Control (continuous forest)	*Alouatta belzebul* *Callicebus moloch* *Cebus apella* *Chiropotes albinasus* *Saimiri sciureus*
Massafra	02°50'06"S, 54°53'14"W	Large (4,500 ha)	*Alouatta belzebul* *Cebus apella*
São Benedito	02°57'05"S, 54°47'41"W	Medium (900 ha)	*Alouatta belzebul* *Cebus apella*
Tabocal	02°37'14"S, 54°43'51"W	Small (30 ha)	*Alouatta belzebul* *Callicebus moloch*

Table 2. Distances between capture sites.

	Massafra	São Benedito	Tabocal
Flona Tapajós	24.5 km	17.5 km	52.7 km
Massafra		17.3 km	28.8 km
São Benedito			37.5 km

3. THE STUDY SPECIES

Mico argentatus (c.f., Rylands et al., 2000) belongs to the "bare-eared" group of marmosets. They are small-bodied monkeys (adult body weight approximately 300 g: Ferrari and Lopes, 1992) characterized by the absence of hairs in the auricular region, in contrast with their "tassel-eared" congenerics (Hershkovitz, 1977; Rowe, 1998). Silvery marmosets present grayish-white pelage on all body parts except the tail, which is black (Figure 1).

The species is endemic to the southern Amazon basin. Its geographic range is limited to the north by the Amazon itself, to the west by the Tapajós River, and to the east by the Tocantins (Hershkovitz, 1977; Ferrari and Lopes, 1990). In the Tocantins-Xingu interfluvium, the species is restricted to the lowland floodplain, being limited to the south by the foothills of the Brazilian shield. This distribution pattern appears to be related to the presence of a second callitrichine species, *Saguinus midas* (Ferrari and Lopes, 1990), which is absent from the Tapajós-Xingu interfluvium, where the present study area is located. In this region, the species ranges much further south (Hershkovitz, 1977; Ferrari et al., this volume).

The ecology of Amazonian marmosets is poorly known, although *M. argentatus* has now been studied in detail at two sites (Veracini, 1997; Albernaz and Magnusson, 1999; Tavares, 1999; Corrêa, in preparation). *Mico argentatus* is ecologically similar to other

Figure 1. Adult female silvery marmoset, *Mico argentatus*. Photo by Paulo E. G. Coutinho.

marmosets and is relatively abundant in disturbed habitat in comparison with primary forest, a pattern also seen in the present study area (Ferrari et al., this volume). In the savannas of Alter do Chão, to the northwest of the study area, silvery marmosets are found in naturally occurring forest fragments (Albernaz and Magnusson, 1999; Corrêa, in preparation), some of which cover less than 10 ha.

Silvery marmosets are found in small family groups of four to 12 individuals, in which a single female reproduces. An additional characteristic of the species, which is also typical of marmosets in general, is the marked predominance of twin births. As in other species, it seems probably that chimerism permits the sharing of a placenta by the normally dizygotic twins (Dixson et al., 1992), creating a problem for the analysis of genetic variability based on blood samples. For this reason, hair follicles were used for genetic analyses, as in the study of Nievergelt et al. (2000).

4. PROCEDURES

4.1. Fieldwork

The abundance of marmosets was estimated at Flona Tapajós, Massafra, and São Benedito using standard line transect surveys (Ferrari et al., this volume). Total transect length was 300 km at Flona Tapajós and 400 km at the other sites. Where sample size permitted, survey results were used for the calculation of estimates of population density based on Fourier series transformations (Ayres et al., 2000). Group densities were converted into population densities using estimates of mean group size derived from the sightings at each site. For this, the absolute minimum size for a social group was defined as four individuals, and records smaller than this value were omitted. At the fourth site (Tabocal), the composition of all resident *M. argentatus* groups was recorded during a behavioral study at the site (A. N. Carvalho, personal communication).

Captures were preceded by a period of monitoring at each site, during which potential trapping sites, such as frequently visited exudate trees, were identified. Following a period of trap habituation, during which traps were baited regularly with bananas, but not set, marmosets were captured using manually operated traps (Figure 2), similar to those used in previous studies of callitrichines (e.g., Digby, 1994). Materials and methods were finalized following field trials using both manual and automatic traps, which demonstrated the method used (Figure 2) to be the most efficient. While they may be more time-consuming in some situations, manual traps have a number of advantages in comparison with automatic devices. In particular, they permit selection of the individuals to be trapped, thereby avoiding both risky captures, such as those of juveniles, pregnant females, or infant carriers, and undesired recaptures.

Captured marmosets were restrained carefully within the traps and sedated with Ketamine Hydrochloride (0.1 mg/kg body weight) before being processed. The animals were weighed and measured, tattooed with their collection number on the inner surface of the right thigh, and fitted with ball-chain collars strung with color-coded plastic beads. Two samples of hair follicles were obtained using sterilized tweezers and stored in plastic pots, with care being taken to avoid contact with any substrate. Samples were maintained at room temperature and transported to the DNA Polymorphism Laboratory at the Federal University in Belém for analysis. Once processed, the marmosets were placed back in the

Figure 2. Trapping site for the capture of silvery marmosets. The traps (left) are triggered from the hide (right) using the nylon lines. Photo by Paulo E. G. Coutinho.

traps in which they were captured and monitored until fully recuperated from the anesthetic, and then they were released.

Extreme care was taken throughout the trapping and processing of animals to ensure the minimization of stress. This appears to be confirmed by the fact that the vast majority of marked animals returned regularly to the traps and could be recaptured. However, a small number of individuals proved impossible to capture, and trapping was especially difficult in the Flona Tapajós, perhaps at least partly due to the low population density of marmosets at this site.

4.2. Genetic Analyses

DNA was extracted from approximately 30 hair follicles selected from one sample from each animal, following the method of Gagneux et al. (1997). Two microsatellite loci, Cj6 and Cj11, isolated in the common marmoset, *Callithrix jacchus*, by Nievergelt (1998), were amplified using Polymerase Chain Reaction (PCR). The reactions took place in volumes of 20 µl containing 10 ng of genomic DNA, 50 mM KCl, 1.5 mM MgCl$_2$, 10 mM Tris-HCL, 50 µM of each dNTP, 0.5 µM of each primer (Cy5 marked forward primers), and a unit of Taq DNA polymerase (Gibco BRL). The reaction was carried out via the following protocol: 4 min at 94°C for denaturation, followed by thirty-five 30 s denaturation cycles at 94°C, association for 30 s at 55°C, extension for 1 min at 72°C, and a final cycle of 5 min at 72°C to ensure complete extension of the PCR products. Analysis of the amplified fragments was carried out in an Alf Express II automatic DNA processor using Allelinks version 1.0 software (Amersham Biosciences).

Intrapopulational variability was quantified in terms of allelic frequencies per locus, the mean number of alleles per locus (A), the proportion of polymorphic loci (P), and observed (H_o) and expected (H_e) heterozygosity under Hardy-Weinberg equilibrium. Mean values were calculated using Pogene32 (Yeh et al., 1997). Differences between H_o and H_e for each population were tested using Student's t. A Marcov chain test (100 batches of 1,000 interactions, with a 1,000-step dememorization process) was used for the calculation of unbiased estimates of exact probability values (P) for the evaluation of possible deviations of Hardy-Weinberg proportions at the two loci in each population. The excess or lack of heterozygotes at each locus was estimated by Wright's (1978) inbreeding coefficient (Fis). Both tests were calculated using Genepop version 3.1 (Raymond and Roussett, 1995). Genetic differences among populations were evaluated pairwise using Nei's (1972) distance, and Wright's fixation index, Fst, using Weir and Cockerham's (1984) variance-based method.

5. RESULTS

5.1. Population Parameters

As expected, population density was much higher in the fragments in comparison with the continuous forest at Flona Tapajós, and there was a clear tendency for increasing density with decreasing fragment size (Table 2). Surveys were conducted at two additional sites within Flona Tapajós, but even so, total sightings were insufficient (19 in 1,700 km surveyed) for the calculation of reliable density estimates. Rates were considerably lower (0.10 and 0.08 per 10 km surveyed) at the other two sites, and the overall scenario is one of both low and variable densities within the protected area. The mean sighting rate for Flona Tapajós is little more than a tenth that recorded at Massafra or São Benedito, which suggests that population density for the National Forest as a whole may be of the order of little more than one individual per square kilometer. Even so, this implies—based on density estimates for the fragments (Table 3)—that the total population would be at least 5,000 individuals, obviously much larger than that of any fragment.

The numbers of sightings recorded at both Massafra and São Benedito can be considered more adequate for the calculation of density estimates, but still fall short of the minimum recommended by Buckland et al. (1993) for reliable estimates based on

Table 3. Sighting rates and population parameters at the four study sites. Values for Tabocal are based on the detailed monitoring of all groups in the forest fragment.

Site	Sightings/rate per 10 km surveyed	Groups per km²	Mean group size ± SD	Population density (individuals per km²)	Estimated total population
Flona Tapajós	6/0.20	<1	5.3 ± 1.3	low	≥5000
Massafra	36/0.90	1.9	5.0 ± 0.9	9.3	418.5
São Benedito	38/0.95	2.3	5.6 ± 2.1	12.6	113.4
Tabocal	-	13.3	7.6 ± 1.2	101.1	30.3

sighting functions. Peres (1993) has suggested pooling data, where possible, as a means of overcoming this problem. This does seem possible in the case of Massafra and São Benedito, not only because of their geographic proximity and overall similarities (which even includes the number of sightings–Table 3), but also because sighting (observer-animal) distances were statistically similar ($t = -1.957$, $p = 0.054$, d.f. = 72). Even so, pooling the data alters density estimates only very slightly, with São Benedito decreasing to 2.2 groups per square kilometer and Massafra increasing to 2.1. Estimates of the total population in each fragment would thus change only negligibly.

In any case, density estimates are only as good as the data from which they are calculated, and their variance is often overlooked in the presentation of results. The 95% confidence limits for the estimates of group density at Massafra and São Benedito are 1.2 to 2.5 and 1.5 to 3.1, respectively, and 1.6 to 2.6 for the pooled data. Given this, it is clear that the real size of the population of each fragment could differ considerably from the values given in Table 3, although it is unlikely that the relative difference between sites— at least, that Massafra has the larger population—would change. Given this, sighting rates may provide less problematic estimates of abundance, especially where differences among sites are the primary focus of a study (Lopes and Ferrari, 2000; Ferrari et al., this volume).

An additional source of bias is the estimate of group size on which the calculation of population density is based. There is a recognized tendency for survey sightings to underestimate group size, and discarding sightings of fewer than four individuals appeared to compensate for this only partially, given that known group sizes at Tabocal, for example, varied between six and nine individuals. It seems likely that population sizes in the larger fragments were underestimated by at least 20%.

5.2. Genetic Variability

A total of 13 alleles were identified for the Cj6 locus, and six for Cj11, with considerable variation in frequencies among populations (Table 4). There is an overall tendency for the mean number of alleles per locus and the proportion of polymorphic loci to increase with increasing sample size.

Average heterozygosity varied from 0.31 to 0.61 (Table 5); whereas, expected values ranged from 0.50 to 0.81 for Flona Tapajós and Tabocal, respectively. The only significant difference between observed and expected heterozygosity was observed at São Benedito (Table 5). However, genotypic frequencies were significantly different from expected according to the Hardy-Weinberg equilibrium for both loci in all populations. Positive values of inbreeding coefficients Fis were recorded in all three populations for which sample size was adequate, in particular for the Cj11 locus (Table 6), indicating the presence of significantly more homozygotes than expected. This suggests that these populations are suffering the effects of either inbreeding or genetic drift.

Despite the limitations of the samples, in terms of the numbers of both loci and individuals, the sum of the evidence indicates clearly that the populations are no longer part of the same gene pool. However, there is no clear indication that population size has a negative effect on genetic diversity, in fact, the smallest population—Tabocal— returned the highest mean heterozygosity (Table 5). This would appear to further support the hypothesis that the populations have suffered the effects of inbreeding or genetic drift, which has reinforced differences among them.

Table 4. Allelic frequencies recorded in the four study populations of *Mico argentatus*.

Locus/allele	Site			
	Flona Tapajós	Massafra	São Benedito	Tabocal
Cj6/A				0.1111
Cj6/B				0.2778
Cj6/C				0.0556
Cj6/D		0.1250		0.0556
Cj6/E	0.2500	0.1250		0.1111
Cj6/F	0.2500	0.2500		0.1111
Cj6/G		0.1250		0.0556
Cj6/H		0.1250		
Cj6/I	0.2500	0.2500		0.0556
Cj6/J	0.2500			0.1111
Cj6/K			0.1667	0.0556
Cj6/L			0.5000	
Cj6/M			0.3333	
Cj11/A			0.1667	
Cj11/B	1.0000	0.4375	0.6667	0.0556
Cj11/C		0.3125	0.1667	0.3889
Cj11/D				0.3333
Cj11/E		0.1250		0.2222
Cj11/F		0.1250		
N	2	8	6	9
Mean number ± SD of alleles per locus	2.50 ± 2.12	5.00 ± 1.41	3.00 ± 0.00	7.00 ± 4.24
Polymorphic loci (%)	50	100	100	100

Table 5. Observed (H_o) and expected (H_e) heterozygosity for the four study populations of *Mico argentatus*, calculated according to Levene (1949).

Locus	Observed (expected) heterozygosity recorded at site			
	Flona Tapajós	Massafra	São Benedito	Tabocal
Cj6	1.000 (1.000)	0.000 (0.867)	1.000 (0.667)	0.889 (0.909)
Cj11	0.000 (0.000)	0.625 (0.725)	0.000 (0.546)	0.333 (0.726)
Mean ± SD	0.500±0.700	0.313±0.440	0.500±0.700	0.611±0.390
	(0.500±0.700)	(0.796±0.100)	(0.606±0.090)	(0.817±0.130)
Student's *t*	0.00 ($P = 1.0$)	- 2.61 ($P > 0.05$)	-0.37 ($P < 0.05$)	-0.29 ($P > 0.05$)

Table 6. Estimates of inbreeding coefficients (Fis) for the Cj6 and Cj11 loci in the four study populations of *Mico argentatus*. Positives values indicate an excess of homozygotes, negative values, a deficit.

Locus	Flona Tapajós	Massafra	São Benedito	Tabocal
Cj6	0.00	+ 1.00	- 0.58	+ 0.02
Cj11	n/a	+ 0.15	+ 1.00	+ 0.56

Estimates of Fst for the two loci were 0.10 (Cj6) and 0.15 (Cj11), once more reinforcing the subdivision among the study populations. This is further emphasized by pairwise comparisons between populations using both Fst and Nei's (1978) genetic distance (Table 7). The smallest genetic distance and the only pair of sites for which the Fst is insignificant (cf., Slatkin, 1995) is Flona Tapajós and Massafra, despite the fact that they are separated by over 20 km. It is interesting to note, however, that while the collecting sites are this far apart, the Massafra fragment is separated from the National Forest only by the Santarém-Cuiabá Highway. Marmosets are generally reluctant to come to the ground, but are known to cross roads (Ferrari, 1988), so they may be able to migrate between these two fragments, and thus maintain gene flow between them.

6. DISCUSSION

Despite the preliminary nature of the data, the results of the present study show that the remnant populations are quite distinct genetically and that they appear to have suffered the effects of inbreeding or genetic drift. Direct evidence of inbreeding depression has been recorded in fragmented populations of golden lion tamarins (Dietz et al., 2000), although no clear indications of this phenomenon were found in the present study. Perhaps equivocally, the population of only 30 animals at Tabocal—far below the minimum viable size by any standards—appeared to be the most variable at both loci. Pope (1998) found high polymorphism and mean heterozygosity in a small, isolated population of muriquis, *Brachyteles arachnoides*, but attributed this to the comparatively recent isolation of the population relative to the generation length of the species. Tabocal, on the other hand, has been isolated from continuous forest for more than two decades, a period that may cover 10 generations of marmosets.

Dietz et al. (2000) modeled the effects of inbreeding depression in remnant populations of golden lion tamarins and found that this factor reduced the predicted long-term probability of survival by a third in small populations (50 individuals), but had no significant effect on larger populations, similar in size to that at Massafra. While lion tamarins are morphologically similar to marmosets, it remains unclear to what extent the same predictions are relevant to the *Mico argentatus* populations studied here. In particular, *M. argentatus* appears to be well adapted to fragmented environments (Albernaz and Magnusson, 1999; Corrêa, in prep.), and may be patchily distributed, even in continuous forest (Veracini, 1997; Tavares, 1999). Given this, it seems likely that both inbreeding and genetic drift have been more important components of the evolutionary

Table 7. Nei's (1978) unbiased genetic distance (in italics) and Wright's (1978) fixation index (in bold script) for pairwise comparisons between the four *Mico argentatus* populations.

	Flona Tapajós	Massafra	São Benedito	Tabocal
Flona Tapajós		*0.199*	*0.385*	*1.503*
Massafra	**-0.069**		*0.592*	*0.583*
São Benedito	**0.125**	**0.116**		*1.668*
Tabocal	**0.202**	**0.056**	**0.225**	

history of *M. argentatus* than for *L. rosalia.* Whether this implies a greater tolerance of these factors, or even a potential for outbreeding depression (Dudash and Fenster, 2000), remains to be seen.

Ongoing analyses include more individuals from both the present study sites and forest fragments within the Alter do Chão savanna, as well as additional loci. The results of the analyses should provide more definitive insights into the genetic variability of these marmoset populations and possible determining factors. Despite evidence for genetic effects of habitat fragmentation, a systematic analysis of complementary ecological factors will be required before definitive guidelines can be drawn up for the management of remnant populations of this species.

7. SUMMARY

The fragmentation of once-continuous populations is likely to have effects such as a loss of genetic variability and inbreeding depression, although little is known about such processes in wild populations of primates. In the present study, the genetic variability of silvery marmosets (*Mico argentatus*) was analyzed at four sites in central Amazonia east of the Tapajós River, representative of different degrees of habitat fragmentation of the original forest cover, ranging from a small (30-ha) fragment to continuous forest (Flona Tapajós). Population size varied from 30 individuals in the smallest fragment to just over 400 in the largest (4,500 ha). In this preliminary study, two microsatellite loci, Cj6 and Cj11, were amplified using PCR and DNA extracted from hair follicles collected from between two and nine marmosets per site. A total of 19 alleles were identified (13 for Cj6 and 6 for Cj11), with considerable variation both within and among populations. Mean heterozygosity was lower than expected in all fragments, and the overall excess of homozygotes indicates that these populations are suffering the effects of either inbreeding or genetic drift. Genetic distances among populations also indicate that they are no longer part of the same gene pool although, equivocally, the smallest population was the most variable at both loci, possibly reflecting random factors in the fragmentation process. Despite their preliminary nature, the results do provide some initial insights into the genetic effects of habitat fragmentation in this Amazonian primate.

8. ACKNOWLEDGMENTS

This study was financed by the National Program for Biodiversity (PRONABIO) of the Brazilian Environment Ministry, through BIRD and CNPq. Additional support was received from FINEP/PRONEX, IBAMA/Santarém and the Kapok Foundation. IBAMA authorized the collection of specimens through special licence 157/2001-DIFAS/DIREC. We are grateful to Robson Mendes and the IBAMA personnel under his supervision, Adenilson Carvalho, Nara, Seubert Jati, Sebastião Lobato, Wellton Costa, Daniela Oliveira, and the many local people from the study area who contributed to the collection of data in their different ways. We also thank Caroline Nievergelt for advice on genetic analyses.

9. REFERENCES

Albernaz, A. L., and Magnusson, W., 1999, Home-range size of the bare-ear marmoset (*Callithrix argentata*) at Alter do Chão, Central Amazonia, Brazil, *Int. J. Primatol.* **20**:665–677.

Ayres, M., Ayres, M., Jr., Ayres, D. L., and Santos, A. S., 2000, *BioEstat 2.0–Aplicações Estatísticas nas Áreas das Ciências Bi ológicas e Médicas,* Sociedade Civil Mamairauá/CNPq, Tefé.

Buckland, S. T., Anderson, D. R., Burnham, K. P., and Laake, J. L., 1993, *Distance Sampling: Estimating Abundance of Biological Populations,* Chapman and Hall, London.

Corrêa, H. K. M., In preparation, Determinantes ecológicos da biologia reprodutiva de *Callithrix argentata,* Doctoral dissertation, Goeldi Museum, Belém.

Dietz, J. M., Baker, A. J., and Ballou, J. A., 2000, Demographic evidence of inbreeding depression in wild golden lion tamarins, in: *Genetics, Demography, and Viability of Fragmented Populations,* A. G. Young and G. M. Clarke, eds., Cambridge University Press, Cambridge, pp. 203–211.

Dixson, A. F., Anzenberger, G., Monteiro da Cruz, M. A. O., Patel, I., and Jeffreys, A. J., 1992, DNA fingerprinting of free-ranging groups of common marmosets (*Callithrix jacchus jacchus*) in NE Brazil, in: *Paternity in Primates: Genetic Tests and Theories,* R. D. Martin, A. F. Dixson, and E. J. Wickings, eds., Karger, Basel, pp. 192–202.

Dudash, M. R., and Fenster, C. B., 2000, Inbreeding and outbreeding depression in fragmented populations, in: *Genetics, Demography, and Viability of Fragmented Populations,* A. G. Young and G. M. Clarke, eds., Cambridge University Press, Cambridge, pp. 35–53.

Ferrari, S. F., and Lopes, M. A., 1990, A survey of primates in central Pará, *Boletim do Museu Paraense Emílio Goeldi, Zoologia* **6**:169–179.

Ferrari, S. F., and Lopes, M. A., 1992, A new species of marmoset, genus *Callithrix* Erxleben 1777 (Callitrichidae, Primates), from western Brazilian Amazonia, *Goeldiana Zoologia* **12**:1–13.

Gagneux, P., Boesch, C., and Woodruff, D. S., 1997, Microsatellite scoring errors associated with noninvasive genotyping based on nuclear DNA amplified from shed hair, *Molec. Ecol.* **6**:861–868.

Hershkovitz, P., 1977, *Living New World Monkeys, Volume 1, with an Introduction to Primates,* Chicago University Press, Chicago.

Lacy, R. C., 1997, Importance of genetic variation to the viability of mammalian populations, *J. Mammal.* **78**:320–335.

Lopes, M. A., and Ferrari, S. F., 2000, Effects of human colonization on the abundance and diversity of mammals in eastern Brazilian Amazonia, *Cons. Biol.* **14**:1,658–1,665.

Nei, M., 1972, Genetic distance between populations, *Am. Nat.* **106**:286–292.

Nievergelt, C. M., Mundy, N. I., and Woodruff, D. S., 1999, Microsatellite primers for genotyping common marmosets (*Callithrix jacchus*) and other callitrichids, *Mol. Ecol.* **5**:1,431–1,439.

Nievergelt, C. M., Digby, L. J., Ramakrishnan, U., and Woodruff, D. S., 2000, Genetic analysis of group composition and breeding system in a wild common marmoset (*Callithrix jacchus*) population, *Int. J. Primatol.* **21**:1–20.

Peres, C. A., 1993, Structure and spatial organization of an Amazonian terra firme forest primate community, *J. Trop. Ecol.* **9**:259–276.

Pope, T. R., 1996, Socioecology, population fragmentation, and patterns of genetic loss in endangered primates, in: *Conservation Genetics: Case Histories from Nature,* J. C. Avise and J. L. Hamrick, eds., Chapman and Hall, New York, pp. 119–159.

Pope, T. R., 1998, Genetic variation in remnant populations of the woolly spider monkey (*Brachyteles arachnoides*), *Int. J. Primatol.* **19**:95–109.

Raymond, M., and Rousset, F., 1995, GENEPOP (Version 3.1): Population genetics software for exact test and ecumenicism, *J. Hered.* **86**:248–249.

Rowe, N., 1998, *A Pictorial Guide to the Primates,* Pongonias Press, New Hampton.

Rylands, A. B., and Keuroghlian, A., 1988, Primate populations in continuous forest and forest fragments in central Amazonia: Preliminary results, *Acta Amaz.* **18**:291–307.

Rylands, A. B., Schneider, H., et al., 2000, An assessment of the diversity of New World primates, *Neotrop. Primates* **8**:61–93.

Slatkin, M. A., 1995, Measure of population subdivision based on microsatellite allele frequencies, *Genetics* **139(1)**:457–462.

Tavares, L. I., 1999, Comportamento de forrageio em *Callithrix argentata (Linnaeus 1771)* na Estação Científica Ferreira Penna, Melgaço–PA, Masters thesis, UFPa, Belém.

Templeton, A. R., 1986, Coadaptation and outbreeding depression, in: *Conservation Biology, The Science of Scarcity and Diversity,* M. E. Soulé, ed., Sinauer Associates, Sunderland, pp. 105–116.

Veracini, C., 1997, O comportamento alimentar de *Callithrix argentata* (Linnaeus 1771) (Primata, Callitrichinae), in: *Caxiuanã*, P. L. B. Lisboa, ed., Museu Paraense Emílio Goeldi, Belém, pp. 437–446.

Weir, B. S., and Cockerham, C. C., 1984, Estimating F-statistics for the analysis of population structure, *Evolution* **38**:1,358–1,370.

Wright, S., 1978, *Evolution and the Genetics of Populations, Vol. 4: Variability Within and Among Natural Populations*, University of Chicago Press, Chicago.

Yeh, F. C., Yang, R. C., and Boylet, T., 1997, *POPGENE (Version 1.32): Software Microsoft Window-based freeware for population genetic analysis*, University of Alberta, Canada.

CHANGES IN DISTRIBUTION OF THE SNUB-NOSED MONKEY IN CHINA

Baoguo Li, Zhiyun Jia, Ruliang Pan, and Baoping Ren[*]

1. INTRODUCTION

Snub-nosed monkeys (*Rhinopithecus*), or golden monkeys, are members of the subfamily Colobinae, family Cercopithecinae. They are very beautiful creatures, but are now distributed in very limited areas. The Sichuan species *(Rhinopithecus roxellana)*, the Yunnan species *(R. bieti)*, and the Guizhou species *(R. brelichi)* are endemic to China (Figure 1). Together with the Vietnamese species *(R. avunculus),* they were originally regarded as one genus (Napier and Napier, 1967). More recently they have been split, the Chinese forms being classified as one subgenus (*R.* [*Rhinopithecus*]) and the Vietnamese as another subgenus (*R.* [*Prebytiscus*]) (Jablonski and Peng, 1993; Jablonski, 1998a). Though limited to isolated regions, they form a graded array of species from *R. avunculus* in subtropical evergreen and deciduous broad-leaf forests at less than 1,000 m altitude to *R. bieti* in temperate, coniferous forests as high as 3,000 to 4,500 m where annual average temperatures hover near freezing (Pan and Yong, 1989; Boonratana and Le, 1998).

The distribution of the snub-nosed monkeys in China has changed from the Pleistocene to the present. The fossil records clearly show that the snub-nosed monkey was once widely spread over south and central China during the Quaternary Period (Gu and Hu, 1991; Gu and Jablonski, 1989; Han, 1982; Jablonski, 1998a; Pan, 1995; Jablonski and Pan, 1988; Wang et al., 1982; Zhao and Li, 1981; Ma and Tang, 1992; IVPP, 1979). Provincial and county annals from the Qing Dynasty to the Republic (1616–1949) document that distribution diminished from south to north and from east to west in China. Most of their populations in south, southwest, east, and central China were extinct (Li and Pan, in press). At present, the Sichuan snub-nosed monkey is found in

[*] Baoguo Li, Department of Biology, College of Life Science, Northwest University, Xi'an, China, 710069 and Field Research Center, Primate Research Institute, Kyoto University, Inuyama, Japan, 484-8506. Zhiyun Jia, Institute of Zoology, Chinese Academy of Sciences, Beijing, China, 100080. Ruliang Pan, Department of Anatomy and Human Biology, The University of Western Australia, Perth, Australia, 6907. Baoping Ren, Department of Biology, College of Life Science, Northwest University, Xi'an, China. Correspondence to B. Li (email: baoguoli@nwu.edu.cn).

Primates in Fragments: Ecology and Conservation
Edited by L. K. Marsh, Kluwer Academic/Plenum Publishers, 2003

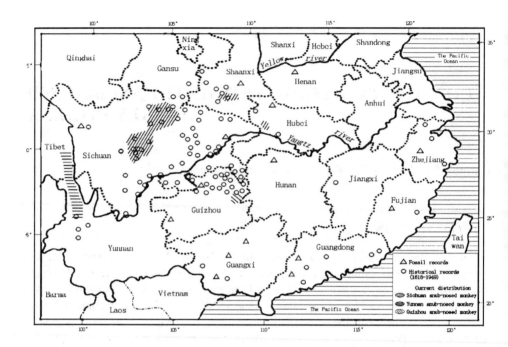

Figure 1. The distribution of snub-nosed monkeys in China from the Pleistocene to 400 years ago and to the present. Fossil locations were taken from Colbert and Hooijer (1953), Zhao and Li (1981), Han (1982), Wang et al. (1982), Pan and Jablonski (1987), Gu and Jablonski (1989), Gu and Hu (1991), Jablonski and Peng (1993), Pan (19950, and Jablonski (1995a and b).

Shaanxi, Sichuan, Hubei, and Gansu Provinces, occurring in subtropical and temperate forests, including deciduous broadleaf and coniferous and deciduous broadleaf mixed plants from 1,400 to 3,300 m above sea level. The Guizhou snub-nosed monkey is found only on Fanjing Mountain, Guizhou Province, in subtropical forest, including evergreen broadleaf, deciduous and evergreen broadleaf mixed plants or deciduous broadleaf plants between 1,200 and 2,100 m. The Yunnan snub-nosed monkey is found in Yunnan Province and Xizang autonomous region, moving in the high mountains within a range of between 3,800 and 4,700 m where snow is present throughout the year. Because of serious deforestation and human activity during the last 400 years, this primate must be considered highly endangered because natural habitats are disappearing or becoming fragmented (Ren et al., 1998; Bleisch et al., 1993; Bleisch 1993; Long et al., 1996; Li et al., 2001).

Even though the Chinese government has carried out strict protection policies, the situation for their survival is still precarious. There are now believed to be only 800 to 1,500 *R. bieti* and *R. brelichi* and 10,000 to 16,000 *R. roxellana* (Ren et al., 1998; Li et al., 2001). Consequently, golden monkeys, like the giant panda, are one of the most endangered species in the world.

Many investigations have been conducted concerning the distribution of the snub-nosed monkeys (Hu et al., 1980; Li et al., 1981; Chen et al., 1982; Bleisch et al., 1993; Li and Liu, 1994; Long et al., 1996; Kirkpatrick, 1998). The monkeys, which are discontinuous in geographical distribution, live in forests in high mountains. As in-depth research continues with field data increasing constantly, investigating the changes in the distribution of these primates is important for their conservation.

2. GEOCHRONOLOGICAL DISTRIBUTION

2.1. Fossil Records and Paleoecology

Fossil records of the snub-nosed monkey can be traced back to the early Pleistocene. This monkey was mainly distributed over the southern provinces, such as Guizhou, Yunnan, Sichuan, and, especially, Guangxi (Hu and Qi, 1978). With regard to current data, all of the monkey fossils were excavated in the Pleistocene and Holocene strata.

Jablonski (1998a) concludes on morphological grounds that the three Chinese species are a monophyletic group with respect to the primitive *R. avunculus* and suggests a divergence during the Pliocene with the three Chinese species originating by mid-Pleistocene. This position for *R. avunculus* is not supported by Zhang and Ryder's molecular data (1998), but their analysis also points to a divergence of *Rhinopithecus* spp. about two million years ago, which is in rough agreement with Jablonski's proposed Pliocene origin for the genus (Jablonski, 1998b).

2.1.1. Fossil Record

Table 1 and Figure 1 summarize distribution of fossil records of Chinese snub-nosed monkeys. We can see that the monkeys were widely distributed in south, southeast,

Table 1. Fossil records of the snub-nosed monkeys in China.

Province	Site	Geochronological scale	Reference
Shaanxi	Gongwangling, Lantian county	Early Pleistocene	Gu and Jablonski, 1989
Henan	Badu Mt., Xinan county	Middle Pleistocene	Gu and Hu, 1991
Sichuan	Yanjinggou, Wanxian county	Mid/Late Pleistocene	IVPP, 1979
Sichuan	Ganzi county	Later Pleistocene	Pan, pers. comm.
Hubei	Yunxian county	Early Pleistocene	Pan, 1995
Hunan	Cili county	Middle Pleistocene	Wang, et al., 1982
Guizhou	Tongzi county	Middle Pleistocene	IVPP, 1979
Guangxi	Heidong, Daxin county	Early Pleistocene	Han, D., 1982
Guangxi	Jiuleng Mt., Du'an county	Middle Pleistocene	Zhao, et al., 1981
Guangdong	Xiashadong, Luoding county	Later Pleistocene	Jablonski, 1998a
Guangdong	Deqing county	Later Pleistocene	Pan, pers. comm.
Guangxi	Luoshanzai, Liucheng county	Early Pleistocene	IVPP, 1979
Fujian	Yong'an county	Early Pleistocene	Pan, pers. comm.
Zhejiang	Shuanglongdong, Jinhua city	Holocene	Ma and Tang, 1992

southwest, and central China, as well as the northwest province of Shaanxi, during the Quaternary period.

2.1.2. Paleoecology

Considering the strata in which golden monkey fossils were excavated, and the accompanying fossil taxa, we conclude that in the early Pleistocene, the monkey mainly inhabited tropical and subtropical forests and was adapted to a warm and damp climate. However, with changes in Quaternary climate, they adapted to the temperate zone. Generally speaking, the elevation of the native habitat of the fossil golden monkeys is low (only several hundred meters). No high-elevation fossils have yet been found. This indicates that they lived in low mountains during the early Pleistocene.

In the middle and late Pleistocene, the habitat of the monkey remained tropical and subtropical forests, as well as forests in warm temperate zones. Although it was colder in the late Pleistocene than in the early Pleistocene (Hu and Qi, 1978), the monkeys could adapt to varied climates during glacial/interglacial periods through migration and vertical movement. Geographically, the three species are assumed to have been separated from each other in the late Pleistocene. Sichuan and Guizhou species were divided by the Yangtze River after they dispersed in two different directions. The Yunnan species was isolated in the Qinghai-Tibet Plateau, far from the other species; its living relatives are only found in southwestern China, a narrow area just between Yangtze and Mekong Rivers.

3. CHANGES AND HISTORICAL DISTRIBUTION

3.1. Appellations in Historical Text

The snub-nosed monkeys not only appeared in ancient literary works, but also were included in historical annals. However, the appellation in ancient books and annals is confusing. In the ancient data, we compared the description of characteristics of the species and examined their appellations in documents from the Qing Dynasty to the Republic (Li et al., 1989). There are a number of names referring to golden monkeys, listed as follows: You, Wei, Guo Ran, Rong, Rou, Jin Xian Rong, Jin Si Yuan, Jin Si Rong, Jin Xian Hou, and Jin Si Hou. It was hard or impossible to determine which of the three species was referred to based on the descriptions in the documents. This may not present a problem if the location where the monkeys were mentioned was available.

3.2. Distribution Records Based on Provincial and County Annals

Table 2 shows literature from 11 provinces in China, including a historic record from the Qing Dynasty in 1616. The distribution of the golden monkeys in the previous 400 years is illustrated in Figure 1. To illustrate dynamic change of the distribution through time is difficult. A dramatic reduction of animal populations has clearly been accelerated in the last two centuries.

Table 2. Literature dealing with the description and distribution of the snub-nosed monkeys from State (Provincial-S) or Local (County-L) governments in China since 1616.

Province	County	Given name in ancient Chinese	Annals	
			Source	Years published
Shaanxi	Qinling Mt. Longshan Mt.	You,Rong,Wei	Shaanxi (S)	Yongzheng 13 years of Qing Dynasty (1887)
	Zhouzhi	Rong	Zhouzhi (L)	Minguo 14 years (1925)
	Longxian (Longzhou)	Rong	Longxian (L)	Kangxi 52 years of Qing Dynasty (1713)
	Xixiang	Rong	Xixiang (L)	Daoguang period of Qing Dynasty (1821–1850)
	Shiquan	Rong	Shiquan (L)	Quangxu 29 years of Qing Dynasty (1903)
	Baoji	Rong	Shaanxi (S)	Qing Dynasty (1664)
Sichuan	Wangyuan	Rong	Wangyuan (L)	Minguo 21 years (1932)
	Wangyuan, Chenkou, Zhongxian	Rong	Taiping (L)	Tianlong 59 years of Qing Ddynasty (1794)
	Fengdu, Dianjian, Liangping	Ji Si Yuan	Zhongzhou (L)	Daoguang period of Qing Dynasty (1821–1850)
	Rongchang, Fushun, Nanxi, Yibin, Pingshan, Mabian, Meigu, Gaoxiang, Xingwen	Rong	Sichuan (S)	Jianqing 15 years of Qing Dynasty (1810)
	Yilong, Xichueng, Nanchueng, Yuechi, Guangan, Linshui, Pengan, Yengshan	Rong	Sichuan (S)	Jiaqing 21 years of Qing Dynasty (1816)
	Wusheng, Hechuan, Tongliang, Bishan, Dazu,Ruengchang, Jiangjin,Jiangbei, Changshou	Ji Si Yuan	Jiangbeiting (L)	Daoguang 24 years of Qing Dynasty (1844)
	Leibe	Rong	Leibeting (L)	Guangxu 19 years of Qing Dynasty (1893)
	Yuexi,Xichang	Rong	Sichaun (S)	Daoguang period of Qing Dynasty (1821–1850)
	Huili	Rong	Huili (L)	Tongzhi 9 years of Qing Dynasty (1870)
	Maoxian, Dayi, Qionglai, Pujiang	Rong	Sichuan (S)	Jiaqing 15 years of Qing Dynasty (1810)
	Wenchuan	Rong, Jin Xian Hou	Sichuan (S)	Jiaqing 10 years of Qing Dynasty (1805)
	Wenchuan	Jin Xian Hou	Wenchuan (L)	Minguo 33 years (1944)
	Baoxing,Yaan, Lushan, Mingshan, Qingxi, Hanyuan, Tianquan, Ruenjing, Xiaojien, Ganzi	Rong	Yazhoufu (L)	Qianlong 4 years of Qing Dynasty (1739)
	Baoxing	Specimen		Collected by Mr. David in 1870
	Kangding	Ji Xian Hou, Rong	Sichuan (S)	Minguo 18 years (1929)
	Guangyuan	You Rong	Guangyuan (L)	Qianlong 50 years of Qing Dynasty (1785)
	Beichuan	Rong	Shiquan (L)	Daoguang period of Qing Dynasty (1821–1850)

Table 2 (continued).

Province	County	Given name in ancient Chinese	Annals	
			Source	Years published
Sichuan (cont.)	Songpan	Rong	Songpan (L)	Minguo period (1912)
	Nanchuan	Guo Ran	Nanchuan (L)	Qianlong 29 years of Qing Dynasty (1764)
Gansu	Wudu, Wenxian	Rong	Shiquan (L)	Daoguang period of Qing Dynasty (1821–1850)
	Kangxian	Ji Si Rong	Kanxiang (L)	Minguo 25 years (1936)
	Tianshui (Qinzhou)	Rong	Qinzhou (L)	Qianlong 29 years of Qing Dynasty (1764)
Hubei	Zhushan	Rong	Zhushan (L)	Tongzhi 4 years of Qing Dynasty (1865)
	Yichang	Wei, Guo Ran	Hubei (S)	Minguo 10 years (1921)
Jiangxi	Yichun	Jin Si Yuan	Yichun (L)	Kangxi 40 years of Qing Dynasty (1701)
Fujian	Yongtai	Wei	Yongtai (L)	Kangxi 40 years of Qing Dynasty (1701)
Zhejiang	Leqing	Jin Si Yuan	Leqing (L)	Guangxu 7 years of Qing Dynasty (1881)
	Shaoxing	You	Shaoxingfu (L)	Qianlong 55 years of Qing Dynasty (1790)
	Huzhou	You	Wuxing (L)	Qianlong 55 years of Qing Dynasty (1790)
Guangdong	Enping	You	Enping (L)	Daoguang 5 years of Qing Dynasty (1825)
	Chaozhou	Ji Si Yuan	Chaozhoufu (L)	Qianlong 27 years of Qing Dynasty (1762)
	Zhaoqing	Ji Si Yuan	Zhaoqinfu (L)	Daoguang 13 years of Qing Dynasty (1833)
	Jieyang	Ji Si Hou	Jieyang (L)	Qianlong 44 years of Qing Dynasty (1799)
	Conghua	Ji Si Yuan	Conghua (L)	Kangxi 40 years of Qing Dynasty (1701)
Guangxi	Fusui (Tongzheng)	Rong	Tongzheng (L)	Minguo 22 years (1933)
	Debao (Zhenanfu)	Jin Sin Yuan	Zhenanfu (L)	Guangxu 18 years of Qing Dynasty (1892)
Guizhou	Sinan, Yenjiang, Dejiang, Wuchuan, Yanhe, Songtao, Fenggang, Jiangkao	Wei	Guizhou (S)	Jiaqing 7 years of Qing Dynasty (1802)
	Tongren, Shiqian	Ji Si Yuan	Tongrenfu (L)	Guangxu 16 years of Qing Dynasty (1890)
	Zhunyi, Suiyang, Tongzi, Xishui, Zhengan, Chishui, Meitan, Renhuai	Wei Rou	Zhunyifu (L)	Daoguang 21 years of Qing Dynasty (1841)
Yunnan	Dali	Rong	Yunnan (S)	Daoguang 15 years of Qing Dynasty (1835)
	Yuengping, Yunlong	Rong	Yunnan (S)	Xuantong 3 years of Qing Dynasty (1911)
	Fengqing	Yu Mian Yuan	Shunningfu (L)	Guangxu 30 years of Qing Dynasty (1904)

Golden monkeys used to be widely distributed in China, including the south, southwest, southeast, and central regions and the two provinces in the northwest (Gansu and Shaanxi). Individuals were abundant around the Yangzi River. Unfortunately, since 1616 increased human pressure forced most of the populations to retreat from the plains and middle mountains or they died out. The extant groups eventually dwelled in isolated mountains with an altitude up to 4,500 m above sea level (Figure 2). A remarkable reduction of the population density occurred in the Sichuan and Guizhou Provinces in the second 200 years. A principal disappearance of residential sites happened in south and southeast China along the coastlines in the first 200 years when there was no trace in Zhejiang, Fujian, and Jiangxi in the southeast and Guangdong and Guangxi in the south. A minimal change was detected in Yunnan in southwest China and in Gansu and Shaanxi in northwest China because these regions are more remote.

4. CURRENT DISTRIBUTION AND HABITATS

Current localities, population estimates, and habitat notes for the Chinese snub-nosed monkeys are summarized below. Refer to Figure 1 for their distribution.

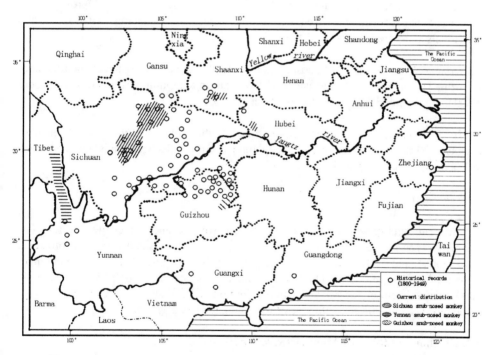

Figure 2. The distribution of snub-nosed monkeys in China about 200 years ago and present.

4.1. Sichuan Snub-Nosed Monkey (*R. roxellana*)

4.1.1. Distribution and Population

R. *roxellana* has the widest distribution of any species of the snub-nosed monkey and is found between longitude 102° to 111° E and latitude 30° to 35° N in the Qunlaishan, Mingshan, Daxiangling, Xiaoxiangling, Qinling, and Daba Mountains (Tan, 1985; Ren et al., 1998). We estimate a population between 15,000 and 16,000 individuals.

4.1.1a. R. r. roxellana. This subspecies is found in the provinces of Sichuan and Gansu. Sichuan habitats include Qingchuan, Pingwu, Songpan, Beichuan, Nanping, Maoxian, Heishui, Wenchuan, Baoxing, Tianquan, Lushan, and Luding in the Qunlaishan Mountains, Mingshan Mountains, and Daxiangling and Xiaoxiangling Mountains. The population is about 10,000 individuals in 100 troops. There are about 6,000 individuals in the Mingshan Mountains, about 3,500 in the Qunlaishan Mountains, and about 500 in the Daxiangling and Xiaoxianling Mountains. (Sichuan Wildlife Resource Investigation Conservation Management Station, 1997).

Gansu habitats include Wenxian in the Mingshan Mountains with about 800 individuals and 8 troops (Zhang, 1995).

4.1.1b. R. r. qinlingensis. This subspecies inhabits the Shaanxi Province, which includes Zhouzhi, Taibai, Foping, Ningshaan, and Yangxian in the Qinling Mountains with about 3,800 to 4,000 individuals in 39 troops (Li et al., 2001).

4.1.1c. R. r. hubeiensis. This subspecies inhabits the Hubei Province, which includes Fangxian, Xingshan, and Batong in Shengnongjia of the Daba Mountains with about 600 to 1,000 individuals in 5 to 6 troops (Ren et al., 1998).

4.1.2. Habitat

Sichuan snub-nosed monkeys live in subtropical and temperate forests. The vegetation type includes deciduous broadleaf forest, coniferous and deciduous broadleaf mixed forest, and coniferous forest with eight formations (Table 3). The elevation ranges from 1,200 to over 3,000 m. The seasonal home range size is about 22.5 km^2 (Li et al., 2000), which is closely correlated with distribution and abundance of resources. A large home range is a unique characteristic of this monkey (Table 4).

4.2. Guizhou Snub-Nosed Monkey (*R. brelichi*)

4.2.1. Distribution and Population

The Guizhou snub-nosed monkey inhabits Guizhou Province, which includes Jiangkou, Songtao, and Yingjiang in Fanjing Mountain of the Wuling Mountains (108° 50' E, 27° 57' N) with about 800 individuals (Bleisch et al., 1993).

Table 3. Vertical distribution of the habitat vegetation form (formation) of the Sichuan snub-nosed monkeys on the south and north slopes in Qinling Mountains.

South Slope (Chaijiaguan area of Ningshaan county)		North Slope (Yuhuangmiao area of Zhouzhi)	
Altitude (m)	Formation		Formation
2900			Abies fargesii
2800			
2700		Betula utilis	
2600			
2500			Betula Pinus armandii
2400	Betula		alba-sinensis Picea wilsonii
2300	alba-sinensis		Betula alba-sinensis
2200	Pinus armandii	Pinus armandii	Q. spinosa
2100	Populus purdomii	Pterocarya hupebensis	
2000	Toxicodendron	Populua purdomii	
1900	vernicifluum	Toxicodendron vernicifluum	
1800			
1700	Quercus aliena		Q. aliena var. acuteserrata
1600	var. Pinus tabuluaefrmis		Q. liaotungensis
1500	acuteserrata Castanea mollissima		
1400	Q. glandulifera var.	Q. variabilis	
1300	breviprtiolata	Q. dentata	
1200	Platycarya strobilacea		

Table 4. Comparison of habitats of Sichuan snub-nosed monkey.

Item	Vegetation types			Altitude (m)	Authors
	Deciduous broadleaf forest	Deciduous and coniferous broadleaf mixed forest	Coniferous forest		
Qinling Mt. in Shaanxi	77.48%[*]	20.78%	1.74%	1,400 to 2,700	Li, 1994
Wolong Reserve in Sichuan	37.2%	51%	11.8%	2,000 to 3,300	Hu et al., 1980
Baishuijiang Reserve in Gansu	Very often in winter	Very often in spring and autumn	Very often in summer	1,600 to 3,200	Zhang, 1995
Sheniongjia Reserve in Hubei	Activities in each type of vegetation			2,000 to 3,100	Hu and Zhang, 1985

* Percent of annual time spent in a particular habitat.

4.2.2. Habitat

The habitat for this monkey is montane subtropical forest, including evergreen broadleaf forest, deciduous and evergreen broadleaf mixed forest, and deciduous broadleaf forest between 1,500 and 2,200 m elevation. Annual rainfall above 1,600 m exceeds 2,000 mm. Temperatures fall below freezing during five months of the year, and snows are common in winter. Monthly mean temperatures are below zero (Celsius). Within the forests used by the monkeys, mono-groves of Asian oaks (*Cyclobanopsis* spp.) and beech (*Fagus longipetiolata*) often occur, while other forest types used by the monkeys do not have a dominant tree species. Common canopy trees include cherries (*Prunus* spp.), maples (*Acer* spp.), *Rhododendron* spp., and birch (*Betula* spp.). Individual monkeys range over more than 10 km^2 of habitat and apparently need a large area of intact forest to survive throughout the year (Bleisch et al., 1993).

4.3. Yunnan Snub-Nosed Monkey (*R. bieti*)

4.3.1. Distribution and Population

This species inhabits Yunnan Province, which includes Deqin, Weixi, Lijiang, Lanping, and Yunlong. The population is about 700 to 1,150 individuals in 11 troops. This species also inhabits Mangkang in Tibet with a population of about 300 to 350 individuals in two troops. Thus, the total population for the Yunnan snub-nosed monkey is about 1,000 to 1,500 individuals in 13 troops (Long et al., 1996). The monkeys are found at 98°37' to 99°41'E and 26°14' to 29°20'N along the Yulong Mountains of southwest China (Long et al., 1996).

4.3.2. Habitat

The monkey is definitely associated with fir forest, with the canopy composed primarily of fir (*Abies* spp.), spruce (*Picea* spp.), evergreen oak (*Quercus* spp.), and *Rhododendron* spp. Temperatures average below zero (Celsius) for several months of the year, and snow can accumulate to over 1 m in depth. Most monkey groups live at elevations of 3,800 to 4,300 m, and they sometimes can go up to a few hundred meters above tree line (4,700 m) and across wide alpine meadows or heath patches on top of snowy mountains. Hence, other than humans, this primate species inhabits the highest elevations (Long et al., 1994, 1996).

5. REASONS FOR CHANGES IN DISTRIBUTION

The change in conditions for survival affects evolution and distribution of organisms. The current biological community in China is the inevitable outcome of long-term adaptation and historical development caused by changes in condition for survival through long-term random or non-random isolation and fragmentation.

5.1. The Effect of Himalayan Orogenesis

In the late Tertiary, the fauna in south and north China had not differentiated from each other. All belonged to the "*Hipparion* fauna." From the Haixi and Yanshan orogenesis up to the early Quaternary Period, the fauna gradually divided into southern and northern elements. In this period, the Qinling Mountains that lie across China were still low and smooth with rather hot and damp climate and flourishing forests, so animals of southern fauna were able to cross over the Qinling Mountains and infiltrate into the north. It is in this period that golden monkeys inhabited the Qinling Mountains (Hu and

Qi, 1978, Gu and Jablonski, 1989). By the middle to late Pleistocene, the Himalayan orogenesis created the Qinghai-Tibet Plateau with the plateau rising as a whole, and the Hengduan Mountains uplifting accordingly. Because of the rapid upheaval, the elevation and topographic characteristics of the Qinling Mountains became similar to the present. As a result, the mountains became a geographical barrier, preventing animals from crossing and also strongly affecting the climatic difference between the south and north. At this stage, the "*Ailuropoda-Stegodon* fauna" in southern Qinling and the "Nihewan fauna" and "*Zhoukoudian* fauna" in northern Qinling gradually replaced the "*Hipparion* fauna" (Xue and Zhang, 1994).

Upheaval of the Qinling and Hengduan Mountains caused changes in rivers, climate, and vegetation, and perhaps, by preventing animals from moving freely, created isolated species distribution patterns. The mountains became natural barriers for the snub-nosed monkey's primary dispersion.

5.2. The Effect of Glacial Climate in the Quaternary Period

In China there were three long ice ages and corresponding interglacials. During glacial advances, some animals migrated from north to south and from high mountains to low. During interglacial periods, those animals returned to their original distribution areas. However, some animals could not return to their original areas. This created a disjunctive distribution pattern of isolated populations, which was similar to some plants (Liu, 1985). This theory has led many to believe that the golden monkey is a relic species, like the giant panda.

Glaciers in China were confined to mountainous regions, and although there were glaciers at the top of mountains, flourishing forests did exist in valley regions with warm climates. With the melting of ice and snow during interglacial periods, the distribution of plants moved into higher elevations, and, accordingly, animals moved upward too. Possibly, the monkeys adapted to glacial climate by vertical migration through heterogeneous valley complexes, becoming the primate species found at the highest elevations (Jablonski, 1998a, 1998b).

5.3. The Effect of Historical Changes in Climate

The climate in China was warm 5,000 years ago, and most regions in north China were subtropical. Carbonized bamboo segments discovered in Longshan Cultural Relics indicates that bamboo was distributed over the Yellow River Valleys. *Rusa* spp. and *Rhizomys sinensis* found in Banpocun Primitive Society Relics near Xi'an demonstrate that the climate in the north was warm and damp. In Yinxu relics (Anyang, Henan Province), bones of several mammals such as *Tapirus sinensis*, *Bubalus* spp., and *Rhizomys sinensis* were excavated; these animals are currently distributed only over tropical and subtropical regions (Zhu, 1986). Because of the warm climate and luxuriant forests, snub-nosed monkeys increased rapidly during this period and became widely distributed. With the following warm-cold interchanging of climate, the warm period became shorter while the cold period became lengthened. The coldest climate occurred between the Ming and Qing dynasties, which greatly affected the structure and distribution of vegetation (Zhu, 1972). As a result, the forests in the north retreated to the south and the climate became drier and colder, which compressed the distribution area of the monkey.

5.4. The Strong Impact from Human Activity in the Last 400 Years

The distribution of the snub-nosed monkey in China has been sharply compressed since the Qing Dynasty and the founding of the Republic only partly because of environmental factors. However, the impact caused by direct or indirect human activity was likely more important. Most of the monkey population disappeared at a great rate in south, southwest, and southeast China during the last 400 years. This sad scenario probably relates to various palpable human activities and social events in recent Chinese history.

5.4.1. Increasing Population Density

The demographic diagram of the regions (Figure 3) related to the snub-nosed monkeys illustrates that humans in the south and southeast areas, such as Zhejiang,

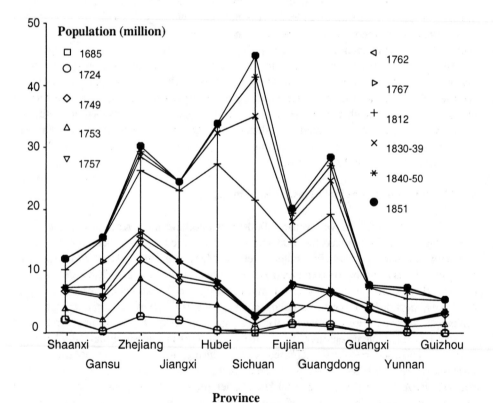

Figure 3. The demography of several provinces where the snub-nosed monkeys were found from 1685 to 1851 (Liang, 1980).

Jiangxi, Hubei, Fujian, and Guangdong, increased rapidly in the 17th and 18th centuries (from 1685 to 1767). This first peak corresponded with the disappearance of monkey populations in the same regions (Figure 2). Another interesting finding is the second peak in the 19th century. Human populations in some regions, such as Hubei and Sichuan, expanded dramatically from 1812 to 1851. This growth rate coincided with the second reduction of the animals' populations in these regions (Liang, 1980). The density of human populations in Sichuan is even more impressive. In 1990 the number was more than 100 million (Shou, et al., 1990), at present it is 110 million (including Chongqing) (Chinese Statistics Bureau, 1993 and 1995). The human population explosion in Sichuan from the 19th century may explain why there has been a dramatic drop in the number of monkeys in this province. Although the human population in Guizhou did not explode as in the other provinces before the last century, this figure grew dramatically after the 1970s. This, as with the case in Sichuan, may be closely associated with the disappearance of most monkey populations in the province. A rapid increase of human density must have resulted in more demands for residential and commercial purposes and an ever deteriorating environment. Animals are always the losers in competition for natural resources. They either die out or are restricted to remote regions that are less favored by humans.

5.4.2. Increasing Economic Activity and Resource Consumption

During the field work, we noticed that most of the local people, especially in remote regions, still maintain traditional ways of earning money by selling logs and firewood in the markets, by trading resources for other living requirements, and by burning wood for cooking and other energy-related purposes. These practices can be traced back to the beginnings of Chinese history. In other words, locals sharing the same environment with animals have to depend on the forest for survival and development. Statistics for the last 10 years indicate that about 40,000 m^3 of logs were removed from the Qinling Mountains annually (communication from the Institute of Forestry Research and Design of Shaanxi). According to information from the Chinese Agriculture Department (1981), "the annual log production before 1950 was 5,670,000 m^3. This figure increases to 50 million m^3 each year, which is close to ten times more than what it was."

Cultivation was another factor causing the disappearance of the forests. Figure 4 illustrates the rapid growth rate of the expansion of land cultivation between 1661 to 1887 in the places containing habitat of the snub-nosed monkeys. Increases in cultivated lands in these regions rapidly accelerated from the 17th to the end of the 19th centuries. In the course of agriculture, as a consequence of human activities, habitat fragmentation caused geographic isolation between various monkey populations (Dong and Tong, 1993). This happened not just between species, but also between populations within species. Limited gene exchange makes new adaptation less likely and decreases population density, especially for the Yunnan species. In Yunnan Province, humans continue to burn down the forest and grass for fertilizing the plantations. This activity obviously destroys the environment inhabited by the monkeys and other animals.

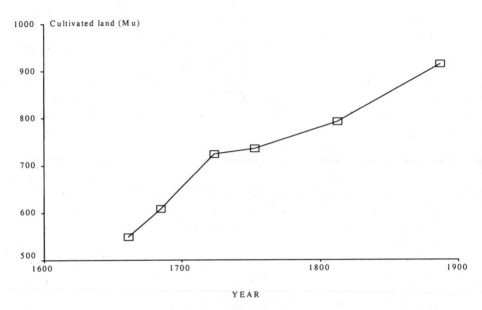

Figure 4. The increasing amount of cultivated land from 1661 to 1887 in the regions where the snub-nosed monkeys used to be found in China (Liang 1980). Mu is a scale used to measure land in China and is equal to 1/15 of a hectare.

5.4.3. Accelerating Deforestation

Deforestation was mainly caused by wars between nations and regional wars between minority groups in recent Chinese history. According to Cou (1998), some big battles, such as those during the Opium Wars between Britain and China from 1840 to 1842 and the Taiping Tianguo (farmer rebellions against the government), which began in 1850 and lasted more than ten years in the regions of southeast China and the areas around Yangtze River, resulted in great losses of humans and animals and in serious environmental damage. During the wars, timber was used frequently to set up the camps (fens and castles) and to make weapons. Fire was a very useful way to defeat enemies in mountainous regions and forests were frequently burned (Dong and Tong, 1993). During the anti-invasion war between China and Japan from 1937–1945 and the civil war between the Communist and National Parties from 1945–1949, modern weapons of destruction (bombs, planes, and heavy cannon) were used in huge battles that occurred in the regions inhabited by the monkeys and other native fauna. These battles caused serious environmental damage and obviously accelerated the disappearance of the animals in south, southeast, and central China.

5.4.4. Hunting

As in the cases of the tiger and other animals used widely in medicine, the snub-nosed monkeys were regarded as 'Rong Gu,' a valuable source of medicinal treatment with wonderful healing qualities. Their bones were considered as valuable as the antlers of deer, which are believed to cure many diseases, for instance, rheumatism, while the skin was used for decoration. It was said that only the officials of the second rank in the federal government were eligible to have a cushion made of monkey's skin (Li et al., 1989).

6. PRESENT STATUS

6.1. Survival of the Sichuan Snub-Nosed Monkey in the Qinling Mountains

The entire population of the Sichuan snub-nosed monkey lives in a restricted area of the five counties of Taibai, Zhouzhi, Foping, Yangxian, and Ningshaan along the main ridge of the Qinling Mountains in the Shaanxi Province of central China (107° 24' to 108° 27'E; 33° 32' to 33° 57'N) (Figure 5). Five nature reserves were established in those counties between 1965 and 1993. The population has been estimated to consist of 3,800 to 4,000 individuals, of whom 2,200 to 2,300 are living within the reserves, and the remaining 1,700 to 1,800 individuals outside the reserves. The population is fragmented into 39 separate troops, each averaging about 100 individuals (Table 5). Of 39 troops, 12 live on the northern slope of the main ridge, while the others live on the southern slope (Chen et al., 1982; Wu et al., 1986; Li and Liu, 1994; Li et al., 1995; Liu and Gao, 1995).

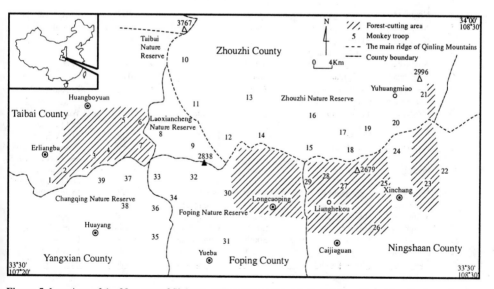

Figure 5. Locations of the 39 troops of Sichuan snub-nosed monkeys in the Qinling Mountains. Each number in the figure shows the location of a troop.

Table 5. The distribution of the Sichuan snub-nosed monkey in the Qinling Mountains of Shaanxi Province.

No.	County	District	North/ South slope	Protected by the nature reserve	Population (estimated)	Source
				Location		
1	Taibai	Erlangbaxiang	South		150	Liu and Gao (1995)
2	Taibai	Erlangbaxiang	South		70	Liu and Gao (1995)
3	Taibai	Erlangbaxiang	South		80	Liu and Gao (1995)
4	Taibai	Huangboyuanxiang	South		150	Liu and Gao (1995)
5	Taibai	Huangboyuanxiang	South		100	Liu and Gao (1995)
6	Taibai	Huangboyuanxiang	South		200	Liu and Gao (1995)
7	Taibai	Huangboyuanxiang	South		100	Liu and Gao (1995)
8	Zhouzhi	Huozhenzixiang	South	Laoxiancheng	150	Li and Liu (1994)
9	Zhouzhi	Huozhenzixiang	South	Laoxiancheng	150	Li and Liu (1994)
10	Zhouzhi	Huozhenzixiang	North	Zhouzhi	100	Li and Liu (1994)
11	Zhouzhi	Huozhenzixiang	North	Zhouzhi	50	Li and Liu (1994)
12	Zhouzhi	Banfangzixiang	North	Zhouzhi	80	Li et al. (2001)
13	Zhouzhi	Banfangzixiang	North	Zhouzhi	120	Li et al. (2001)
14	Zhouzhi	Banfangzixiang	North	Zhouzhi	150	Li et al. (2001)
15	Zhouzhi	Anjiaqixiang	North	Zhouzhi	100	Li et al. (2001)
16	Zhouzhi	Anjiaqixiang	North	Zhouzhi	150	Li et al. (2001)
17	Zhouzhi	Xiaowangjianxiang	North	Zhouzhi	120	Li et al. (2001)
18	Zhouzhi	Xiaowangjianxiang	North	Zhouzhi	100	Li et al. (2001)
19	Zhouzhi	Xiaowangjianxiang	North	Zhouzhi	110	Li et al. (2001)
20	Zhouzhi	Shuangmiaozixiang	North	Zhouzhi	90	Li et al. (1999)
21	Zhouzhi	Shuangmiaozixiang	North		120	Li et al. (1999)
22	Ningshaan	Lianghexiang	South		70 to 80	Li et al. (2001)
23	Ningshaan	Xinchang and Lianghexiang	South		100	Li et al. (2001)
24	Ningshaan	Xinchangxiang	South		100	Li et al. (2001)
25	Ningshaan	Xinchangxiang	South		70 to 80	Li et al. (2001)
26	Ningshaan	Xinchang and Caijiaguanxiang	South		70 to 80	Li et al. (2001)
27	Ningshaan	Caijiaguanxiang	South		100	Li et al. (2001)
28	Ningshaan	Caijiaguanxiang	South		60 to 70	Li et al. (2001)
29	Ningshaan and Foping	Caijiaguan and Longcaopingxiang	South		100	Li et al. (2001)
30	Foping	Yuebaxiang	South	Foping	60	Wu et al. (1986) and Li et al. (1995)
31	Foping	Yuebaxiang	South	Foping	100	Li et al. (1995)
32	Foping	Yuebaxiang	South	Foping	50	Wu et al. (1986) and Li et al. (1995)
33	Foping	Yuebaxiang	South	Foping	80	Wu et al. (1986) and Li et al. (1995)
34	Foping, Yangxian	Yueba and Huayangxiang	South	Foping, Chongqing	200	Li et al. (2001)
35	Yangxian	Huayangxiang	South	Chongqing	100	Li et al. (1995)
36	Yangxian	Huayangxiang	South	Chongqing	150	Li et al. (1995)
37	Yangxian	Huayangxiang	South	Chongqing	50 to 60	Li et al. (1995)
38	Yangxian	Huayangxiang	South	Chongqing	80	Li et al. (1995)
39	Yangxian	Huayangxiang	South	Chongqing	100	Li et al. (1995)

Throughout the range of the species, local people have been educated about the significance of this species and the necessity for its protection by means of public lectures, posters, and leaflets provided by the local government. Since about 1980, five nature reserves covering 13% of the total area of the five counties have been established (Table 6) (Ma, 1997). Despite these actions, the population of *R. roxellana* remains threatened by various human impacts.

6.1.2. Habitat Decrease Outside of the Nature Reserves

Sixteen troops with 1,700 to 1,800 members currently inhabit unrestricted areas outside of the nature reserves. Since the 1960s, the government has instituted four forest bureaus for the exploitation of forests, including commercial logging (Table 7). This policy has resulted in large-scale forest destruction and serious habitat loss for wildlife. *R. roxellana* have experienced habitat loss and declining food availability, forcing the monkeys out of their natural habitat to an early death (Li et al.,1999). In 1993, the Changqing Forestry Bureau was urged to stop commercial wood harvesting and to adopt a nature conservation policy, including establishing a nature reserve to protect monkey species as well as giant panda and other rare mammals. Unfortunately, because of the poor state of the local economy, the main source of revenue is commercial logging, which is still being conducted outside of the nature reserves.

6.1.3. Isolation of Monkey Troops

A 12-year field survey of two troops in the Yuhuangmiao region has revealed that no animals migrated between the troops. This was most likely attributed to the fact that the two troops are isolated by human-made barriers, such as farms, bush, villages, and rivers (Li and Liu, 1994). Of the 39 troops, many are isolated by unfavorable habitat, by farmland, by roads, and by mines, all of which serve to fragment natural habitats and present barriers that the monkeys are unlikely, or even unable, to cross. As a result of commercial logging, some forest habitat has been reduced to bush. These areas are usually dominated by *Lespedeza formosa, Carpinus turczaninowii C. cordata* var. *mollis, Rosa sertata, Euonymus giraldii, Philadelphus incanus,* and *Abelia macrotera,* all species that arrived and spread freely after cutting.

Table 6. The distribution of nature reserves established for the protection of the Sichuan snub-nosed monkey in the Qinling Mountains, Shaanxi Province.

Nature reserve	Location (county)	Established year	Area (ha)	Main wildlife species
Zhouzhi	Zhouzhi	1985	52,931	Golden monkey, Takin, and their habitats
Foping	Foping	1980	29,240	Giant panda, Golden monkey, Takin, and their habitats
Laoxiancheng	Zhouzhi	1993	12,380	Giant panda, Golden monkey, and their habitats
Changqing	Yangxian	1993	29,906	Giant panda, Golden monkey, and their habitats
Taibai	Taibai, Zhouzhi, Meixian	1965	56,325	Montane forest ecosystem and wildlife

Table 7. The commercial logging information from four forestry bureaus of Shaanxi Province (to 1995).

Forestry bureau	Establishe d year	Managing area (ha)	Building driveway (km)	Cutting woods (m³)	Forest cutting area (ha)
Taibai	1963	128,220	605.60	2,920,800	23,240
Ningxi	1966	82,720	393.60	3,085,300	13,330
Changqing	1968	29,910	286.67	1,770,500	4,306
Longcaoping	1980	20,040	52.60	490,800	1,914

6.1.4. Poaching

Hunting protected wildlife is prohibited by the "Wildlife Protection Law of the People's Republic of China." Unfortunately, this law has rarely been enforced in remote areas. Even though the local people know that the monkey is strictly protected, they sometimes kill them in the course of hunting unprotected animals for provisions, medicine, and fur. In addition, poachers hunt and kill the monkeys for money. For example, two monkeys were snared by poachers in Zhouzhi Nature Reserve in 1992. Since 1980, an estimated 50 to 70 individuals have been killed in the Qinling Mountains.

6.1.5. Impact of Tourism

The Qinling Mountains are an important resource for nature-related tourism because of the attractive landscape, historical heritage, and biodiversity. During the 1980s and 1990s, visitor numbers (mostly domestic and occasionally from abroad, including Japan, USA, and Germany) increased annually. Five thousand tourists visited the region in 1985; by 1995, that number had increased to 50,000 each year (Zheng, 1997). Tourists have contributed significantly to the local economy, and, as a result, several counties and nature reserves have embarked on developing tourism programs that have included the rushed construction of roads and modern hotels. Considerable quantities of rubbish are produced by the tourist industry and much of it is dumped locally. Taibai Mountain tour, for example, has caused serious environmental pollution, which may affect the fragile forests that wildlife is dependent upon.

6.2. The Main Threat to the Guizhou Snub-Nosed Monkey

The habitat for this animal is montane subtropical forest between 1,200 and 2,100 m elevation. Individual monkeys range over more than 10 km² of habitat and apparently need a large area of intact forest to survive throughout the year. While hunting with guns is no longer a serious threat to these animals, snares set for other wildlife in the reserve may still kill monkeys. However, the main threat to the survival of the Guizhou snub-nosed monkey is the continued destruction and fragmentation of their forest habitat within the reserve (Bleisch, 1993).

Fanjing Mountain is located in an area of China with a very low per capita income. The park is home to over 1,000 families and is surrounded by villages on all sides. Many local people obtain most of their cash by producing charcoal within the reserve's buffer zone, by collecting medicinal plants, and, sometimes, by poaching protected timber or wildlife. With limited resources and training and spotty cooperation from local authorities, the management of the reserve has not been able to cope with these pressures. Destruction of forest in Fanjing Mountain has now reached a crisis level (Bleisch, 1993).

In fall and winter, Guizhou golden monkeys appear to favor magnolia flower buds, although *Magnolia* spp. are not common in the forest. This suggests that magnolia flower buds are a critical component of the diet of the monkeys in winter, a season when monkeys may face a scarcity of nutritious food resources. Unfortunately, these buds are also sought by local collectors, who collect them for market value as a source of medicinal oils. If magnolia flower buds are a critical resource for the monkeys, the new competition could have a detrimental effect on the monkeys' ability to meet their energy needs at a time of energy shortage. Lopping and felling of magnolias by bud collectors means that there may be a lasting shortage of magnolia buds in years to come (Bleisch, 1993).

Habitat destruction by local people is the most serious threat to survival of this species. In most areas of the buffer zone, collection of firewood, timber, and wood for charcoal production is occurring at unsustainable rates. As wood is removed from the forest faster than the remaining trees can regenerate, the forest is degraded. Eventually, only a scrub dominated by dwarf bamboo, coarse grasses, and bracken remains. Monkeys cannot survive in these degraded habitats (Bleisch, 1993).

Forest at lower elevations of Fanjing Mountains is under the most severe threat from deforestation, since these forests are most accessible to villagers who live below. Unfortunately, these lower forests may also be critically important for the survival of the Guizhou snub-nosed monkeys. However, the monkeys do not use low-elevation habitats after severe winter storms. If these special refuges are destroyed by tree cutting, the population may be severely affected in the future (Bleisch, 1993).

6.3. Survival of the Yunnan Snub-Nosed Monkey

Of the 13 troops, only home ranges of two troops (Yiyong and Cikatong in Deqin) are totally within Baimaxueshan Nature Reserve, the only formal nature reserve set up for protecting the monkey (Table 8). Some of the Wuyapuya (in Deqin) troop's home range and most of the Guomorong troop's home range (in the border area between Deqin and Weixi) were outside of the reserve. Xiaochangdu and Milaka troops are within the Honglaxueshan Protected Area, an informal nature reserve in Mangkang. Jinsichang and Dapingzi troops are within the Jinsichang-Laojunshan Protected Area, an informal nature reserve in Lijiang. The other five troops (Bamei and Adong troops in Deqin, Heishan and Neidaqin troops in Lanping, and Longmashan troop in the border area between Lanping and Yunlong) are outside of the nature reserves and have received little protection (Long et al., 1996).

6.3.1. Poaching

Poaching is the greatest threat to the survival of the species. The golden monkey might be able to survive fragmentation if poaching of both the resources and monkeys

Table 8. Locations of the 13 troops of the Yunnan snub-nosed monkey (Long et al., 1996).

No.	Site	Latitude and Longitude	Population	Elevation (m)
1	Xiaochangdu	29°20′N, 98°37′E	>200	3800 to 4300
2	Milaka	29°06′N, 98°45′E	<150	3700 to 4200
3	Bamei	28°54′N, 98°46′E	<50	3700 to 4300
4	Adong	28°43′N, 98°55′E	<50	3600 to 4200
5	Wuyapuya	28°29′N, 99°06′E	>200	3500 to 4400
6	Yiyong	28°20′N, 99°07′E	<50	3400 to 4200
7	Cikatong	28°02′N, 99°03′E	<150	3200 to 4100
8	Guomorong	27°40′N, 99°15′E	<200	3300 to 4000
9	Jinsichang	26°52′N, 99°34′E	<150	3300 to 4000
10	Heishan	26°42′N, 99°19′E	<50	3100 to 3600
11	Dapingzi	26°40′N, 99°41′E	<50	3200 to 3900
12	Neidaqin	26°26′N, 99°21′E	<100	3000 to 3400
13	Longmashan	26°14′N, 99°15′E	<100	3000 to 3500

were not a factor. Local people in the monkey's distribution area are composed of Tibetan, Lishu, Yi, Naxi, Bai, and Pumi minorities. All these ethnic groups traditionally hunt. Since the 1960s, various modern hunting weapons give the local poachers more power to kill the animals in the forest. Once the habitat is destroyed, monkeys have to face the hungry poachers' smoky guns and are doomed. Since the home range of the monkey is extensive and human transportation is difficult in the area, staff members of the reserves or local forestry departments have problems with law enforcement. Thus, poaching is difficult to detect and to control (Long et al., 1996).

6.3.2. Habitat Fragmentation

Although birth control has been the basic national policy in China for about two decades, the local people population in the monkey's distribution area is still increasing (1.21% per year) primarily because the birth control policy is difficult to implement in remote areas. Currently, the population is over 300,000 in the snub-nosed monkey range. The population increase certainly results in accelerated consumption of various forest resources and, in turn, results in more deforestation (Long et al., 1996).

Because local economy is poor, local government has to depend on commercial logging as the main revenue source. This is the reason for much of the deforestation. The fir-forest is vulnerable since fir grows very slow due to the high elevation and low temperature. Its natural afforestation typically requires more than 100 years. Thus, it is believed that the forest might vanish forever once it is cut away (Long et al., 1996).

6.3.3. Local Peoples' Awareness of the Monkey

Even today, very few people know about the snub-nosed monkey because of few reports by public media. Many poachers do not know that they are eating the most beautiful animals in the world. That only a few staff members of the local forestry department, who are supposed to implement the conservation duties in the area, know about the monkeys makes the situation very serious (Long et al., 1996).

7. SUMMARY

The distribution of the snub-nosed monkey is changing by shifting ranges from low mountains to high mountains, from tropical and subtropical forest to temperate forest, and from eastern and southern China to western and northwestern China. This is attributed to historic, orogenesis, climate change, and especially human activity impacts. If the current rate of fragmentation does not slow down in the near future, there is a strong possibility that the snub-nosed monkeys will be further reduced in China and will probably entirely disappear. The most threatened species are those in Guizhou and Yunnan. Their dramatic shrinking populations have resulted in less genetic diversity and reduced breeding. This obviously limits population increase and weakens resistance to disease. Furthermore, because of the increasing impact of human activities such as poaching, monkeys are now confined to resource limited habitat (Bleisch et al., 1993; Kirkpatrick et al., 2001; Li et al., 2000). Guizhou species may disappear first, followed by the Yunnan species. The Sichuan species may survive a little longer since they have relatively large populations and a wide distribution.

8. ACKNOWLEDGMENTS

We extend our sincere thanks to Professor Jim Moore for revising the early version of the manuscript, Doctors Noel Rowe and Ardith A. Eudey and Professors Kazuo Wada and Kunio Watanabe for supporting the project. We also wish to thank Professor Charles Oxnard and Doctor Len Freedman for kind support on the study and Doctor Zhang Y. for his help in fossil references. The study was supported by a grant from the National Natural Science Foundation of China (1999, No.39970116), Primate Conservation Inc., USA (1998), and Shaanxi Education Commission Fund of China (2001).

9. REFERENCES

Allen, G. M., 1938, *The Mammals of China and Mongolia*, American Museum of Natural History, New York.
Bleisch, W. V., Cheng, A., Re, X., and Xie, J., 1993, Preliminary results from a field study of wild Guizhou snub-nosed monkeys (*Rhinopithecus brelichi*), *Folia Primatol.* **60**:72–82.
Bleisch, W. V., 1993, Management recommendations for Fanjing Mountain Nature Reserve, unpublished manuscript.
Boonratana, R., and Le, X. C., 1998, Preliminary observations of the ecology and behavior of the Tonkin snub-nosed monkeys (*Rhinopithecus [Presbytiscus] avunculus*) in Northern Vietnam, in: *The Natural History of the Doucs and Snub-nosed Monkeys*, N. G. Jablonski, ed., World Scientific Publishing, Singapore, pp. 207–215.
Chen, F. G., Min, Z. L., Gan, G. F., Luo, S. Y., and Xie, W. Z., 1982, The resource and conservation of golden monkey, *Chinese Wildlife* **2**:7–10.
Chinese Agriculture Department, 1981, Annual report of the Chinese agriculture, Agriculture Publisher, pp. 15.

Colbert, E. H., and Hooijer, D. A., 1953, Pleistocene mammals from the limestone fissures of Szechuan, China. Bulletin of the Museum of Natural History **102**:1,024.

Cou, Y. Z., 1998, *Historic Events in Chinese History*, Zhejiang Education Publisher, pp. 864–1,057.

Dong, Z. Y., and Tong, X. F., 1993, *Chronological Change of the Vegetation in China*, Forest History Association, Chinese Forest Council, Jiling News Press Publishing Bureau, pp. 222–225.

Gu, Y., and Jablonski, N. G., 1989, A reassessment of *Megamacaca lantiangensis* of Gongwangling, Shaanxi Province, *Acta Anthropology Sinica* **8**:343–346.

Gu, Y., and Hu, C., 1991, A fossil cranium of *Rhinopithecus* found in Xinan, Henan Province, *Vert. Palasiatica* **29**:55–58.

Han, D., 1982, Mammalian fossils from Daxin county, Guangxi Province, *Vert. Palasiatica* **20**:58–64.

Hu, C., and Qi, T., 1978, Gongwangling Pleistocene mammalian fauna of Lantian, Shaanxi, *Palaeontol. Sin., New Ser. C.* **21**:1–64.

Hu, J., Deng, Q., Yu, Z., Zhou, S., and Tian, Z., 1980, A study on the ecological biology of Giant panda, Golden monkey, and Takin, *J. Nanchong Normal College* **2**:1–19.

IVPP (Institute of Vertebrate Paleoantology and Paleoanthropology), 1979, *Handbook of Chinese Paleovertebrate Fossils*, Science Press, Beijing.

Jablonksi, N. G., 1998a, The evolution of the doucs and snub-nosed monkeys and the question of the phyletic unity of the odd-nosed colobines, in: *The Natural History of the Doucs and Snub-nosed Monkeys*, N. G. Jablonski, ed., World Scientific Publishing, Singapore pp. 13–52.

Jablonski, N. G., 1998b, The response of catarrhine primates to Pleistocene environmental fluctuations in East Asia, *Primates* **39**:29–37.

Jablonski, N. G., and Pan, Y. R., 1988, The evolution and paleobiogeography of monkeys in China, in: *The Paleoenvironment of East Asia from the Mid-Tertiary*, J. S. Aigner, N. G. Jablonski, G. Taylor, D. Walker, and W. Pinxian, eds., Centre of Asian Studies, University of Hong Kong, pp. 849–867.

Jablonski, N. G., and Peng, Y. Z., 1993, The phylogenetic relationships and classification of the doucs and snub-nosed langurs of China and Vietnam, *Folia Primatol.* **60**:36–55.

Kirkpatrick, R. C., 1998, Toward a gazetteer of the snub-nosed and douc langurs, in: *The Natural History of the Doucs and Snub-nosed Monkeys*, N. G. Jablonski ed., World Scientific Publishing, Singapore. pp. 337–372.

Kirkpatrick, R. C., Zou, R. J., and Zhou, H. W., 2001, Digestion of selected foods by Yunnan snub-nosed monkey *Rhinopithecus bieti* (Colobinae), *Am. J. Phys. Anthrop.* **114**:156–162.

Li, B., Chen, C., Ji, W., and Ren, B., 2000, Seasonal home range changes of the Sichuan snub-nosed monkey (*Rhinopithecus roxellana*) in the Qinling Mountains of China, *Folia Primatol.* **71**:375–386.

Li, B., He, P., Yang, X., Wei, W., Ren, B., Yang, J., Si, K., and Liu, Y., 2001, The present status of the Sichuan snub-nosed monkey in the Qinling Mountains of China and a proposed conservation strategy for the species, *Biosphere Cons.* **3**:107–114.

Li, B., and Liu, 1994, On the Sichuan snub-nosed monkey in the north slope of Qinling Mountains, *Monkey* **386**:3–10.

Li, B., and Pan, R. L., 2002, Extinction of the snub-nosed monkey (*Rhinopithecus*) in China during the past 400 years, *Int. J. Primatol.* **23(6)** (in press).

Li, B., Ren, B., and Gao, Y., 1999, A change in the summer home range of Sichuan golden monkeys in Yuhuangmiao, Qinling Mountains, *Folia Primatol.* **70**:269–273.

Li, B., Shen, W., Chen, F., 1989, Textual criticism of historical name of the Golden monkey, in: *Progress in the Studies of the Golden Monkey*, F. Chen, ed., Northwest University Publishing Press, Xi'an, pp. 30–32.

Li, B., Xiong, C., and Li, Z., 1995, The distribution patterns of Sichuan golden monkey and Rhesus monkey in Shaanxi, China, *J. of Pro Natura* **4**:125–137.

Li, Z., Ma, S., Hua, C., and Wang, Y., 1981, Distribution and habitat of golden monkey in Yunnan, *Zool. Research* **21**:9–18.

Liang., F. Z., 1980, *The Statistic Material of Ancient Chinese Population and Land*, Shanghai People's Publisher pp. 258–380.

Liu, S., 1985, Climate change and plant migration, Thesis, Science Press, Beijing, pp. 65–73.

Liu, S. F., and Gao, Y. F., 1995, Analysis of the distribution and changing numbers of golden monkey in the Qinling Mountains, in: *Primate Research and Conservation*, W. P. Xia and Y. Z. Zhang, eds., China Forestry Publishing Press, Beijing, pp. 191–196.

Long, Y., Kirkpatrick, C., Zhong, T., and Xiao, L., 1994, Report on the distribution, population, and ecology of the Yunnan snub-nosed monkey, *Primates* **35**:241–250.

Long, Y., Kirkpatrick, C., Zhong, T., and Xiao, L., 1996, Status and conservation strategy of the Yunnan snub-nosed monkey, *Chinese Biodiversity* **4**:145–152.

Ma, A., and Tang, H., 1992, On discovery and significance of a Holocene *Ailuropoda-Stegodon* fauna from Jinhua, Zhejiang, *Vert. Palasiatica* **30**:295–312.

Ma, N. X., 1997, The features of biodiversity and the construction of nature reserve in Shaanxi Province, in: *Studies on the Resource, Environment, and City*, S. Yie, ed., Northwest University Publishing Press, Xi'an, pp. 30–35.

Napier, J. R., and Napier, P. H., 1967, *A Handbook of Living Primates: Morphology, Ecology, and Behavior of Nonhuman Primates*, Academic Press, New York.

Pan, Y. R., 1995, Fossil primates discovered in China, in: *Primate Research and Conservation*, W. Xia and Y. Zhang, eds., China Forestry Publishing House, Beijing, pp. 99–105.

Pan, Y. R., and Jablonski, N. G., 1987, The age and geographical distribution of fossil cercopithecids in China, *Hum. Evol.* **2**:59–69.

Pan, W. S., and Yong, Y., 1989, The biology of the golden monkey, in: *Progress in the Studies of the Golden Monkey*, F. Chen, ed., Northwest University Publishing Press, Xi'an, pp. 3–7.

Ren, R. M., Kirkpartrick, C., Jablonski, N. G., Bleisch, W. V., and Le Xuan, C., 1998, Conservation status and prospects of the snub-nosed langurs (Colobinae: *Rhinopithecus*), in: *The Natural History of the Doucs and Snub-nosed Monkeys*, N. G. Jablonski, ed., World Scientific Publishing, Singapore, pp. 301–314.

Sichuan Wildlife Resource Investigation Conservation Management Station, 1997, Status and conservation strategy of the Sichuan golden monkey in Sichuan province, (personal communication).

Shou, X. H., Li, X. F., and Suan, Z. Y., 1990, *The Handbook of Social Material in China*, Reference Publisher in Social Science, pp. 480–1,304.

State Statistic Bureau of China, 1993, *The Demography in China*, Statistic Publisher of China, Beijing, pp. 354.

Statistical Bureau of China, 1995, *The Yearbook of Statistic Material of China*, Statistical Publishing Press, Beijing, pp. 46–48.

Tan, B. J., 1985, The status of primates in China, *Primate Cons.* **5**:63–77.

Wang, L., Lin, Y., Chang, S., and Yuan, J., 1982, Mammalian fossils found in the northwest part of Hunan Province and their significance, *Vert. Palasiatica* **20**:350–358.

Wu, J. Y., Han, Y. P., Yong, Y. G., and Zhao, J., 1986, The mammals in Foping Nature Reserve, *Chinese Wildlife* **3**:1–4.

Xue, X., and Zhang, Y., 1994, Zoogeographical divisions of Quaternary mammalian faunas in China, *Acta Theriologica Sinica* **14**:15–23.

Zhang, T., 1995, Population and conservation of the snub-nosed monkey on the northern slope of the Qin Rause, Gansu, China, in: *Primate Research and Conservation*, W. Xia and Y. Zhang, eds., China Forestry Publishing House, Beijing, pp. 138–142.

Zhang, Y., and Ryder, O. A., 1998, Mitochondrial cytochrome b gene sequences of langurs: Evolutionary inference and conservation relevance, in: *The Natural History of the Doucs and Snub-nosed Monkeys*, N. G. Jablonski, ed., World Scientific Publishing, Singapore, pp. 65–71.

Zheng, S. W., 1997, Diversity of wildlife species and conservation management of rare and endangered animals in northwest China, in: *Wildlife and Conservation*, M. T. Song, ed., Shaanxi Scientific and Technological Press, Xi'an, pp. 95–102.

Zhao, Z., and Li, X., 1981, Human fossils and associated fauna of Jiulengshan hill, Guangxi, *Vert. Palasiatica* **19**:45–54.

Zhu, K., 1972, Preliminary study on the climatic change in China in recent 5,000 years, *Acta Archaeologica Sinica* **1**:15–38.

Zhu, Z., 1986, *The Natural History of Shaanxi Agriculture*, Shaanxi Science Publishing Press, Xi'an, pp. 366–448.

ANALYSIS OF THE HYPOTHETICAL POPULATION STRUCTURE OF THE SQUIRREL MONKEY (*SAIMIRI OERSTEDII*) IN PANAMÁ

Ariel R. Rodríguez-Vargas[*]

1. INTRODUCTION

1.1. Taxonomy, Distribution, and Natural History

The Central American squirrel monkey (*Saimiri oerstedii*) is an endemic species from the coast of the Pacific Ocean to southeast Costa Rica and southwest Panamá (Hershkovitz, 1984). Three common species (*S. ustus, S. sciureus,* and *S. boliviensis*) are found in South America and are distributed in the tropical forests of the Guayanas, the Amazon basin, the high Orinoco, and the high Magdalena (Hershkovitz, 1984). Nine taxas are distinguished at the level of subspecies, of which two correspond to the Central American species (*Saimiri o. oerstedii* and *Saimiri o. citrinellus*) (Hershkovitz 1984). In Panamá only the subspecies *S. o. oerstedii* is found. The squirrel monkey is exclusively arboreal and uses a great variety of habitat (Figure 1). It mainly exploits the lower and middle forest canopy. This primate forms the largest and most cohesive groups unlike any other Neotropical primate (Kinsey, 1997). They are mainly insectivorous and frugivorous (Kinsey, 1997).

Panamá, in spite of being a Central American nation with eight primate species, lacks information on the original and current distribution of its primates. The majority of the maps published with the distributions of the taxas of interest in Mesoamerica are considered inexact because they lack up-to-date field data (Rodríguez-Luna et al., 1996a and b). The exact distribution of the squirrel monkey was described for the first time by Rodríguez (1999a) in the southwest part of the Republic of Panamá. Bangs (1902) determined the presence of the squirrel monkey on "the lowest slopes of the Talamanca Mountain Range." Other authors, such as Bennett (1968), Baldwin and Baldwin (1971, 1972, 1976, 1981), and Rodríguez (1996b), among others, have given information about

[*] University of Panamá, Center for Studies of Biotic Resources (CEREB), Exact and Technology Natural Sciences Faculty, 223-0212 Panamá City, Panamá, email: arielrdrz@yahoo.com.

Primates in Fragments: Ecology and Conservation
Edited by L. K. Marsh, Kluwer Academic/Plenum Publishers, 2003

Figure 1. Juvenile squirrel monkey of Panamá. Photo by Ariel R. Rodríguez-Vargas.

the presence of the squirrel monkey up to the most recent and complete data by Rodríguez (1999a and b).

Since 1970, the squirrel monkey has been included in the Threatened and Endangered species list by the U.S. Fish and Wildlife Sercvice (USDI, 1984). In 1980, the National Environmental Authority for the Republic of Panamá included this species under the category "Danger of Extinction" (RENARE, 1980). The International Union for Nature Conservation (IUCN, 1995) included it under the category of "Endangered." Since 1975 the International Agreement on Threatened Forest Species of Fauna and Flora (CITES I) has regulated the international commerce of this species (CITES, 1991). The latest revision of conservation subcategories of the IUCN (1996) for *S. o. oerstedii* in Panamá determined the status as "Endangered (EN)/B1+2abcde, C2a)," meaning that the taxa is confronting a high risk of extinction in their wild state in the near future. This status is characterized by their area of occupation of less than 500 km^2 and their area of occurrence of less than 5,000 km^2, a severely fragmented population, a notable reduction of the area and quality of the habitat, a reduction of the number of subpopulations, and a reduction of the number of reproducing individuals.

Since squirrel monkey populations are poorly studied in the region, the general objective of this chapter is to analyze, interpret, and discuss the reasons for the assignation of hypothetical population structures and their implication in strategies for the conservation and management of the squirrel monkey.

2. STUDY AREA

The area includes all the extreme southeast of the Province of Chiriquí. It covers an area of approximately 3,500 km². It includes all the territory of the following districts: Alanje, Barú, David, Renacimiento, Dolega, and Boquete and part of Bugaba (Figure 2).

3. METHODOLOGY

3.1. Determination of the Metapopulations and Subpopulations

Based on the definitions of subpopulation and metapopulation (Table 1), I used empirical evidence of movement patterns and predictions of increased movement patterns under enviromental conditions of scarce resources, and applied them to squirrel monkey's living in Chiriquí. I wish to clarify that we are dealing with population models and they should be viewed from that perspective, especially because in this study I have not been able to prove that there is an effective flux of individuals between the subpopulations

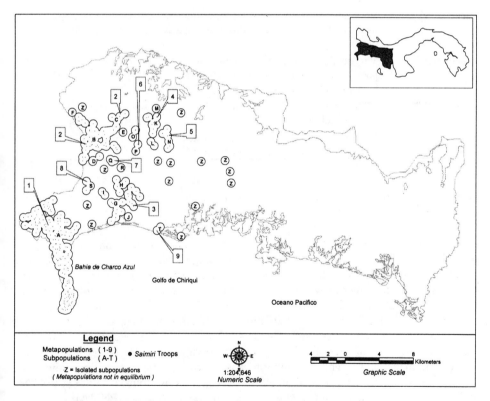

Figure 2. Location and identification of the metapopulations and subpopulations of the squirrel monkey. The distribution of this species is limited to the extreme southwest of the Isthmus of Panamá. See Table 2 to identify names of metapopulations shown here.

Table 1. Definitions of theoretical metapopulation classes.

Metapopulation classes	Definition	Reference
Classic	Population grouped in patches where there is continuous recolonization and extinction.	Levins, 1970; Harrison, 1991
Continent	Highly connected, high population density.	Harrison, 1991
Continent-Continent	Two or more large populations with high density but only slightly connected.	Stith et al., 1996
Disjunt	Two or more populations, small sized populations, but poorly connected.	Stith et al., 1996
Continent-Island	Large area with a primary population with small patches around it. The population is maintained through dispersal.	Harrison, 1991
Subcontinent-Island	Population nucleus significant, but not large enough to contribute to immigration for a long term.	New definition based on this study (see discussion section)
Subcontinent	Population large and stable but virtually isolated.	Stith et al., 1996
Island	Very small and highly isolated population (in the case of squirrel monkey this means at least two troops).	New definition based on this study (c.f. Section 5)
Island-Island	Very small and highly isolated population, poorly connected to another very small and highly isolated population.	New definition based on this study (c.f. Section 5)
Semi-Island	Small isolated population, similar to a "true" island condition.	New definition based on this study (c.f. Section 5)
Non-equilibrium	Very tiny populations completely isolated with high risk of extinction in the immediate future.	Harrison, 1991

within a metapopulation. Thus, I calculate the probable population conformation of the squirrel monkey in Panamá.

Through the use of the program ArcView, Version 3.0 (ESRI, 1990), I determined the probable maximum radius for the dispersal of individuals in each one of the Chiriquí troops. I used 257 georeferenced points corresponding to the local groups and calculated the model with the "buffer area" command. The maximum radius of range for all squirrel monkeys in Central America was used as a reference. According to a long-term study in Parque Nacional Corcovado (PNC), Costa Rica, Boinski (1987) found that squirrel monkeys in a continuous forest area utilized a maximum range of 176 hectares. I increased the radius of dispersion by 35% (238 hectares) to delimit the expected maximum range of dispersion of a troop or individuals for the conformation of a subpopulation structure. This assumption was based on the data on maximum and minimum perimeters of range determined for this species by Boinski (1987) at the PNC and by Wong (1990a and b) at the Parque Nacional Manuel Antonio (PNMA).

If a squirrel monkey troop increased their range by an average of 35%, then given the current conditions of discontinuous and disturbed forest cover, I hypothesize Panamá lacks two-thirds of the vital resources for their survival with respect to the areas of PNC and PNMA. Depending on the severity of fragmentation, a troop may not be able to cohesively recolonize available patches of habitat. If even one individual from a troop can move between subpopulations, even sporadically, the hypothesized principle would be achieved for the metapopulation theory of sporadic immigrations (Levins, 1970; Hasting and Harrison, 1994; Hanski 1996).

The total squirrel monkey population in Panamá is approximately 4,755 individuals wihtin 2,613.41 km^2. The mean troop size for squirrel monkeys in Panamá is 18.5 individuals, with a range between 4 to 40 (Rodríguez, 1999b). Wong (1990b) reported a mean of 41 individuals per troop in Pacific Central, Costa Rica. However, Bionski and Sirot (1997) a minimal critical value for squirrel monkey troop size is 15 individuals. To make estimates of population density for subpopulations and metapopulations, the average for combined troops in Panamá was used. Rodriguez 1999a found sex and age structurs on average to be 33% male (5.8 individuals), 45% female (7.9), 14% juveniles (2.4), and 9% infants (1.5). For this study, however, I followed Boinski and Sirot (1997) who suggest a normal troop should have 10% males, 40% females, 25% juveniles, and 25% infants.

4. RESULTS

Based on the spatial and distance distributions among troops, 23 metapopulations were distinguished, of which 14 corresponded to isolated tropos. The metapopulations ranged between two to 139 troops (Figure 2). In the same way, a total of 34 subpopulations were identified, including the 14 isolated troops. The three main subpopulations ("A," "B," and "C") are, at the same time, the nucleus of the three main metapopulations of Chiriquí (Figure 2).

The metapopulations with the largest number of troops are those in Burica, Renacimiento, Alanje, Boquerón, and David, respectively. Of these, only the first two have 800 or more adult, juvenile, and infant individuals (Table 2).

Table 2. Characterization of the squirrel monkey metapopulations in Panamá.

Metapopulation*	Structure †	Subpopulation	No. troops	Population	Density (ind./km^2)
Burica (1)	Continent	1	139	2572	8.37
Renacimiento (2)	Continent-Island	5	48	888	4.3
Alanje (3)	Subcontinent-Island	4	25	463	3.33
Boquerón (4)	Subcontinent-Island	3	11	204	3.46
David (5)	Subcontinent	1	6	111	3.29
Concepción (6)	Island-Island	2	5	93	3.18
Gariché (7)	Island-Island	2	4	74	3.58
Jacú (8)	Semi-island	1	3	56	2.81
Guarumal (9)	Island	1	2	36	3.9
Isolated Troops	Non-equilibrium	1	14	259	-

* The bold numbers identify each metapopulation in Figure 2. † See text definitions and references in Table 1. Subpopulation and No. troops indicated the total of geo-referenced troops within each metapopulation buffer area.

With respect to the estimated number of adult individuals for each metapopulation, only that of Burica was larger than 1,000 individuals. The rest of the metapopulations did not exceed 500 adult individuals. The metapopulation with the greatest population density is Burica with 8.37 individuals/km^2, followed by Renacimiento (4.30), Guarumal (3.90), Gariché (3.58), Boquerón (3.46), Alanje (3.33), David (3.29), Concepción (3.18), and Jacú (2.81). The rest of the metapopulations correspond to isolated troops with a density of 2.60 individuals/km^2. As for the subpopulations density, only seven exceed 3.50 individuals/km^2, these are: A (9.49), B (4.88), Q (4.11), T (3.90), K (3.84), O (3.75), and G (3.55) (Figure 2).

5. DISCUSSION

The application of the concept of metapopulation is important for this study according to the criteria by Wiens (1996) because it includes two ecological realities: (1) the fragmentation of the habitat system occupied and the subdivided spatial pattern of the population and (2) the prediction of a metapopulation structure that can increase the persistence of the population as a whole, in spite of the fact that many subpopulations may have disappeared. From the point of view of conservation biology, the application of the metapopulation concept should be applied to reduced populations so that their management will ensure the probability of surviving stochastic events (Gilpin and Soulé, 1986; Gilpin, 1987; Hanski, 1991). Metapopulation management strategies should involve the management of the subpopulations to maximize the probability of survival of the species (Hanski, 1991). For this reason, I firmly believe that metapopulations and subpopulations, because of their characteristics and population size, should be considered as units for management and conservation.

According to the subpopulation and metapopulation models described in the methodology, the population of the squirrel monkey in Panamá is made up of subpopulations and metapopulations where three zones stand out as priorities for the conservation of the species: the Península of Burica, Renacimiento, and Alanje, in spite of the fact that this study did not determine the real possibilities of migration between subpopulations.

Under this focus, the metapopulation of Burica, identified as "continent metapopulation," with a large number of troops is a population that may have better opportunities of survival in the long run in the Panamanian territory, if a plan for conservation for the short- or medium-term is established in this zone. Given the pattern of grouping as a single unit, I have defined it as a continent metapopulation, which means that the population is large and is grouped together in a compact way with one or no peripheral subpopulations.

The subpopulations at Renacimiento are in danger in the short-term of suffering from increased fragmentation of the population. There is a strong tendency for isolation of the peripheral groups within the subpopulation nucleus "B" (Figure 2). Therefore, it is important that population dynamic studies of the squirrel monkey in this zone be established. Based on the theory, these subpopulations seem to behave themselves as a typical continent-island metapopulation, defined by Harrison (1991) as a system of habitat fragments located within a dispersion distance of a large habitat fragment (continent) where the local populations are never extinguished. In this system we could include the theory of "drain-source" where the habitat (Pulliam, 1988), populations

(Howe et al., 1991), and continent fragment (Stith et al., 1996) has individuals arriving at the peripheral fragments, in spite of the fact that these may be constantly extinguished.

The conformation of the metapopulation of Alanje, although it seems compact, is a population highly affected by the intensive use of the soil that has provoked the almost complete disappearance of the forests that would allow a dynamic interaction of the troops present in the area. It is probable that there is little or no interaction among troops; thus, it would be important to establish immediate studies and plans for monitoring to try to save this population in Chiriquí. These troops could function as a metapopulation that I have called "subcontinent-island" by considering that the subpopulation nucleus is significant, but not sufficiently large enough to maintain a contribution in the long term of individuals immigrating toward the peripheral subpopulations. The metapopulation of Boquerón I have also considered a subcontinent-island. The metapopulation of Davíd is considered as only a "subcontinent" (Stith et al., 1996), similar to that of Boquerón but lacking in peripheral subpopulations. The metapopulation of Concepción is called "island-island" because it behaves like two isolated subpopulations, very close to each other but with a moderate probability of interchange of individuals.

For its part, the metapopulation of Jacú has been called "semi-island," since, in spite of being made up of three troops, it is a virtually isolated population. A more detailed examination of this population indicates that a high probability of an absence of metapopulation dynamics exists since the troops find themselves isolated from each other by the Río Chiriquí Viejo and Río Jacú. In spite of that, it can be called metapopulation according to the empirical criteria of Harrison (1994); a group of conspecific populations, possibly, but not necessarily, interconnected. This concept is significant because the effects of fragmentation may not diminish viability as compared to originally continuous population (Harrison, 1994). Finally, the metapopulation of Guarumal can be described as an "island" according to my criteria, based on the exclusively arboreal behavior of the squirrel monkey. This is a reduced population limited to a small area, since the landscape corresponds to an inadequate matrix for displacement and interconnection with other populations. This population has a high risk of extinction and will increase the gap between occidental and oriental populations of this species in Chiriquí.

All isolated troops I have considered are unstable metapopulations according to the criteria of Harrison (1991). Unstable populations are defined as those populations whose rate of extinction, over the long run, exceeds the rate of colonization or vice versa. An extreme case is where the local populations are so far from each other that there are no migrations between them and there are no possibilities for recolonization (Hanski and Simberloff, 1997). Brown (1987) argues that the rate of recolonization is inversely proportional to the distance that separates the isolated populations. In Chiriquí, this criteria effectively operates, given the considerable distance that separates the unstable metapopulations.

Shaffer (1981) argues that the additional problems confronting the isolated troops is an effect of the aggressive matrix of pastures and human settlements. For squirrel monkeys this means internal changes to the squirrel monkey population and high sensitivity to external pressures, leading the squirrel monkey population to a potential local extinction. These points from Shaffer should be taken seriously, particularly when deciding priorities for the relocation of troops.

Determining the locations and characteristics of the squirrel monkey population in Chiriquí may allow for a more efficient management system for forest and primate. However, under the current ecological and socioenviromental context of the province, it is not feasible to include the total population and its distribution within a conservation plan. Even if the plan was ideal, it is not always the most practical, given the variety of interests and economic resources required for a project of this magnitude. I propose that the biological conservation criteria of Wiens (1996) is adopted for the region. He established that the concept of metapopulation is useful for the design and management of reserves because it emphasizes explicit spatial relations, the importance of dispersion, and structure of the landscape. For reasons of sociopolitical nature, it is not always possible to apply these concepts to the "real world," but one should make the attempt.

Although this chapter determines metapopulations which may be the most vulnerable over time, it is important to continue research those populations while establishing conservation and management plans. In particular, research into the amount of connectedness and population migration would be a priority. Modeling lays the groundwork for field investigation, but without further data on population interactions, conservation priorities for even isolated populations becomes difficult.

6. CONCLUSION

The total squirrel monkey population in Panamá is minimal with just under 4,755 individuals. Its density reflects a notable shrinking of the population and a probable extinction of the greater part of their small distribution in Panamá. If the extrinsic adverse factors associated with human activities continue to operate, we will be faced with the gradual and continuous disappearance of the forest remnants in the region. The survival of the squirrel monkey in Panamá depends on management practices that may include restoring the habitat, relocations, and the administration of the isolated and highly reduced populations, based on the habitat requirements and on demographic characteristics of this species. Unlike what Boinski et al. (1998) suggest, in Panamá the efforts of management should have as a minimum scale of work, a metapopulation and not a local population. Burica and Renacimiento metapopulations showed a conservation priority and good management. Environmental education efforts are required. Conservation efforts in other identified metapopulations in this study should be carefully analized.

7. SUMMARY

This work discusses hypothetical population structures and their implication in strategies for the conservation and management of the squirrel monkey in Panamá. I used ArcView 3.0, where I found a total of 23 metapopulations and 34 hypothetical subpopulations. The most numerous in individuals were as follows: Burica, with one subpopulation, Renacimiento, with five subpopulations, Alanje, with four subpopulations, Boquerón, with three subpopulations, and David, with one subpopulation. The rest of the population consists of isolated troops. From these above-mentioned populations, only the first ones have more than 800 individuals (adults, juveniles, and infants), but only the population of Burica has more than 1,000 adult individuals. All the other hypothetical population structures have less than 500 adult individuals. The density of the hypothetical metapopulation structures are the following (individuals/km^2): Burica—8.37, Renacimiento—4.30, Guarumal—3.90, Gariché—3.58,

Boquerón—3.46, Alanje—3.33, David—3.29, Concepción—3.18, and Jacú—2.81. In agreement with the theoretical analysis of the probable structure and dynamics of hypothetical metapopulations identified in Panamá, I believe that the population of Burica behaves as would a typical "continental" metapopulation, Renacimiento as a "continent island," Alanje and Boquerón as "subcontinent islands," David as a "subcontinent," Concepción as an "island island," Gariché as an "island island," Jacú as a "semi-island," Guarumal as an "island," and the rest of the isolated troops as "non-equilibrium." The development of this population model allows us to describe and prioritize conservation efforts and management of the squirrel monkey in Panamá.

8. ACKNOWLEDGMENTS

My thanks to G. Wong (PRMVS), E. Rodríguez-Luna (IPS), H. Chaves (PRMVS), E. Vargas (PRMVS), R. Samudio (STRI), A. Taymes (U. P.), N. Zamora (DAAD), W. VanSickle (IDEA WILD), W. Quintero (UNACHI), M. Araya (UNA), M. McCoy (PRMVS), R. Montagne (PRMVS), and the IX PRMVS Team for your help and advice. I also thank Conservation International, DAAD, IDEA WILD, Universidad de Panamá, International Primatological Society, and World Preservation Trust International for their financial support to squirrel monkey research in Panamá. Finally I thank L. Marsh and the editorial team of "Primates in Fragments" book for your comments and edits to the manuscript.

9. REFERENCES

Baldwin, J. D., and Baldwin, J. I., 1971, Squirrel monkeys (*Saimiri*) in natural habitats in Panamá, Colombia, Brasil, and Peru, *Primates* **12**(1):45–61.

Baldwin, J. D., and Baldwin, J. I., 1972, The ecology and behavior of squirrel monkeys (*Saimiri oerstedi*) in a natural forest in western Panama, *Folia Primatol.* **18**:161–184.

Baldwin, J. D., and Baldwin, J. I., 1976, Primate population in Chiriquí, Panamá, in: *Neotropical Primates Field Studies and Conservation*, R. W. Thorington and P. G. Heltne, eds., National Academy of Sciences, Washington, D.C., pp. 20–31.

Baldwin, J. D., and Baldwin, J. I., 1981, The squirrel monkeys, genus *Saimiri*, in: *Ecology and Behavior of Neotropical Primates, Volume I*, A. Coimbra-Filho and R. A. Mittermeier, eds., Academia Brasileira de Ciencias, Rio de Janeiro, Brasil, pp. 277–330.

Bangs, O., 1902, Chiriqui mammalia, *Bull. Mus. Comp. Zoo.* **39**(2):15–51.

Bennett, C., 1968, Human influences on the zoogeography of Panama, *Iberoamericana* **51**:1–121.

Boinski, S., 1987, Habitat use by squirrel monkey (*Saimiri oerstedi*) in Costa Rica, *Folia Primatol.* **49**:151–167.

Boinski, S., and Sirot, L., 1997, Uncertain conservation status of squirrel monkeys in Costa Rica, *Saimiri oerstedi oerstedi* and *Saimiri oerstedi citrinellus*, *Folia Primatol.* **68**:181–193.

Boinski, S., Jack, K., Lamarsh, C., and Coltrane, J. A., 1998, Squirrel monkeys in Costa Rica: drifting to extinction, *Oryx* **32**(1):45–58.

Brown, A., 1987, Aplicación de estudios ecológicos en la cría y conservación de primates, *Boletín Primatológico Argentino* **5**(1):133–145.

CITES, 1991, Species listed on CITES appendices summary, Convention on International Trade in Endangered Species of Wild Fauna and Flora, Compiled by P. Biber, Secretary of Convention, Basel, Suiza.

ESRI, 1990, Understanding SIG, Enviromental Systems Research Institute, Redlands, California, USA.

Gilpin, M. E., 1987, Spatial structure and population vulnerability, in: *Viable Populations for Conservation*, M. E. Soulé, ed., Cambridge University Press, Cambridge, pp. 125–139.

Gilpin, M. E., and Soulé, M. E., 1986, Minimun viable populations: Processes of species extinction, in: *Conservation Biology: the Science of Scarcity and Diversity*, M. E. Soulé, ed., Sinauer Associates Press, MA, USA, pp. 19–34.

Hanski, I., 1991, Single-species metapopulation dynamics: Concepts, models, and observations, *Biological J. Linn. Soc.* **42**:17–38.

Hanski, I., 1996, Metapopulation ecology, in: *Population Dynamics in Ecological Space and Time,* Q. E. Rhodes, Jr., R. K. Cheser, and M. H. Smith, eds., University of Chicago Press, Chicago, USA, pp. 13–43.

Hanski, I., and Simberloff, D., 1997, The metapopulation approach, its history, conceptual domain, and application to conservation, in: *Metapopulation Biology, Ecology, Genetics, and Evolution,* I. Hanski and M. E. Gilpin, eds., Academic Press, San Diego, CA, USA, pp. 5–26.

Harrison, S., 1991, Local extinction in a metapopulation context: An empirical evalutation, *Biol. J. Linn. Soc.* **42**:73–88.

Harrison, S., 1994, Metapopulations and conservation, in: *Large-Scale Ecology and Conservation Biology,* P. J. Edwards, R. M. May, and N. R. Webb, eds., Oxford Blackwell, USA, pp. 111–128.

Hasting, A., and Harrison, S., 1994, Metapopulation dynamics and genetics, *Annual Review and Ecology Systematic* **25**:167–188.

Hershkovitz, P., 1984., Taxonomy of squirrel monkeys genus *Saimiri* (Cebidae, Platyrrhini): A preliminary report with description of a hitherto unnamed form, *Am. J. Primatol.* **7**:155–210.

Howe, R. W., Davis, G. J., and Mosez, V., 1991, The demographic significance of "sink population," *Biol. Cons.* **57**:239–255.

IUCN, 1995, 1994 IUCN red list of threatened animals, WCC, IUCN Species Survival Commission and Bird Life International, B. Groombridge, ed.

IUCN, 1996, IUCN red list of threatened animals, Unión Internacional para la Conservación de la Naturaleza, Gland, Suiza, J. Baillie and B. Groombridge, eds., 368 pp.

Kinsey, W. G., 1997, *New World Primates: Ecology, Evolution, and Behavior,* Aldine de Gruyter, New York, 436 pp.

Levins, R., 1970, Extinction, in: *Some Mathematical Problems in Biology,* M. Gerstenhaber, ed., American Mathematical Society, Providence, USA, pp. 75–107.

Pulliam, R. H., 1988, Sources, sinks, and population regulation, *Am. Nat.* **132**:652–661.

RENARE, 1980, Lista de fauna en peligro de extinción en la República de Panamá, Ministerio de Desarrollo Agropecuario, Dirección Nacional de Recursos Naturales, Panamá.

Rodríguez, A. R., 1996, Los mamíferos amenazados de Panamá, *El Universal de Panamá* (A-7, 5 Septiembre), Panamá.

Rodríguez, A., 1999a, Estatus de la población y hábitat del mono tití, *Saimiri oerstedii,* en Panamá, Tesis de Maestría, PRMVS, Universidad Nacional, Costa Rica.

Rodríguez, A., 1999b, Modelos poblacionales del mono ardilla en Panamá como herramientas de conservación, Resumen, XVII Congreso Científico Nacional (4-8 Octubre 1999), Universidad de Panamá, Panamá.

Rodríguez-Luna, E., Cortés-Ortiz, L., Mittermeier, R., Rylands, A., Carrillo, E., Wong, G., Matamoros, F., and Nuñez Motta, J., 1996a, Hacia un plan de acción para los primates mesoamericanos, *Neotrop. Primates* **4(Suplemento)**:119–133.

Rodríguez-Luna, E., Cortez-Ortiz, L., Mittermeier R., and Rylands, A., 1996b, Plan de Acción para los Primates Mesoamericanos, Grupo Especialista en Primates-Sección Neotropical. Xalapa, Veracruz, México. Borrador de Trabajo. 102 pp. + 19 de anexos.

Shaffer, M. L., 1981, Minimum population sizes for species conservation, *Bioscience* **31**:131–134.

Stith, B. M., Fitzpatrick, J. W., Woolfenden, G. E., and Pranty, B., 1996, Classification and conservation of metapopulations: A case study of the Florida scrub jay, in: *Metapopulations and Wildlife Conservation,* D. R. McCullough, ed., Island Press, Washington, USA, pp. 187–215.

U. S. Department of the Interior, 1984, *Endangered and Threatened Wildlife and Plants,* U. S. Goverment Printing Office, Washington, USA. 24 pp.

Wiens, J. A., 1996, Wildlife in patchy environments: metapopulations, mosaics, and management, in: *Metapopulations and Wildlife Conservation,* D. R. MacCullough, ed., Islands Press, Washington D.C., USA, pp. 53–84.

Wong, G., 1990a, Uso del hábitat, estimación de la composición y densidad poblacional del mono tití (*Saimiri oerstedi citrinellus*) en la zona de Manuel Antonio, Quepos, Costa Rica, Tesis de Maestría, Programa Regional en Manejo de Vida Silvestre, Universidad Nacional, Heredia, Costa Rica, 78 pp.

Wong, G., 1990b, Ecología del mono tití (*Saimiri oerstedi citrinellus*) en el Parque Nacional Manuel Antonio, Costa Rica, Tesis de Grado, Universidad Nacional, Heredia, Costa Rica, 57 pp.

PRIMATE SURVIVAL IN COMMUNITY-OWNED FOREST FRAGMENTS:
Are Metapopulation Models Useful Amidst Intensive Use?

Colin A. Chapman, Michael J. Lawes, Lisa Naughton-Treves, and Thomas Gillespie[*]

1. INTRODUCTION

Human modification of ecosystems is threatening biodiversity on a global scale (Cowlishaw, 1999; Cowlishaw and Dunbar, 2000; Chapman and Peres, 2001). A recent Food and Agriculture Organization report (FAO, 1999) indicates that tropical countries are losing 127,300 km^2 of forest annually, and this does not consider the vast area being selectively logged (approximately 55,000 km^2; FAO, 1990). The extent of tropical forests burning each year is highly variable and difficult to measure precisely (FAO, 1999; Nepstad et al., 1999), however, the forests of Southeast Asia (Kinnaird and O'Brien, 1999) and the Brazilian Amazon (Nepstad et al., 1999) are especially impacted by the combination of droughts from El Niño and burning for agriculture (FAO, 1999). In 1997 and 1998 an area of 2 million ha of forest burned in Brazil and 4 million ha burned in Indonesia (FAO, 1999).

These modifications to tropical forests do not just result in the forest being uniformly reduced in size, they also result in forest being fragmented. To understand the conservation value of these fragments is critical, because they may represent opportunities to make important conservation gains. The reason fragments become important for conservation is related to the fact that, today, less than 5% of tropical forests are legally protected from human exploitation, and many of these legally protected areas are subjected to illegal exploitation (Redford, 1992; Oates, 1996). Furthermore, many tropical species are locally endemic or are rare and patchily

---

[*] Colin A. Chapman, Department of Zoology, University of Florida, Gainesville, Florida, 32611 and Wildlife Conservation Society, Bronx, New York, 10460. Michael J. Lawes, Forest Biodiversity Programme, School of Botany and Zoology, University of Natal, Bag X01, Scottsville, 3209, South Africa. Lisa Naughton-Treves, Department of Geography, University of Wisconsin, Madison, Madison, Wisconsin, 53706 and Center for Applied Biodiversity Science, Conservation International, Washington, DC, 20036. Thomas Gillespie, Department of Zoology, University of Florida, Gainesville, Florida, 32611. Correspondence to C. A. Chapman (email: cachapman@zoo.ufl.edu).

Primates in Fragments: Ecology and Conservation
Edited by L. K. Marsh, Kluwer Academic/Plenum Publishers, 2003

distributed (Struhsaker, 1975; Richards, 1996). Such restricted distributions predispose many tropical forest species to an increased risk of extinction when habitats are modified (Terborgh, 1992), because national parks and reserves, even if effectively protected, will fail to conserve species whose ranges do not fall within a protected area. As a result, conservation of many tropical forest species will depend on the capacity of fragmented forests to support their populations. Primates are valuable species to study the effects of fragmentation because they are relatively easy to census, and there is often a large body of information on their behavior from intact forests. Furthermore, because many species are locally endemic and endangered or threatened, it is critical to formulate informed management plans.

Ecological research over the last decade reveals that conserving animals in forest fragments is difficult given the unpredictable and complex interactions between species experiencing rapid habitat change (Laurance and Bierregaard, 1997). With respect to primates, this difficulty arises from three sources. First, a number of simple logical predictions made by some of the first researchers studying primates in forest fragments have not proven to be general. For example, home range size is frequently cited as influencing a species ability to survive in a fragment (Lovejoy et al., 1986; Estrada and Coates-Estrada, 1996). However, Onderdonk and Chapman (2000), found no relationship between home range size and the ability to live in fragments for a community of primates in Western Uganda. Similarly, it has been suggested that a highly frugivorous diet may limit the ability of a species to live in fragments (Lovejoy et al., 1986; Estrada and Coates-Estrada, 1996). However, Tutin et al. (1997) found that several frugivorous species were at higher or similar densities in forest fragments than in the intact forest of Lopé Reserve, Gabon (Tutin, 1999; Onderdonk and Chapman, 2000). The complexity of the issue is illustrated by redtail monkeys frequently moving between forest fragments near Kibale, using available forest corridors and crossing agricultural areas; whereas, blue monkeys, which have a similar diet and social organization, do not use these fragments or corridors (Chapman and Onderdonk, 1998; Onderdonk and Chapman, 2000). In contrast, blue monkeys near Budongo Forest Reserve, Uganda, often reside in fragments and travel through agricultural land (Fairgrieve, 1995).

Secondly, studies of fragments and their primate populations often involve attempts to understand dynamic systems. Typically, studies are conducted in areas where the long-term history of the fragments is not well known. One does not know if a study population in the fragment is at equilibrium or not. Many years may pass after isolation before a population will respond numerically to fragmentation. For example, Struhsaker (1976) documented that it was nearly 10 years after the loss of approximately 90% of a major food resource that a statistically significant decline in vervet monkeys (*Chlorocebus aethiops*) at Amboseli, Kenya, could be detected. Furthermore, the human use of fragments and resulting ecological impacts are too often ignored. Most fragments are not protected; they are on private land and are used by local landowners. Thus, fragments change structure and composition as landowners use the forest for grazing or to extract timber or fuelwood or allow fallow land to regenerate. This fact has not been fully appreciated, probably because a number of previous studies have been conducted in forest fragments that are protected (i.e., they are within a protective reserve; Lovejoy et al., 1986; Tutin et al., 1997; Tutin 1999). While these studies in protected reserves have provided us with many insights, they are not typical of most fragments, and they may have biased our perception of the value of forest fragments to primates.

Third, although the theoretical effects of habitat isolation and fragment size are well known (Hanski, 1994; Hanski and Gilpin, 1997), their effects on individual species are rarely studied in detail (Harrison et al., 1988; Thomas et al., 1992). Where they are, it is generally acknowledged that fragmentation of once continuous habitat has had a detrimental effect on species' persistence (Laurance and Bierregaard, 1997), and there is hope that species may persist in metapopulations. The dynamics and persistence of such metapopulations are governed by the interaction between the life history of species, which determines the rates of local extinction and colonization, and landscape properties (e.g., area of the fragment, distance to other fragments, Hanski, 1994). However, robust and predictive metapopulation models of a species persistence in fragmented forests demand data from a large number of fragments (>40 fragments, Hanski and Gilpin, 1997; Lawes et al., 2000). The difficulties of surveying a large number of fragments frequently limits data to presence-absence records that provide limited information on life-history constraints (e.g., density, diet, fecundity) or species persistence. These difficulties do not easily reveal the processes responsible for observed patterns of persistence and distribution. As a result, insights are difficult to gain from the theoretical models that can be accurately applied to management objectives.

The objectives of this study were to (1) document the changes in forest structure of a series of forest fragments outside of Kibale National Park, Uganda, over a 5-year period, (2) describe the persistence of primates in those fragments over that period, (3) quantify changes in the size and structure of black-and-white colobus (*Colobus guereza*) populations, and 4) consider the value of metapopulation models, particularly incidence-function models, to the management of this community-owned forest fragment system. The area in which these fragments are located is a matrix of small-scale agriculture, grazing land, and tea plantations. Local residents are using all fragments for multiple purposes, including fuelwood collection and charcoal manufacture. Although we have a relatively small data set relative to what is needed for metapopulation modeling, we use a simple incidence function metapopulation model to investigate patterns of primate fragment occupancy. We also incorporate the life-histories of the primate species and the landscape properties in our analyses.

2. METHODS

2.1. Study Site

The primates in 20 forest fragments were censused from May to August 1995 (Onderdonk and Chapman, 2000), and 19 were recensused in May to August 2000. These forest fragments are neighboring Kibale National Park, Uganda (766 km²), located in western Uganda (0 13' - 0 41' N and 30 19' - 30 32' E) near the foothills of the Rwenzori Mountains. Kibale is a mid-altitude moist evergreen forest that receives approximately 1,750 mm (1990 to 1999) that falls primarily in two rainy seasons (Chapman et al., 1997; Struhsaker, 1997; Chapman and Lambert, 2000). Before clearing for agriculture, there was likely continuous forest throughout the study region and it was directly connected to what is now Kibale forest. The forest in the fragments was probably similar to the forest within the national park, but it has been largely deforested and is now dominated by smallholder agriculture.

The forests and wildlife of western Uganda have long been influenced by human activities, but these activities have dramatically intensified over the past 50 years (Howard, 1991, Naughton-Treves, 1999). Pollen records suggest that forest clearing began in Uganda at least 1,000 years ago with the introduction of agriculture and iron making (Hamilton, 1974, 1984). Until the 20th century, the forests of western Uganda were sparsely settled by Bakonjo and Baamba hunter-gatherers (Taylor, 1962; Steinhart, 1971). War and epidemics likely caused forests to expand at the end of the 19th century (Osmaston, 1959; Paterson, 1991). Shortly thereafter, Batoro herders and agriculturalists arrived in the region from the north, displaced the Bakonjo and Baamba, and began a lengthy period of deforestation. By the end of the 20th century, nearly all forest outside of officially protected areas has been converted to farms, grazing areas, or tea plantations (Naughton-Treves, 1997). Only small pockets of forest remain in areas unsuitable for agriculture. Thus, the forest fragments that we studied were either forested areas associated with swampy valley bottoms or on the steep forested rims of crater lakes (Table 1). While the precise timing of isolation of these forest remnants is not known, local elders describe them as 'ancestral forests' (Naughton-Treves, unpublished data). Aerial photographs taken in 1959 indicate that most fragments have been isolated from Kibale at least since that time, although many have decreased in size. Oral histories suggest that they have been present for decades.

Human population surrounding Kibale has increased seven-fold since 1920, surpassing 272 individuals per km^2 at Kibale's western edge (versus 92 per km^2 for the District; NEMA, 1997). Population growth rate varies among parishes, but is typically between 3% and 4%. (In the parishes with the majority of the fragments growth rate

Table 1. Characteristics of forest fragments outside of Kibale National Park, Uganda[a].

Fragment	Area (ha)	Fragment Type	Distance to Kibale (km)	Nearest Fragment (m)	Redtail	Red Colobus	B&W Colobus	Chimp	Forest Status
Rutoma #3	0.8	HS	2.2	100	1/0	0/0	0/0	1/0	Deforested
Dry Lake	1.2	HS	6.1	153	1/0	0/0	1/0	0/0	Deforested
Rutoma #1	1.2	HS	2.4	80	1/1	0/1	1/1	1/0	Remaining
Kiko #4	1.2	VB	1.1	70	0/0	1/1	1/1	1/0	Remaining
Durama	1.4	HS	1.1	60	1/0	0/0	0/0	0/0	Deforested
Kiko #3	1.7	VB	1.1	70	1/0	1/1	1/1	0/0	Remaining
Rutoma #4	2.0	HS	2.1	80	1/0	0/0	1/0	0/0	Deforested
Lake Nyanswiga	2.2	CL	6.0	155	1/1	0/0	1/1	1/1	Remaining
Kyaibombo	2.3	VB	1.1	162	1/0	0/0	1/0	0/0	Deforested
Ruihamba	2.4	VB	4.1	300	0/1	0/1	1/1	0/1	Remaining
Nkuruba - Fish Pond	2.8	VB	3.7	70	1/1	1/1	1/1	1/1	Remaining
Lake Nyaherya	4.6	CL	6.1	300	1/1	0/1	1/1	0/1	Remaining
Rutoma #2	4.9	HS	3.0	150	1/0	0/1	1/1	1/0	Remaining
Rusenyi	4.9	VB	1.1	50	1/1	0/1	1/1	0/0	Remaining
Kiko #2	5.0	VB	1.8	125	1/0	1/1	1/1	0/0	Remaining
Kiko #1	6.2	VB	2.0	50	1/0	1/1	1/1	0/0	Remaining
Nkuruba Lake	6.4	CL	3.6	70	1/1	0/1	1/1	1/1	Remaining
CK's Durama	8.7	VB/HS	0.2	150	1/0	1/1	1/1	1/0	Remaining
Lake Mwamba	28.7	CL	7.2	100	1/?	0/0	0/0	0/0	Remaining

[a] The presence (1) and absence (0) of each species in 1995 and 2000 are indicated (95/00). Forest Status is labeled as deforested when it is viewed by the researchers to have insufficient trees remaining to support resident primate populations and no residents were seen in the 2000 survey. If a solitary individual was in a fragment, it was not assumed that the fragment could support the species (HS = Hillside, VB = Valley Bottom, CL = Crater Lake).

averages 3.87%; NEMA, 1997.) Batoro farmers remain the dominant local ethnic group in the area (~52% of population). However, waves of other immigrants into the area (e.g., Bakiga) have intensified the demand for agricultural land and forest products (NEMA, 1997; MFEP, 1992).

Of Kibale's 12 primate species (chimpanzees–*Pan troglodytes*, gray-cheeked mangabey–*Lophocebus albigena*, red colobus–*Procolobus badius*, black-and-white colobus–*Colobus guereza*, red-tailed monkeys–*Cercopithecus ascanius*, blue monkeys– *C. mitis*, l'hoest's monkey–*C. lhoesti*, vervets–*Chlorocebus aethiops*, olive baboon– *Papio anubis*, potto–*Perodictus potto*, Matschie's bush babies–*Galago matshiei*, and Thomas's bush babies–*Galagoides thomasi*), only six have been recorded in the fragments (red-tailed monkeys, chimpanzees, baboons, red colobus, black-and-white colobus, and vervets; fragments have not been sampled for the three nocturnal primates).

2.2. Surveying Primate Fragment Occupancy

In 1995, forest fragments were selected if they had a fairly clearly defined boundary, were isolated from other fragments or tracts of forest by ≥ 50 m, and were small enough to count all black-and-white colobus groups. Twenty fragments were visited in the first survey. One large fragment was surveyed in 1995, but was not resurveyed. In the first survey the following parameters were measured in each fragment: primate species present, black-and-white colobus group size and composition, tree species richness, area of the fragment, and distance to the nearest fragment (see Onderdonk and Chapman, 2000 for details of these methods). We determined which primate species were present by observations made over a 2- to 4-day period. Ideally, species abundance rather than presence-absence would be used as an index of success in a fragment, but these data were only possible to obtain for the black-and-white colobus. For each group of black-and-white colobus encountered we determined size and composition (age/sex classes follow Oates, 1974). To obtain reliable estimates of group counts an observer would often stay with a group for up to a day and wait for members to make a coordinated movement crossing an opening. Since many of the fragments were on the slopes of the crater lakes, we were often able to get above the group. In such instances, animals were highly visible.

In the survey conducted in 2000, the same parameters were measured, with the exception of fragment size and distance to the next fragment, although changes in the condition of the fragments were noted. In addition, in the second survey the composition of red colobus groups was determined. Detailed descriptions of how the fragments were being used by the local people were also recorded. From long-term research at one fragment (Lake Nkuruba, Chapman et al., 1998), we know that redtail monkeys and chimpanzees frequently move among fragments (i.e., they use multiple fragments in a week). In contrast, the colobines are much more site tenacious and rarely move among fragments (i.e., to colonize a new fragment). As a result, when contrasting presence-absence data between time periods, we focus on colobines.

2.3. Environmental Factors and Forest Use by Local Landowners

The forest fragments we studied provide multiple resources to local citizens, including medicinal plants, foodstuffs, fodder, building materials, and, most importantly, fuelwood (Table 2). Over 98% of residents neighboring Kibale rely exclusively on

Table 2. Patterns of land use of 16 of the 19 fragments used to assess the long-term viability of primate populations in forest fragments near Kibale National Park, Uganda[a].

Forest Fragment	Area (ha)	Households	Ethnicity (Dominant)	Tenure	Brew Beer	Distill Gin	Charcoal	Cattle/ Goats	Woodlot
Rutoma I	1.2	8	Mixed[b]	V	0	25	0	0	13
Kiko #4	1.2	5	Mixed (K)	T	0	0	0	0	100
Durama	1.4	4	Mixed (T)	V	50	0	0	0	75
Kiko #3	1.7	3	Mixed[b]	T	0	0	0	0	33
Rutoma IV	2.0	8	Mixed (T)	V	25	0	13	38	38
Lake Nyanswiga	2.2	4	Mixed (T)	V	0	0	50	25	25
Kyaibombo	2.3	7	Toro	C	29	29	14	43	86
Rwaihamba	2.4	8	Mixed (T)	V	25	38	0	50	88
Nkuruba - FishPond	2.8	2	Other	C	0	0	0	50	100
Lake Nyaherya	4.6	8	Mixed (T)	V	0	0	63	13	13
Rusenyi	4.9	11	Toro	V	9	9	27	0	27
Kiko #2	5.0	6	Mixed (K)	T	0	0	50	0	50
Kiko #1	6.2	9	Mixed (K)	T	0	0	0	11	44
Nkaruba Lake	6.4	2	Other	C	0	0	0	0	50
Ck'sDurama	8.7	16	Mixed (K)	V	6	31	0	12.5	50

[a] Ethnicity includes Batoro (T), Bakiga (K), and Other (Mzungu, Munyankole, Catholic Church, etc.), mixed indicates that households from a number of ethnic groups were using the fragment. For the other parameters reported we indicate the percentage of the households that were engaged in the indicated activity (e.g., brewing beer). Land tenure types – V = customary claim by village, T = property of tea company, C = property of Catholic Church. [b] no dominant ethnicity.

fuelwood or charcoal for energy; one of the highest levels in the world (Bradley, 1991; Government of Uganda, 1992). Rapid population growth, expanded commercial charcoal and brick production, industrial fuelwood demands, and technological change are fundamentally altering the relationship between forests and forest users. Furthermore, the demand for forest products has intensified in a context of insecure property rights, resulting in rapid deforestation.

The land and tree tenure arrangements governing local access to resources in these forest fragments are complex and rooted in customary systems. Traditionally, clans governed land use and allocated plots to individual members who then carefully demarcate their property by planting living fences or clearing fields. This system has largely persisted, despite the nationalization of all land in Uganda in 1975 (Place and Otsuka, 2000), and is the de facto tenure system for a majority of the fragments we studied. Seven of the forest fragments are formally held under semi-permanent leaseholds by the Catholic Church or tea company, but both the Church and the tea company permit local residents to use the forests according to customary systems. Traditionally, individuals managing plots of forest allowed kin to freely harvest firewood, drinking water, and medicinals from their property, although certain species, hardwoods in particular, required special permission to harvest (Kaipiriri, 1997). Many villagers today complain that controlling access to their forests has become difficult, given increasing scarcity and value of forest resources. As is typical of much of the tropics, they often resort to deforestation as a means of securing land ownership (Sjaastad and Bromley, 1997). Others are bribed or coerced by charcoal manufacturers to allow them to produce charcoal in their forest.

To better understand forest use by local citizens and its potential impact on primate survival, we collected detailed data on how fragments were being used by the local people, as well as general socioeconomic parameters for each fragment. For example, for 16 of the 19 fragments, we determined the number of households that owned land directly adjacent to the forest. Fragments are considered to be property of these households and thus the number of households should represent an index of pressure on the fragment. We also noted the ethnicity of households (Bakiga immigrants are thought to use forests more intensively than Batoro residents, Kaipiriri, 1997) and whether they had eucalyptus woodlots on their land. Woodlots may take pressure off fragments, because landowners would have access to alternative fuelwood sources. Alternatively, plantations may indicate that fuelwood resources in the fragments are being depleted and farmers are now planting trees on their own land to have access to fuel in the future. Finally, for each household we interviewed residents and determined if they were using fuelwood from the fragment to brew beer, distill gin, and/or produce charcoal. We also determined if each household had goats or cattle since these animals are often allowed to graze in fragments.

2.4. The Incidence-Function Model

A population that consists of several subpopulations linked together by immigration and emigration is regarded as a metapopulation. The fraction of suitable habitat fragments occupied at any given time (incidence) represents a balance of the rate at which subpopulations go extinct in occupied fragments and the rate of colonization of empty fragments. However, the measurement of colonization and extinction rates is very time-consuming and thus the practical application of metapopulation models can be difficult. Hanski (1994) has argued that incidence functions, based on relatively easily collected presence-absence data from a large number of fragments, can provide relative or absolute rates of extinction and colonization at low cost.

Here we used an incidence function to model the presence or absence of a primate species in any given fragment. Incidence functions are discrete-time stochastic fragment models and a metapopulation-level extension of a first-order Markov chain model for an individual fragment (Hanski, 1994). However, to generate state transition probabilities from a species-incidence curve derived from one survey, we assume that occupancy of the system at the time of the survey is at equilibrium (Hanski, 1994; Thomas, 1994). As a means of determining whether or not primate metapopulations are indeed in a steady-state, we compare species fragment occupancy models between two time periods. The difference in fragment occupancy between time periods enables a prediction probability of future persistence and consequences of habitat loss to a suite of forest primates.

Hanski (1994) showed that if fragment i is currently empty, it has the probability, C_i, of becoming recolonized in unit time, and if fragment i is currently occupied, it has a constant probability, E_i, of becoming empty (local extinction). This elementary model describes the incidence of a species in fragment i as the stationary probability of fragment occupancy (J_i):

$$J_i = C_i / (C_i + E_i) \qquad (1)$$

To account for the generally observed trends of increasing occupancy with increasing area and decreasing isolation, an incidence function model based on a metapopulation consisting of a 'mainland' (whose population is invulnerable to extinction) with small forest fragments around it (i.e., mainland-island incidence-function model) requires the following assumptions (Hanski 1994; Lawes et al., 2000): 1) the colonization probability, C_i, is a negative exponential function of distance from the mainland; 2) the relatively week dependence of colonization on fragment area, A_i, is ignored; and 3) the extinction probability, E_i, strongly depends on fragment area but not on isolation. Hanski (1994) provides an elementary mainland-island incidence-function model that combines these assumptions into one function. The colonization probability model is given by

$$C_i = e^{-\beta D_i}$$
(2)

where D_i is the distance of the ith fragment from the mainland, in this case Kibale National Park, and β is a constant. The extinction probability model is

$$E_i = \frac{c}{A_i^x},$$
(3)

where A_i is the area and c and x are constants. Substituting (Eq. 2) and (Eq. 3) into (Eq. 1) yields the incidence function for the mainland-island metapopulation model as

$$J_i = \frac{1}{1 + \dfrac{ce^{\beta D_i}}{A_i^x}},$$
(4)

We fitted this model (Eq. 4) to our data on fragment occupancy for each primate species and survey period and estimated parameter values (i.e., c, β, and x), using the non-linear regression routines in SPSS. In addition, we calculated all values of A_i and D_i for $J_i = 0.9$ and $J_i = 0.5$ and displayed these incidence lines in the log$A$ on log$D$ scatterplot summary of each species' occupancy. The main trends in primate fragment occupancy were derived from these graphic summaries, mainly because sample size (number of forests) was small and the parameters in the model had large margins of error, diminishing their usefulness.

3. RESULTS

3.1. Forest Use

On average 6.7 households had access to the resources in any given fragment (range 2 to 16). Although there is a correlation between the size of the fragment and the number of households that they support, the strength of the relationship is not strong ($r = 0.54$, $p = 0.040$). For example, the largest fragment did support the most households ($N = 16$), but the second largest fragment only supported two households. Given the fact that the average household in this area contains 4.8 people (NEMA, 1997), these fragments are supporting the fuelwood needs of an average of 32 people. Evidence suggests that the fragments also support a significant share of the neighboring families up to three farms away (>120 families or >576 people, Naughton-Treves, unpub. data).

Estimates of domestic fuel use by people around Kibale indicate that a typical family uses 8.4 kg of fuelwood each day for cooking (Wallmo and Jacobson, 1998). Thus, in one year a local family would use 3,066 kg. (This value is slightly lower than average for non-liquid propane gas users in East Africa; Kammen, 1995.) However, most of the fuelwood gathered for cooking is comprised of fast-growing secondary species like *Acanthus pubescens* and *Vernonia* spp. that are relatively abundant on fallow land (Naughton-Treves and Chapman, 2002).

Commercial uses of the forest involves the extraction of higher volumes of wood and greater long-term impacts given the selection for slow-growing hardwood logs. Extensive clearing of fragments often occurred when neighboring households were engaged in beer brewing (an average of 9.6% of the households were engaged in beer production–8.45 kg per episode, on average 19 times a year), gin (8.8%–875 kg per episode, on average 16 times a year), or charcoal production (14.5%–5935.9 kg per episode, on average 18 times a year). On average 16.2% of the households had cattle or goats.

Just over half of the households (52.8%) that were adjacent to the fragments also had woodlots. We did not have the impression that the fragments where most households did not have woodlots were more degraded than fragments with a number of adjacent woodlots.

3.2. Primate Population Change

Of the 16 fragments that we studied in 1995 that supported resident populations of black-and-white or red colobus, three had been cleared to the extent that primates were no longer present in 2000. The fragments that were cleared had supported five groups of black-and-white colobus (31 individuals total) and were also used by redtail monkeys. For all fragments in 1995 we counted 165 black-and-white colobus, while in 2000 only 118 animals were seen (Figure 1). During the first census, there were 0.405 infants for every adult female, while in the recensus there was only 0.026 infant for every female.

For red colobus the situation was very different. There were seven fragments with red colobus groups in 1995, and none of these fragments were cleared by the time of the 2000 census. In the 2000 census, red colobus groups were found in these original seven fragments and in four additional fragments. In the 2000 census, 159 red colobus were counted and the ratio of infants to adult females was 0.25.

Redtail monkeys were seen or reported by local landowners to be in 18 of the 20 fragments that were surveyed in 1995 and in seven of the 19 surveyed in 2000. They were in fragments that were largely cleared and no longer supported either of the colobine species. Redtail monkeys are known to move between forest fragments and are notorious crop raiders (Naughton-Treves 1997, 1998, Naughton-Treves et al., 1998), thus it seems likely that they used the last few trees of highly degraded fragments to move throughout the landscape while feeding mainly outside of the fragments. Chimpanzees were seen once in 1995 and once in 2000. Evidence of chimpanzees, such as nests, dung, or wadges, was found in nine fragments in 1995 and in five fragments in 2000. Chimpanzees are reported throughout this region and are frequent crop raiders. Blue monkeys and mangabeys were not seen in any of the fragments during either of the surveys. We asked people living near each fragment if they had ever seen or heard blue monkey or mangabeys, both of which have very loud distinct calls, and no one reported them in the area. In fact, while there are Rutoro names for most of Kibale's primates,

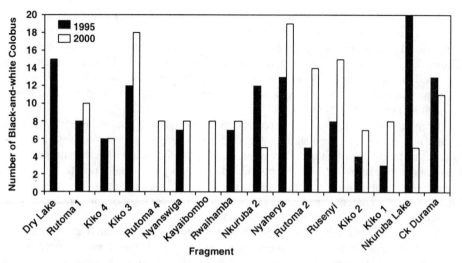

Figure 1. The number of black-and-white colobus found in forest fragments neighboring Kibale National Park, Uganda, in a 1995 and 2000 census.

there is no local name for blue monkeys or mangabeys, suggesting that they have very little contact with them.

3.3. Mainland-Island Incidence-Function Model

Using nonlinear regression we fitted the mainland-island incidence-function model (Eq. 4) to the data. In all cases the ratio of the 'model' sum of squares to the 'error' sum of squares, which is analogous to the F-ratio in linear regression procedures, was greater than four. Thus in all cases the model accounted for a substantial amount of the variance. However, the absolute values of the ratios of the parameter estimates to their standard errors were small (<2), indicating not much confidence in the coefficient values. Thus the curve fits are at best an indication of the trends in occupancy pattern of each primate species and in each survey (Figure 2). Interpretation of the trends is as follows:

Red colobus can persist in fragments that are just over a hectare in size. Occupancy increased with increasing fragment areas, but distance of the fragment from the mainland did not critically affect occupancy. There was little difference in the curves generated from the 1995 and 2000 surveys, with the exception of how the presence of red colobus in distant fragments influenced the curve. Black-and-white colobus are much like the red colobus in being able to persist in very small fragments; however, they are in more fragments than the red colobus. There is very little evidence of a distance effect, yet they are found occupying some of the most distant fragments.

Redtail monkeys exhibited a different pattern. They were found in nearly all the fragments in the 1995 survey and their occupancy was not strongly limited by area or distance. In fact, there were no probability curves for 1995. The curve generated from the 2000 survey suggests that redtail incidence was relatively independent of area and generally decreased with increasing distance from Kibale. This probability curve was in a direction opposite to what would be predicted; that is, incidence is greater for small

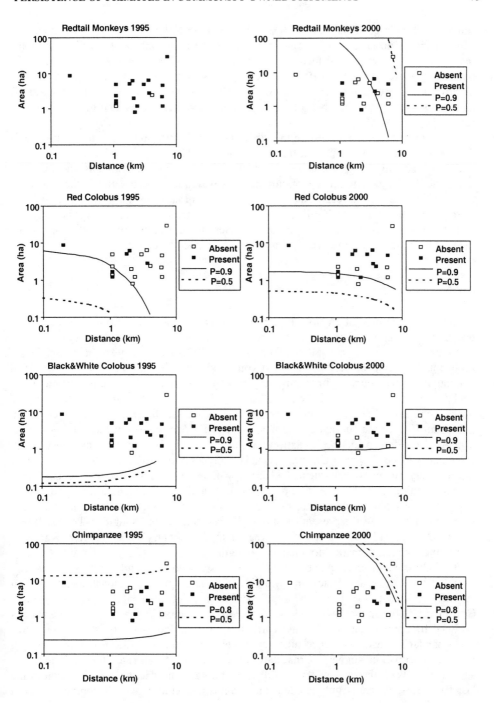

Figure 2. The log-log relationship between area and isolation distance of occupied (solid) and unoccupied (open) forest fragments for four species of primates living in fragments neighboring Kibale National Park, Uganda. The solid and dashed lines provide fitted percent occupancies from logistic equations at different *p* values.

fragments than large. This may because they are easier to locate in a small fragment and can hide from observers in a large fragment. Redtails are known to readily cross the matrix, and they frequently feed on crops in the agricultural land. Comparison of the occupancy of fragments from 1995 and 2000, shows a decline. However, since redtails so readily move among fragments, this may not represent a true decline in numbers. A detailed study of the number of animals and movement rates among fragments is required.

Like redtails, chimpanzees are thought to be very mobile and often change fragments from week to week, as they track fruiting phenology. The differences between surveys is not likely meaningful because of the dynamic nature of their movements and fragment occupancy. A single community of chimpanzees would likely use many of the fragments that we studied. As a result, they are not ideal animals for the application of this metapopulation model. This emphasizes the importance of understanding the behavior and life-history characteristics of the species in question for developing sound management plans for conserving primates in fragmented landscapes.

4. DISCUSSION

Much of the previous work on primates living in fragmented habitats has involved the study of fragments in protected areas or has largely ignored the human use of fragments (Lovejoy et al., 1986; Tutin et al. 1997; Tutin, 1999). In reality, most fragments are not protected; they are on land held by private citizens or local communities who depend on them for fuelwood, medicinals, or game. While these studies in protected reserves have provided us with many insights, they may have biased our perception of the value of forest fragments to primates. Our study documented that in five years 19% of the surveyed fragments were cleared to the extent that they no longer supported resident primate populations. If this level of clearing remains constant, all fragments will be cleared in 26 years. This is likely an optimistic projection since current development suggests that rates of clearing will increase. There is little unused land remaining to be farmed by the growing population, and people are increasingly turning to what was considered marginal lands (e.g., draining swamps and farming on the steep sides of hills) that are the present locations of forest fragments (Naughton-Treves, 1999). A paved road will soon connect the region to the country's capital city, Kampala, increasing the ease to market for charcoal and gin.

The different primate species we examined responded differently to the changes seen over the last five years. Black-and-white colobus is a species demonstrated to exhibit considerable dietary flexibility and to prosper in many degraded habitats (Onderdonk and Chapman, 2000; Oates, 1977, 1996). Yet, their population in the forest fragments has declined by approximately 30% and currently their reproductive rate is extremely low. The mainland-island incidence-function model indicated no distance constraint on fragment occupation, suggesting that they can survive in a fragmented forested landscape such as this. With the current level of reproduction and the decline we have seen in the last five years, it seems unlikely that this fragmented landscape will continue to be a viable habitat for this species.

In contrast to the black-and-white colobus, red colobus were found in more fragments in the 2000 census in comparison to the 1995 census, and their adult female to infant ratio was much higher than black-and-white colobus. Since the fragments were being degraded, it is difficult to envision a reason why their numbers would increase between the two time periods (as indicated by the increased number of fragments they occupy). It is unlikely that they are just now colonizing the fragments that have been there for decades. It is possible that the fragments are operating as a sink and excess animals from the Kibale National Park emigrate to the fragments only to subsequently do poorly in the long term (Pulliam, 1988). However, the ratio of adult females to infants does not support the idea that the fragments are poor-quality sink habitats. With the decline in black-and-white colobus, small groups of red colobus may be more successful in the small fragments. This is supported by 43.2% dietary overlap between a red colobus group range of the red colobus group.

Both redtail monkeys and chimpanzees have similar patterns of use in the fragments and similar mainland-island incidence-functions. We believe that these species are highly mobile, moving readily among fragments, likely tracking the availability of fruit resources. Given this, we are unable to state if the population size of these species or the number of fragments they are occupying have changed over the timeframe of our study. Attempting to apply metapopulation models does not likely provide us any valuable insights into the conservation of these species, but it does highlight the importance of understanding animal behavior and their life-history characteristics for developing sound management plans for conserving primates in fragmented landscapes. These two species are responding to changing conditions at a landscape level and the deforestation of one fragment may have little effect, because animals can travel to alternative fragments. However, if there is general degradation of all fragments and destruction of some, the animals may be negatively affected.

Metapopulation theories are based on the idea that random fluctuations in local populations cause local extinctions and thus unoccupied fragments are available for recolonization (Hanski, 1994; Hanski and Gilpin, 1997). In the forest fragment system we studied, extinctions appear to be driven by increased levels of deforestation that degrade the habitat and make it unsuitable for the primates. Once the fragments are deforested they are used for agriculture and thus are not available for recolonization. Such limitations of metapopulation models have been previously recognized (Thomas, 1994; Lawes et al., 2000). However, despite these limitations, incidence-function metapopulation models provide useful information for the management of animal populations within such fragmented landscapes.

The habitat in this fragmented forest landscape is deteriorating. For black-and-white colobus, evidence indicates that this deterioration is leading to a decline in their population size and poor birth rates. If the situation does not change more fragments will be cleared and the remaining ones will become further degraded. To reverse the present trends would require a major conservation effort, on a scale and of a nature that is not typically done. To stop the fragments from being cleared would require the cooperation of the local people, since this is their land. Alternative sources of income would have to be found (e.g., ecotourism), fuelwood supplies from elsewhere would have to be made available (e.g., a large scale woodlot project), and a great deal of effort would have to be placed in education and outreach to obtain the willing support of all the communities. In all reality, it is unlikely that a project of this magnitude will be initiated. If it is not, it is inevitable that the animals in this fragmented landscape will be lost and the habitat destroyed.

5. SUMMARY

As deforestation and habitat fragmentation accelerate throughout the tropics, the survival of many forest primates depends largely on their ability to cope with such changes. In 1995 we censused 20 forest fragments near Kibale National Park, Uganda, that had existed for several decades. For each fragment we determined the presence or absence of all diurnal primate species and population sizes of black-and-white colobus (*Colobus guereza*). Five years later, we recensused the same fragments and discovered that of the 16 fragments inhabited by primates in 1995, three had been largely cleared and resident primate populations were no longer present. Population declines and lowered fertility rates in the remaining fragments were documented for some species. For example, the black-and-white colobus populations declined from 165 in 1995 to 118 animals in 2000, and there were 0.405 infants per adult black-and-white colobus female in 1995 versus 0.026 infants per female in 2000. For red colobus (*Procolobus badius*) the situation was very different. Red colobus groups occupied seven fragments in 1995, and they were found in an additional four fragments in 2000. In the 2000 census, 159 red colobus were counted and the ratio of infants to adult females was 0.25. The extent of forest clearing in the fragments was documented and factors encouraging clearing are considered. Treating the primates in these forest fragments as putative species metapopulations, we consider whether or not metapopulation principles are useful in conservation planning. In addition, we consider the susceptibility of predictions of species persistence derived from metapopulation principles, and used in management plans, to further human disturbance of forests.

6. ACKNOWLEDGMENTS

Funding for this research was provided by the Wildlife Conservation Society and National Science Foundation (grant number SBR-9617664, SBR-990899). Permission to conduct this research was given by the Office of the President, Uganda, the National Council for Science and Technology, the Uganda Wildlife Authority, and the Ugandan Forest Department. ML would like to thank the National Research Foundation of South Africa for funding support. We would like to thank Lauren Chapman, Steven Piper, and Adrian Treves for helpful comments on this work.

7. REFERENCES

Bradley, P. N., 1991, *Women, Woodfuel, and Woodlots*, Macmillan Ltd., London.
Chapman, C. A., Chapman, L. J., Wrangham, R., Isabirye-Basuta, G., and Ben-David, K., 1997, Spatial and temporal variability in the structure of a tropical forest, *Afr. J. Ecol.* **35**:287–302.
Chapman, C. A., and Lambert, J. E., 2000, Habitat alteration and the conservation of African primates: A case study of Kibale National Park, Uganda, *Am. J. Primatol.* **50**:169–186.
Chapman, C. A., and Onderdonk, D. A., 1998, Forests without primates: Primate/plant codependency, *Am. J. Primatol.* **45**:127–141.
Chapman, C. A., and Peres, C., 2001, Primate conservation in the new millennium: The role of scientists, *Evol. Anthro.* **10**:16–33.
Chapman, L. J., Chapman, C. A., Crisman, T. L., and Nordlie, F. G., 1998, Dissolved oxygen and thermal regimes of a Ugandan crater lake, *Hydrobiol.* **385**:201–221.

Cowlishaw, G., 1999, Predicting the decline of African primate biodiversity: An extinction debt from historical deforestation, *Cons. Biol.* **13**:1,183–1,193.

Cowlishaw, G., and Dunbar, R., 2000, *Primate Conservation Biology*, University of Chicago Press, Chicago.

Estrada, A., and Coates-Estrada, R., 1996, Tropical rainforest fragmentation and wild populations of primates at Los Tuxtlas, Mexico, *Int. J. Primatol.* **17**:759–783.

Fairgrieve, C., 1995, The comparative ecology of blue monkeys (*Cercopithecus mitis stuhlmanni*) in logged and unlogged forest, Budongo Forest Reserve, Uganda: The effects of logging on habitat and population density, Ph.D. Dissertation, University of Edinburgh, Edinburgh, Scotland.

FAO, 1990, Forest resources assessment 1990–Tropical countries, Food and Agriculture Organization of the United Nations, Forestry Paper 112, Rome.

FAO, 1999, State of the world's forests, Food and Agriculture Organization of the United Nations, Rome.

Government of Uganda, 1992, 1991 national housing and rural settlement census, Kampala, Uganda.

Hamilton, A. C., 1974, Distribution patterns of forest trees in Uganda and their historical significance, *Vegetation* **29**:218–228.

Hamilton, A. C., 1984, *Deforestation in Uganda*, Oxford University Press, Nairobi.

Hanski, I., 1994, Patch-occupancy dynamics in fragmented landscapes, *TREE* **9**:131–135.

Hanski, I., and Gilpin, M. E., 1997, *Metapopulation Biology: Ecology, Genetics, and Evolution*, Academic Press, San Diego.

Harrison, S., Murphy, D. D., and Ehrlich, P. R., 1988, Distribution of bay checkerspot butterfly, *Euphydryas editha bayensis*: Evidence for metapopulation model, *Am. Nat.* **132**:360–382.

Howard, P. C., 1991, *Nature Conservation in Uganda's Tropical Forest Reserves*, IUCN, Gland, Switzerland.

Kaipiriri, M., 1997, Local use of non-timber forest products at Kibale Forest, Uganda, Master's Thesis, Makerere University, Kampala, Uganda.

Kammen, D., 1995, Cookstoves for the developing world, *Sci. Amer.* **273**:64–67.

Kinnaird, M. F., and O'Brien, T. O., 1999, Ecological effects of wildfire on lowland rainforest in Sumatra, *Con. Biol.* **12**:954–956.

Laurance, W. F., and Bierregaard, R. O., Jr., 1997, *Tropical Forest Remnants: Ecology, Management, and Conservation of Fragmented Communities*, University of Chicago Press, Chicago.

Lawes, M. J., Mealin, P. E., and Piper, S. E., 2000, Patch occupancy and potential metapopulation dynamics of three forest mammals in fragmented afromontane forest in South Africa, *Con. Biol.* **14**:1,088–1,098.

Lovejoy, T. E., Bierregaard, R. O., Jr., Rylands, A. B., Malcolm, J. R., Quintela, C. E., Harper, L. J., Brown, K. S., Powell, A. H., Powell, G. V. N., Schubart, H. O. R., and Hays, M. B., 1986, Edge and other effects of isolation on Amazon forest fragments, in: *Conservation Biology: The Science of Scarcity and Diversity*, M. E. Soulé, ed., Sinauer Associates, Sunderland, pp. 257–285.

MFEP, 1992, Population and housing census report, Ministry of Finance and Economic Planning, Government of Uganda, Kampala, Uganda.

Naughton-Treves, L., 1997, Farming the forest edge: Vulnerable places and people around Kibale National Park, Uganda, *The Geographical Review* **87**:27–46.

Naughton-Treves, L., 1998, Predicting patterns of crop damage by wildlife around Kibale National Park, Uganda, *Con. Biol.* **12**:156–168.

Naughton-Treves, L., 1999, Whose animals? A history of property rights to wildlife in Toro, western Uganda, *Land Degradation and Development* **10**:311–328.

Naughton-Treves, L., and Chapman, C. A., 2002, Fuelwood resources and forest regeneration on fallow land in Uganda, *J. Sustain Forestry* **14**:19–32.

Naughton-Treves, L., Treves, A., Chapman, C. A., and Wrangham, R. W., 1998, Temporal patterns of crop raiding by primates: Linking food availability in croplands and adjacent forest, *J. Appl. Ecol.* **35**:596–606.

NEMA (National Environment Management Authority), 1997, Kabarole District environment profile, NEMA, Kampala, Uganda.

Nepstad, D. C., Veríssimo, A., Alencar, A., Nobre, C., Lima, E., Lefebvre, P., Schlesinger, P., Potter, C., Moutinho, P., Mendoza, E., Cochrane, M., and Brooks, V., 1999, Large-scale impoverishment of Amazonian forests by logging and fire, *Nature* **398**:505–508.

Oates, J. F., 1974, The ecology and behaviour of black-and-white colobus monkeys (*Colobus guereza* Ruppell) in East Africa, Ph.D. Thesis, University of London, London, UK.

Oates, J. F., 1977, The guereza and its food, in: *Primate Ecology*, T. H. Clutton-Brock, ed., Academic Press, London, pp. 275–321.

Oates, J. F., 1996, Habitat alteration, hunting, and the conservation of folivorous primates in African forests, *Austral. J. Ecol.* **21**:1–9.

Onderdonk, D. A., and Chapman, C. A., 2000, Coping with forest fragmentation: The primates of Kibale National Park, Uganda, *Int. J. Primatol.* **21**:587–611.

Osmaston, H. A., 1959, Working plan for the Kibale and Itwara Central Forest Reserves, Toro District, W. Province, Uganda, Forest Department, Uganda.

Paterson, J. D., 1991, The ecology and history of Uganda's Budongo Forest, *Forest and Conservation History* **35**:179–187.

Place, F., and Otsuka, K., 2000, Population pressure, land tenure, and tree resource management in Uganda, *Land Economics* **76**:233–251.

Pulliam, R., 1988, Sources, sinks, and population regulation, *Am. Nat.* **132**:652–661.

Redford, K. H., 1992, The empty forest, *Bioscience* **42**:412–422.

Richards, P. W., 1996, *The Tropical Rain Forest*, (2nd edition), Cambridge University Press, Cambridge.

Sjaastad, E., and Bromley, D. W., 1997, Indigenous land rights in Sub-Saharan Africa: Appropriation, security, and investment demand, *World Development* **25**:549–562.

Steinhart, E. I., 1971, Transition in Western Uganda: 1891-1901, Ph.D. Dissertation. Northwestern University, Chicago.

Struhsaker, T. T., 1975, *The Red Colobus Monkey*, University of Chicago Press, Chicago.

Struhsaker, T. T., 1976, A further decline in numbers of Amboseli vervet monkeys, *Biotropica* **8**:211–214.

Struhsaker, T. T., 1997, *Ecology of an African Rain Forest: Logging in Kibale and the Conflict between Conservation and Exploitation*, The University Press of Florida, Gainesville, Florida.

Taylor, B. K., 1962, *The Western Lacustrine Bantu*, Sidney Press Ltd., London, UK.

Terborgh, J., 1992, *Diversity and the Tropical Rain Forest*, Scientific American Library, New York.

Thomas, C. D., 1994, Extinction, colonization, and metapopulations: Environmental tracking by rare species, *Con. Biol.* **8**:373–378.

Thomas, C. D., Thomas, J. A., and Warren, M. S., 1992, Distributions of occupied and vacant butterfly habitats in fragmented landscapes, *Oecologia* **92**:563–567.

Tutin, C. E. G., 1999, Fragmented living: Behavioural ecology of primates in a forest fragment in the Lopé Reserve Gabon, *Primates* **40**:249–265.

Tutin, C. E. G., White, L. J. T., and Mackanga-Missandzou, A., 1997, The use of rainforest mammals of natural forest fragments in an equatorial African savanna, *Con. Biol.* **11**:1,190–1,203.

Wallmo, K., and Jacobson, S. K., 1998, A social and environmental evaluation of fuel-efficient cook stoves and conservation in Uganda, *Env. Conser.* **25**:99–108.

RELATIONSHIPS BETWEEN FOREST FRAGMENTS AND HOWLER MONKEYS (*ALOUATTA PALLIATA MEXICANA*) IN SOUTHERN VERACRUZ, MEXICO

Erika M. Rodriguez-Toledo, Salvador Mandujano, and Francisco Garcia-Orduña*

1. INTRODUCTION

Los Tuxtlas, a region located in the state of Veracruz, Mexico, is the most northern geographic distribution of tropical rain forest in America (Dirzo and Miranda, 1991). In this region there are two species of primates, the howler monkey (*Alouatta palliata mexicana*) and spider monkey (*Ateles geoffroyi vellorosus*). Populations of both species have declined 90% (Estrada and Coates-Estrada, 1995). For this reason, these species are considered to be vulnerable according to NOM-059-ECOL-1994 and endangered according to Appendix I of CITES (Rylands et al., 1995). In Mexico, hunting or capturing these primates is illegal, however, the practice continues in the region (Rodriguez-Luna et al., 1987). Nevertheless, the main reason of reduction in distribution and abundance is the dramatic deforestation of tropical rain forest (Silva-Lopez et al., 1988; Estrada and Coates-Estrada, 1995). Loss in 1986 alone was estimated at 84% of the original forest (Dirzo and Garcia, 1992), but the percentage is even higher today. As a consequence, there is relatively intact habitat found only in the higher parts of the region's volcanoes, while in the lower areas many *Alouatta* and *Ateles* groups have been forced to live in forest fragments (Garcia-Orduña, 1995). This is, unfortunately, the most common scenario for the monkeys, thus creating urgency for the development of management plans designed to conserve the primates not only in relatively intact reserves, but also in the disturbed landscape.

---

* Erika M. Rodriguez-Toledo, Posgrado en Manejo de Fauna Silvestre, Instituto de Ecología A. C., km2.5 Carret. Antigua a Coatepec No.351, Congregación el Haya, CP91070, Xalapa, Veracruz, México. Salvador Mandujano, Departamento de Ecología y Comportamiento Animal, Instituto de Ecología A. C., km2.5 Carret. Antigua a Coatepec No.351, Congregación el Haya, CP91070, Xalapa, Verzcruz, México. Francisco Garcia-Orduña, Instituto de Neuroetología, Universidad Veracruzana, Ave. Dos Vistas s/n, km2.5 Carret. Xalapa-Veracruz, CP91000, Xalapa, Veracruz, México. Correspondence to S. Mandujano (email: mandujan@ecologia.edu.mx).

Primates in Fragments: Ecology and Conservation
Edited by L. K. Marsh, Kluwer Academic/Plenum Publishers, 2003

In the last 30 years, data about *Alouatta* and *Ateles* ecology in Los Tuxtlas has been gathered (Estrada, 1982; Estrada and Coates-Estrada, 1995; Garcia-Orduña, 1995; Juan-Solano, 2000; Silva-Lopez et al., 1988). Specifically, studies have provided evidence that the fragmentation influences certain demographic and behavioral factors (Jimenez, 1992, Silva-Lopez et al., 1993; Garcia-Orduña 1996, Estrada et al., 1999; Juan-Solano, 2000). Studies on the relationships between forest fragment characteristics and *Alouatta* group sizes, have been done in this region (Rodriguez-Luna et al., 1987; Silva-Lopez et al., 1988; Estrada and Coates-Estrada, 1995; Estrada et al., 1999). The main result of these studies is that as a fragment increases in surface and quality, the size of the *Alouatta* groups also increases. Evidently, this result provides us with initial knowledge of the possible consequences of forest fragmentation on primate species. In this chapter we present results on the relationships between fragment characteristics and howler monkeys in this region but in another highly altered landscape. If similar results are found in many fragments within Los Tuxtlas, then it may be possible to make some generalizations about the impact of fragmentation on this primate species and, from this, management implications.

We think that further research is needed to understand the functional relationship between landscape structure and primate population dynamics at a regional scale. In this respect, metapopulation theory could provide a feasible, conceptual, and methodological framework for the development of this kind of research. A demonstration of the relationships between forest fragment characteristics and group size of monkeys is the first step in understanding how fragmentation affects population dynamics and extinction-survival probabilities at a regional scale. In conservation applications, the usual question of interest is how much the viability of a species is affected by the number and size of fragments within a region and by the rates of movement of individuals among fragments (Hoopes and Harrison, 1998).

2. PROJECT OBJECTIVES

The principal objective is to analyze the possible effects of forest fragmentation on demography and behavior of the *Alouatta* and *Ateles* metapopulations in the southern buffer zone of the Special Biosphere Reserve Los Tuxtlas. An *in situ* conservation program could be possible if we consider the spatial and temporal metapopulation dynamics at a landscape level. We will thus be able to define local population viability and effective population size. These measures will lead to the establishment of certain fragments as priority areas that need increased connectivity through corridors. Therefore, this project is based on the theory and methods of landscape ecology, population and metapopulation ecology, and conservation biology. We intend to 1) develop actual and potential digital landscape models; 2) analyze the local primate groups (hereafter "subpopulations" or "local populations") and metapopulation dynamics to estimate the local subpopulation extinction probability and regional metapopulation persistence; 3) know the behavioral variation of foraging, fragment use, and movements; and 4) propose management strategies to increase the persistence probability at a regional scale.

3. METAPOPULATION BACKGROUND

Metapopulation theory has been used recently to address the conservation problem of animals in fragments (Harrison, 1994). The principal paradigm of the metapopulation concept for conservation is to maintain equilibrium between the extinction-colonization processes. Theoretically, the persistence of a metapopulation at both temporal and spatial scales depends on the equilibrium between extinction and the rate of colonization of vacant fragments (Hanski, 1999). Metapopulations exist as various local populations of the same species within a fragmented system surrounded by an inhospitable matrix (McCullough, 1996). Under certain conditions, a local population may go extinct or may colonize previously uninhabited fragments. Thus, metapopulations are maintained over a region (Wiens, 1996).

The extinction rate depends on factors associated with local populations and the landscape. The main factor is the structure and abundance of the local population, because at a small size, events such as stochastic demography and environment, allelic effect, and genetic depletion, could increase the probability of extinction (Gotelli, 1998). There are other important factors that affect the survival of the local population within the landscape, such as connectivity among fragments, number, size, quality and age of fragments, the number, length, and quality of corridors, and the matrix characteristics surrounding each fragment (Burkey, 1989). In general, it is expected that as size and quality increase and isolation decreases, the probability of extinction diminishes and colonization increases. Thus, a metapopulation within a landscape of small size, low quality, and high isolation of fragments, has a greater probability of extinction because of a negative balance of colonization-extinction rates (Hanski, 1999).

First we need to define metapopulation. A metapopulation could be "natural" or the result of the segmentation of an original population (Hanski, 1999). Therefore, some fragments could have a fraction of individuals of the original population. This group of individuals may form a new local population that, throughout time, could have its own demographic process. The viability of this new local population will depend on intrinsic factors such as the number of individuals and sex, age, and genetic composition of the fraction of the original segmented population. As a consequence, the extinction and persistence probability of each new local population will depend on its demographic dynamic and the movements of individuals among populations. Thus, a new non-natural metapopulation could be formed if three conditions are met: density dependence of local population dynamics, spatial asynchrony in local population dynamics, and limited dispersion linking the local populations (Hanski, 1991).

In the original metapopulation model (Levins, 1970; cited in Hanski, 1999), all fragments are of equal quality, equally spaced, and dispersed. However, few metapopulations meet this simplified definition (Harrison, 1994). Actually, there are other concepts of metapopulation where a finite number of local populations, landscape structures, and non-random movements are considered. For details of these models, see Hanski (1999). Although most populations and habitats are patchy at some spatial scales, this does not always imply the type of dynamics portrayed in metapopulation models (Hoopes and Harrison, 1998). Important metapopulation attributes may be expected to arise when habitats are fragmented at a spatial scale comparable to a species' long-distance dispersal capabilities. Alternatively, habitats may be fragmented so finely those populations are not really subdivided, or so coarsely that populations are completely isolated from one another. In these cases, we have a patchy population and a non-equilibrium metapopulation, respectively. Also, very uneven fragmentation may lead to mainland-island dynamics in which only the largest populations and habitat fragments really matter for persistence. Thus, initial information about landscape structure, fragment occupation, population dynamics, and individual movements are needed to determine if local populations are structured as metapopulations. From this, it will be possible to analyze the information using metapopulation models.

4. POSSIBLE METAPOPULATION SCENARIO IN LOS TUXTLAS

Originally, howler and spider monkeys were throughout the region of Los Tuxtlas. The area of this region was estimated at 155,122 ha (Diario Oficial de la Federación, 1988), and was dominated by different types of tropical forest, principally tropical rain forest (Rzedowski, 1978). An important aspect is that an abrupt discontinuation of the original vegetation did not exist. This region was inhabited by one pandemic or genetically intermixed population of each primate species. However, the high rate of deforestation as a consequence of cattle breeding and crop production (Guevara et al., 1997) has drastically reduced the original vegetation (Dirzo and Garcia, 1992). Thus the original populations were segmented and the primates were forced into fragmented habitat. As a consequence, the original populations of *Alouatta* and *Ateles* in Los Tuxtlas could be described as a "non-natural" metapopulation. Regardless, Estrada and Coates-Estrada (1995) describe the need for conservation of these primates at landscape and metapopulation scale.

According to predictions based on metapopulation theory at a regional scale, we expect a higher probability of persistence for *Alouatta* and *Ateles* metapopulations as long as more fragments are occupied and group sizes increase. Occupation would depend on the probability of fragment colonization, which in turn is determined by the size, quality, and connectivity of fragments. The largest well connected fragments and those with the best habitat conditions would have a greater likelihood of being occupied and/or colonized by these primates. Furthermore, the larger a monkey group is, the lower the likelihood of its local extinction due to effects such as stochasticity and/or demography. If the landscape has been highly transformed due to deforestation so that the number, size, quality, and connectivity of fragments are quite low, the probability of persistence on a regional scale will decrease due to limited fragment occupation and less colonization of empty fragments. This type of scenario clearly indicates the pressing need for a proposal of measures to mitigate and reverse fragmentation. Thus, an initial step is to describe the landscape structure and occupation of fragments by primates to know how the fragment characteristics are associated with monkey presence/absence and group size. In this chapter we present an analysis of these aspects only for *Alouatta*. The *Ateles* results are beyond the scope of this chapter.

5. STUDY AREA

Los Tuxtlas is naturally divided into two parts: the north volcano "San Martin Tuxtlas," and the south volcano "San Martin Pajapan." This region was declared a Special Biosphere Reserve in 1988 (Figures 1 and 2). The elevation in this region goes from sea level to 1,780 m. The climate is temperate to warm with a mean annual temperature of 24 to 26°C, with a minimum of 16°C and maximum of 35°C (García, 1981). Annual precipitation ranges between 1,850 and 4,600 mm. The principal vegetation type is tropical evergreen rain forest according to the Miranda System (Miranda and Hernandez, 1963). Research was conducted in basins originating from the "Sierra Santa Marta" and "San Martin Pajapan" volcanoes in a buffer zone south of the Los Tuxtlas Biosphere Reserve (18°18' to 18°26' N, and 94°45' to 94°55' W). Specifically, the study area included eight *ejidos* from the council of Tatahuicapan de Juarez. The *ejidos* are Mirador Pilapa, Magallanes, López Arias, Venustiano Carranza, Tecuanapa, Úrsulo Galván, Piedra Labrada, and Guadalupe Victoria. The study area was defined as the landscape formed by two rivers (Tecuanapa and Pilapa) and their corresponding watersheds (Figure 1). The total study area was 3,987 ha and was a representative habitat for the Los Tuxtlas area. But the width of these rivers could be in many parts a natural barrier that limits the constant movement of individuals outside the landscape study area. Thus, for study proposes we considered the landscape a closed area. Future studies of primates' movements will show if this is true.

6. METHODOLOGY

6.1. Landscape Description

A digital map of the zone was developed using ArcView. The program was used to calculate fragment areas and perimeters as well as distances to the continuous forest edge, the nearest fragment, and the nearest town. Aerial photographs were taken (INEGI, 1999) to create a map at 1:20,000 scale. Photo interpretation that took into account the tone and texture of the photos facilitated the identification and demarcation of evergreen rain forest vegetation and secondary vegetation fragments. According to Glander (1975), we defined "fragment" as dominant vegetation having an arboreal height >10 m. In some cases, an area had secondary vegetation but it was <10 m. If these secondary vegetation areas were the surrounding matrix, then they were not considered a fragment. But if the secondary vegetation was close to primary vegetation, we considered both as one fragment. Also, a differentiation was made between relatively undisturbed fragments, those characterized by trees height >15 m, and significantly disturbed fragments, which featured secondary vegetation under 15 m. The classification of fragment in undisturbed and disturbed was confirmed during field work.

In order to ascertain the degree of fragment isolation, three distances were taken into account: the shortest measurement from each fragment to the continuous forest at Sierra Santa Marta or San Martin Pajapan (considered possible "sources"), the shortest distance from each fragment to the nearest town, and the straight-line distance between fragments.

a)

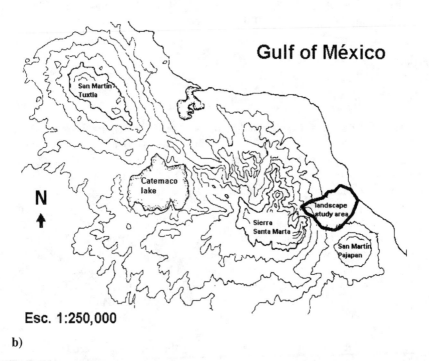

b)

Figure 1. Geographic location of Veracruz state and the region of Los Tuxtlas (a), and the Los Tuxtlas Special Biosphere Reserve, showing the landscape study are, circled (b).

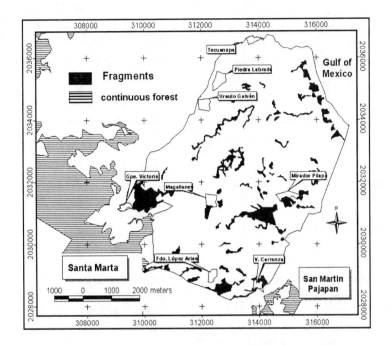

Figure 2. The landscape study area showing towns, fragments (black), and continuous forest (stripe).

To estimate fragment shape, we used a shape index (SI) that takes into consideration fragment perimeter (P) and area (A):

$$SI = \frac{P}{2\sqrt{\pi A}} \tag{1}$$

This index evaluated the complexity of fragment shape by comparing it to a standard circular fragment of the same surface area (Mas and Correa-Sandoval, 2000). When the index value is exactly equal to the standard, the fragment is circular in shape. As the index increases, fragment shape becomes more complex or irregular.

6.2. *Alouatta* Sampling

Two field studies were conducted, one from January to March of 2001 and the other from February to April of 2002. During this period the presence and location of each fragment were corroborated and rectified. A census of all fragments identified in the study area was also conducted and the presence or absence of *Alouatta* groups was noted for each. In order to determine the presence of monkeys, two to three people surrounded the fragment and made observations with binoculars and vocalizations over a period of four to five hours from 0800 to 1800 hours. This procedure was followed for each of the fragments smaller than 10 ha. In fragments larger than 10 ha, samples were taken in one day. Every fragment was checked from four to six times during the study. A total of 1,000 hours of observation were sampled over all fragments. When a group was spotted, an attempt was made to count the total number of individuals and in the greatest possible detail to determine group composition (sex and age).

We estimate the population density (D_{pop}) as

$$Dpop = \frac{n}{A} \quad , \tag{2}$$

where n = the sum of individuals from all groups inhabiting the occupied fragments and A = the total surface of complete landscape.

Also, we estimated the specific ecological density (Ds_{ecol}) of fragments as

$$DSecol = \frac{n_i}{Ah_i} \quad , \tag{3}$$

where n_i = the number of individuals from each group and Ah_i = the surface of the fragment where the group is found. In this case, we estimated ecological density considering both occupied and unoccupied fragments. This procedure diminished the possible bias of considering only occupied fragments for estimating ecological density. In the two cases, density were expressed as individuals per hectare.

We estimated the distance of each group to the nearest-neighbor group and also the isolation of each group with respect to all groups in the landscape using ArcView. These distances are an indirect approximation of the movements of individuals among groups. If there is minor distance between fragments, there is greater probability of dispersal by the primates throughout the landscape.

In order to ascertain the relationship between monkey presence or absence and group size with certain fragment characteristics such as size, shape, isolating distance, and vegetation height, independent tests of Chi squared were applied to presence or absence data, and simple linear correlation tests were applied to group size data. Non-significant results indicated that presence and size were independent from fragment characteristics.

7. RESULTS AND DISCUSSION

7.1. Fragment Characterizations

The landscape in the study area has been dramatically altered since 1980. Originally, there were 3,987 ha of forest for primates. Now only 335 ha (8.4%) remain. The surrounding matrix is formed principally by crops like maize and beans and livestock pastures. Unfortunately, economic practices, poor technological development, and low financial support have led to an extensive conversion of the original tropical forest within the last two decades.

We identified 64 fragments in this study (Figure 3). The range of elevation is from sea level to 700 m. All fragments were located on the shores of the rivers and streams and just a few of them in the top of some inaccessible hills. The slope of the fragments is very steep (>30°). Some of the fragments near the sea coast had less severe slope but had permanent inundation. The inaccessibility of these fragments was the principal cause for the lack of deforestation. Of the total fragments, 81% are less than 5 ha, and only five fragments (8%) are over 10 ha (Table 1). The biggest fragment is 67 ha. Therefore, the actual landscape is characterized as a small habitat area for primates, divided into smaller fragments.

Using canopy height as an indicator of the original vegetation, disturbance might be considered a weak habitat quality indicator; however, there are some interesting comparisons found in this study. For instance, 27 fragments had relatively well-preserved canopy (trees taller than 15 m), while 37 fragments had disturbed vegetation (trees <10 m) (Figure 3). When compared to fragment size, all of the fragments larger than 10 ha had a well-preserved canopy (>10 m height), while 58% of the fragments smaller than 10 ha had suffered disturbance (<10 m height). This result was statistically significant (X^2= 5.1, d.f. = 1, p = 0.02). Thus, the landscape is characterized by many small fragments with some grade of disturbance in the arboreal stratum. There were no obvious relationships between fragment disturbance and distance to towns, since both well-preserved and disturbed fragments were equally spaced from the towns.

Many of the fragments are narrow and long and follow streams and rivers. Taking this into account, 64 fragments (45%) had a relatively irregular shape (Table 1). A more irregular fragment might increase the relationship of surface to perimeter and could increase the edge effect. It was found that the disturbance of the fragments was dependent of their shape (X^2 = 16.4, d.f. = 3, p = 0.001) and their size (X^2 = 8.4, d.f. = 1, p = 0.005). The biggest fragments had the most irregular shape (Table 1). To better understand this relationship in terms of primate survival, we would need to know the floristic composition and to be able to evaluate the edge effect.

Relatively well-preserved and continuous forest is found at the top of the "Sierra Santa Marta" and "San Martin Pajapan" volcanoes. In theory, the closer the fragment was to the source of continuous habitat the more likely the fragment would be colonized. The majority of fragments are relatively far away from a potential source habitat. At all fragments, 85% were greater than 1,000 m from continuous forest with an average of 2,800 m (Table 1).

Conversely, 85% of the fragments were less than 200 m from the nearest fragment (Table 1); the average distance from the nearest fragment was 123 m. However, this distance is a straight line from fragment to fragment and does not necessarily mean that there is a real corridor between them. In this site, there is a lack of data about primates walking at ground level moving from one fragment to another. If the mean distance between fragments is beyond the natural movements of the primates in a place with no corridors, then the groups inhabiting the fragments are relatively isolated from the nearest fragment. In contrast, if the actual distance among fragments is within the dispersal capability of the primates, then the dynamics of the population depends only on the groups inhabiting fragments and not on the continuous forest.

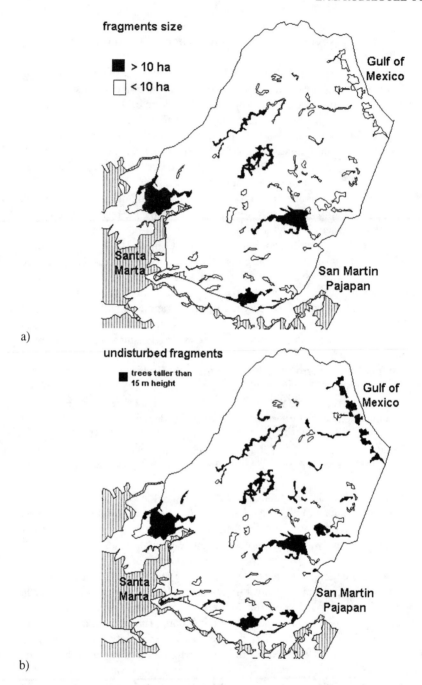

Figure 3. Location of the 64 fragments in the landscape study area. Map a shows the size of the fragments, and Map b shows the relatively well-preserved (tree taller than 15 m height) fragments.

Table 1. Number of fragments in the landscape (total 3,987 ha) and the number of fragments occupied by at least one individual of *Alouatta palliata mexicana* in the study area at Tatahuicapan de Juarez, Los Tuxtlas, Veracruz, Mexico.

Fragment Characteristics	Number of Fragments in the Landscape	Number of Fragments Occupied by *Alouatta*
Size		
<10 ha	59	11
11 to 70 ha	5	5
Canopy Disturbance		
Well-preserved		
<10 ha	22	7
11 to 70 ha	5	5
Poor-preserved		
<10 ha	37	4
11 to 70 ha	0	0
Shape Index (SI)		
Regular (SI= 1.1 to 2.0)		
<10 ha	35	8
11 to 70 ha	0	0
Moderate (SI= 2.1 to 3.0)		
<10 ha	21	3
11 to 70 ha	2	2
Irregular (SI= 3.1 to 5.0)		
<10 ha	3	0
11 to 70 ha	3	3
Distance to Continuous Forest (m)		
<1,000	10	1
1,001 to 2,000	12	4
2,001 to 7,000	42	11
Distance to Closest Fragment (m)		
<200	56	14
201 to 400	7	2
401 to 600	1	0
Distance to Towns (m)		
<900	37	7
901 to 1,800	17	2
1,801 to 2,700	10	7

Another characteristic of the study area is that fragments were relatively close to towns (Figure 2). In fact, 58% of fragments were less than 1,000 m from the nearest town (Table 1). The fragments that were particularly close include Magallanes, Guadalupe Victoria, Mirador Pilapa, and Lopez Arias. In this case, the inaccessibility is not the principal reason for the conservation of these fragments. Local people recognize that fragments represent a source of water for their crops and personal needs. This is an important aspect because it opens the possibility for a management plan for the conservation of primates through the understanding of the beneficial and ecological preservation of fragments by the people.

7.2. Howler Population

In the study area there were 16 fragments, 13 with more than one individual of *Alouatta* and three with isolated individuals (Figure 4). Therefore, only 25% of the 64 fragments in the region were occupied (Table 1). This distribution is very small compared to the available number of fragments, which suggests that the primate survival in this landscape may be problematic.

A total of 80 individuals of *Alouatta* were counted in the occupied fragments. The size of the groups ranged from 2 to 15 individuals; while the average group size was 6 ±1 (S.E.) individuals. Three isolated males of *Alouatta* inhabited separate fragments. If these individuals were in the process of dispersal or lived permanently in these fragments is unknown. Because fragments are small and isolated, it is important to determine the

Figure 4. Distribution (black fragments) of *Alouatta* groups in the study landscape. The shaded area represents the continuos forest.

probability of local extinction and metapopulation persistence in this landscape. We predict two possible sceneries in the near future: 1) a high risk of local extinction of many groups simply by stochastic demography and environment and 2) an overpopulation of some fragments because of the small fragment size and low connectivity with other fragments that limits the dispersal of individuals.

The population density (total number of individuals per entire landscape area) was estimated as 0.02 individuals/ha. This estimate is very low because it includes the entire landscape area. However, considering the "ecological density" (total number of individuals per fragment area) the estimate was 0.28 ± 0.09 individuals/ha. This estimate is more realistic because the abundance is divided by habitat. It is very interesting to note that as fragment size decreases, the ecological density increases (Figure 5); and bigger fragments show a low ecological density. This result means that overpopulation exists in some very small fragments throughout the study area. Thus, potential inter-specific competition among individuals could exist because of the low carrying capacity in small fragments. The effect on survival and fecundity rates is unknown and requires urgent study.

A very important result is that a relationship was observed between fragment size and *Alouatta* presence in the fragments ($X^2 = 4.46$, d.f. = 1, $p = 0.04$). Of the fragments that measured under 10 ha, only 19% were occupied; in contrast, in those fragments larger than 10 ha, groups were present in 100% (Table 1). Group presence was also associated with canopy height ($X^2 = 8.78$, d.f. = 1, $p = 0.005$), as 75% of the groups were found in fragments with a well-preserved canopy (Table 1). In contrast, the presence of monkeys was

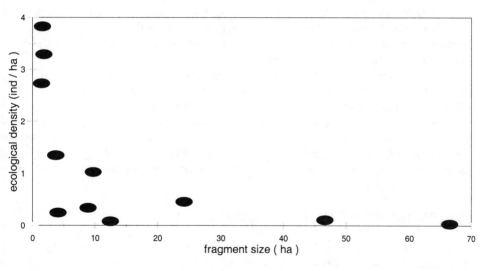

Figure 5. Relationships between fragment size and ecological density (number of individuals per fragment size).

independent from fragment shape ($X^2 = 0.104$, d.f. $= 1$, $p = 0.75$). A similar number of groups were found in regular- and irregular-shaped fragments (Table 1). There is a positive association between monkeys who live in large fragments and those that have a well preserved canopy, which shows that size and distribution do have an effect over population. Therefore, howlers show a preference for the characteristics of fragments. An important question is if this preference is the result of an active selection of these fragments by howlers or simply the result of historic processes of isolation for each group.

Group presence was not significant as compared to the distance to the continuous forest (Table 1, $X^2 = 1.4$, d.f. $= 2$, $p = 0.51$). This result is important because it implies that howlers do not depend on continuous forest as source/mainland area. Therefore, the persistence of local groups depends on demographic structure and on the movements of individuals among fragments. In continuous forest there is a high abundance of primates. So, creating corridors between fragments would possibly increase the isolated subpopulations.

The average distance of the nearest group was 393 ± 328 m; while the average distance of a particular group to all other groups was $2,731 \pm 1,735$ m. This result shows the critical situation of primates, since the isolation among groups is quite large. There is a negative consequence to individuals and groups as a result of such long distances. The long distance to the nearest group could be an isolation factor preventing some individuals from dispersal. If dispersal were attempted, there is a probability of mortality during the movement from one group to another. There is also a chance that individuals remain isolated. For example, we observed three males that inhabited an isolated fragment. In either case, the consequence is a reduction of primates in this landscape.

Group presence to the nearest town was not significant (Table 1, $X^2 = 2.1$, d.f. $= 2$, $p = 0.35$). A group's proximity to towns did not influence a group's presence. Some groups inhabit fragments very close to people, such as in Mirador Pilapa, Magallanes, Lopez Arias, and Guadalupe Victoria. A possible reason for this might be that the illegal hunting and capture of monkeys has recently decreased. In some towns we noted a slight positive modification of the people's perception of primates.

Finally, there was no relationship between group size and fragment size ($r = -0.10$, $F = 0.09$, d.f. $= 1$, 10, $p = 0.77$), fragment shape ($r = 0.15$, $F = 0.20$, d.f. $= 1$, 10, $p = 0.66$), distance to continuous forest ($r = 0.001$, $F = 0.0002$, d.f. $= 1$, 10, $p = 0.99$), to the nearest town ($r = -0.27$, $F = 0.72$, d.f. $= 1$, 10, $p = 0.42$), and distance to the nearest fragment ($r = 0.31$, $F = 0.96$, d.f. $= 1$, 10, $p = 0.35$). These results may be due to the low number of individuals that inhabit bigger fragments. In fact, we had hoped to find a greater number of individuals in larger fragments, as had been found in other studies (see Section 8 below).

8. COMPARISON WITH OTHERS STUDIES

It is interesting to compare our results with data reported by Estrada et al. (1999) for a landscape north of Los Tuxtlas. These authors analyzed the group size of *A. p. mexicana* in relation to fragment characteristics such as size and isolation. The main landscape characteristics and the situation of the howler monkey in the north of Los Tuxtlas are compared with our results on a similar region in the south (Table 2). The primary difference between the sites is the northern sites have larger forest area and a minor fragmentation level

Table 2. Comparison of the howler monkey in two different landscapes in Los Tuxtlas. The data for the north landscape were obtained from Estrada et al. (1999). South data are this study.

Characteristics	South	North
Landscape total area	3,987 ha	3,200 ha
Number of fragments	64	38
Fragments total area	335 ha	838 ha
Fragments area percentage respect to the landscape total area	8%	26%
Number of fragments under each category size		
<10 ha	92%	71%
11 to 70 ha	8%	18%
71 to 150 ha	*	11%
Average isolation distance between near fragments	123 m	<200 m

* This size class is not in the study fragments.

as compared to our southern study sites. Differences in distribution frequency of fragment size for the two landscapes are reflected in *Alouatta* abundance for the two sites. In the north landscape there are 157 individuals and 27 troops. In contrast, the south has 80 individuals and 13 troops. Likewise, in the northern landscape the howler number increases as the area of fragments increases. In our study, we did not find this relationship.

Based on the data in Estrada et al. (1999), we calculated a linear regression equation between the number of individuals (N) and fragment size (S), such an equation results as N = 2.33 + (0.085 × S), which was statistically significant (r = 0.58, F = 17.9, d.f. = 1, 37, p = 0.001). On this basis, we estimated the number of individuals expected to inhabit each fragment in our study, assuming a similar relationship between fragment area and group size. Results indicated that in our landscape, the number of observed individuals is significantly different from that expected (Figure 6, X^2 = 55.3, d.f. = 10, p = 0.001). More individuals than expected were observed on 6 of the 11 occupied fragments. Specifically, there were fragments (<5 ha) with a higher number of monkeys. Meanwhile, two of the largest fragments (between 10 and 70 ha) had less individuals than the expected number. Apparently, the study area features more individuals than the expected, based on the small size of fragments. This explains why estimation of ecological density in our research is higher, 0.28 individuals/ha, than the 0.029 to 0.036 individuals/ha reported by Estrada and Coates-Estrada (1995).

These apparently contradictory results, that is, observation of relatively large groups of primates in small fragments and small groups in large fragments, may be explained with the combination of several factors: 1) the effect of fragment structure (e.g., difference in size, shape, isolation, quality), 2) the process by which the natural forest became fragmented and how groups of monkeys were left in each fragment, 3) the structure of groups (sex, age, number of individuals) when they were initially isolated, 4) demographic dynamics (birth and death rates, migration rates) for each group during the period when fragments were being

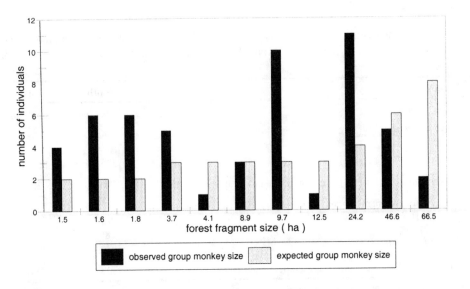

Figure 6. Number of *Alouatta* individuals observed on each fragment where their presence was observed in the study area, number of individuals expected on each fragment as a function of the regression equation obtained from information found in Tables 2 and 3 of Estrada et al.(1999). See text for further details.

occupied, and 5) hunting that took place in previous years. For the moment, our data do not permit the exact determination of which of these factors influence the finding that group size does not increase in fragments that are larger and closer together. Nevertheless, the fact that most of the fragments have a surface area of under 10 ha and feature more disturbed vegetation supports the suggestion of Silva-Lopez et al. (1993) that these factors are key to the presence or absence of monkeys. This does not, however, explain why there are no larger groups of *Alouatta* on less disturbed fragments of substantial size. It is therefore probable that other factors such as primate hunting in previous years (Rodriguez-Luna et al. 1987) and the fragmentation history of the particular site play a role.

9. HOWLERS AND FRAGMENT CHARACTERISTICS

For this research, both fragment size and canopy height were demonstrated to be the most important variables explaining *Alouatta* presence. A well-preserved canopy height means better habitat quality, since there is a greater availability of food resources for howlers (Jimenez, 1992; Silva-Lopez et al., 1993). In fact, fragment quality has been proven to exert a strong influence on survival and abundance of populations of other mammals (Saunders et al., 1991). The research of Estrada and Coates-Estrada (1995) is pertinent, as they found a greater primate presence in fragments where vegetation was taller than in those fragments where canopy height was smaller than 10 m. We found that with a decrease in fragment area, canopy height deteriorates. The way that locals use these areas could be influential (Rodriguez-Luna et al., 1987), although another explanation could be that an edge effect takes place. In large, narrow fragments, trees may fall from strong winds that blow from the Gulf of Mexico in several months of the year, damaging fragment structure.

Alouatta presence and group size do not depend on the distance from fragments to the continuous forest. At this moment, areas with extensive and relatively undisturbed vegetation are found in the high and intermediate parts of the San Martin Pajapan and Sierra Santa Marta volcanoes (Rodriguez-Luna et al., 1987). It might be assumed that these areas are functioning as sources for primates and that the fragments are sinks. Although this aspect requires detailed analysis, our data do not support this explanation, as greater proximity to continuous forest did not mean a larger number of occupied fragments. Therefore, source areas should be seriously considered in management planning. In the study area, howler groups are isolated and there is a high probability that they will disappear due to catastrophes, stochastic events, genetic depletion, or social dysfunction (Lande and Barrowclough, 1987). Hanski (1999) provides corroboration for the concept that with an increase in fragment size, connectivity, and habitat quality, the areas are more likely to be occupied and/or colonized. It is therefore necessary to preserve as much original vegetation in fragments as is possible, giving priority to those larger than 10 ha. Concerning smaller fragments, priority should be given to those that have a canopy height of over 15 m. There are also unoccupied fragments that based on their characteristics could function as potentially good habitats for groups of monkeys. It is therefore necessary to formulate proposals for the creation of corridors that permit individual movement between fragments, thus facilitating fragment recolonization. As the number of occupied fragments increases, so will the probability of persistence for monkey groups in this landscape, and this in turn will assure the conservation of *Alouatta* populations in Mexico.

10. SUMMARY

The relationship between certain fragment characteristics (size, shape, isolation, and vegetation disturbance level), *Alouatta palliata mexicana* presence, and the number of individuals per group was analyzed in a section of the buffer zone found on the Los Tuxtlas Special Biosphere Reserve in the state of Veracruz, Mexico. Fieldwork was carried out in early 2001 and 2002. Sixty-four fragments were identified, representing 8.4% of the 3,987 ha that make up the study area. Of these, 92% had a surface area of under 10 ha, while only 8% exceeded this size. In the largest fragment measured (67 ha) only 42% had a relatively well-preserved canopy height (exceeding 15 m). Specifically, 100% of the fragments measuring over 10 ha had well-preserved canopy, while of the total fragments under 10 ha in size, 58% showed signs of disturbance and only 19% were occupied by howlers. Forty-five percent had a relatively regular shape and were found principally in riparian zones. The fragments were far from continuous forest, which could be a source/mainland of individuals to fragments. Fragments were near to other fragments and to towns. Groups of *Alouatta* were found only in 25% of the 64 fragments. Thirteen troops and three isolated individuals totaled 80 individuals. Average group size was 6 ± 1 individuals (ranging from 2 to 11). Ecological and population density were estimated at 0.28 and 0.02 individuals/ha, respectively. Some very small fragments are overpopulated. Group presence was independent of fragment shape, straight-line distance between fragments, distance to the continuous forest, and distance to the nearest town. The actual distance between groups is very large. Eighty-two percent of the groups were found in fragments with well-preserved canopy height. Clearly, management proposals are urgently needed to increase population size and the number of fragments with primates in this landscape. A management plan is needed where regeneration and reforestation of fragments and corridors are a priority. It is urgent to involve local people in conservation plans that consider their social and economics issues.

11. ACKNOWLEDGMENTS

The authors would like to thank physicist Rosario Landgrave and biologist Norma Corona for their assistance in the analysis of Geographic Information Systems as well as the family of Mateo Gutierrez, especially Ruben, for his hospitality and invaluable help with fieldwork. Also to Odette Brunell for her field assistance. The first author received a grant from the National Council of Science and Technology (CONACYT) for the pursuit of postgraduate research. We would also like to thank the U.S. Fish and Wildlife Service, which provides assistance to the Natural Protection through Conservation Program and the Wildlife Management Masters Program that form part of the Latin American Alliance of Wildlife Management Training. Finally, we are grateful to the Department of Ecologia y Comportamiento Animal of the Instituto de Ecologia A.C., for providing the support necessary for the completion of this research. Actually, the American Society of Primatology brings a grant to help in some aspects of this study.

12. REFERENCES

Burkey, T. V., 1989, Extintion in nature reserves: The effect of fragmentation and the importance of migration between reserve fragments, *Oikos* **55**:75–81.

Diario Oficial de la Federación, 1994, NOM-059-ECOL-1994, Organo del Gobierno Constitucional de los Estados Unidos Mexicanos, Tomo CDLXXXVIII, 10, México, D.F.

Dirzo, R., and Garcia, M. C., 1992, Rates of deforestation in Los Tuxtlas, a Neotropical area in Veracruz, Mexico, *Conserv. Biol.* **6**:84–90.

Dirzo, R., and Miranda, A., 1991, El límite boreal de la selva tropical húmeda en el continente americano: Contracción de la selva y solución de una controversia, *Interciencia* **16**:240–247.

Estrada, A., 1982, Survey and census of howler monkeys (*Alouatta palliata*) in the rain forest of Los Tuxtlas, Veracruz, México, *Am. J. Primatol.* **2**:363–372.

Estrada, A., and Coates-Estrada, R., 1995, La contracción y fragmentación de las selvas y las poblaciones de primates silvestres: El caso de los Tuxtlas, Veracruz, in: *Estudios Primatológicos en México*, E. Rodríguez-Luna, L. Cortés, and J. Martínez-Contreras, eds., Universidad Veracruzana, Xalapa, Veracruz, México, pp. 25–60.

Estrada, A., Anzures, A., and Coates-Estrada, R., 1999, Tropical rain forest fragmentation, howler monkeys (*Alouatta palliata*), and dung beetles at Los Tuxtlas, Mexico, *Am. J. Primatol.* **48**:253–262.

Forman, R. T. T., and Godron, M., 1986, *Landscape Ecology*, John Wiley and Sons, New York.

García, E., 1981, *Modificaciones al sistema de clasificación climática de Koeppen*, Instituto de Geografía, UNAM, México, D.F.

García-Orduña, F., 1995, Fragmentación del hábitat y demografía de primates en la región de los Tuxtlas, Veracruz, in: *Estudios Primatológicos en México*, E. Rodríguez-Luna, L. Cortés, and J. Martínez-Contreras, eds., Veracruz, México, pp. 61–80.

García-Orduña, F., 1996, Distribución y abundancia del mono aullador *Alouatta palliata* y el mono araña *Ateles geoffroyi* en fragmentos de selva del municipio de San Pedro Soteapan, Veracruz, Undergraduate thesis, Universidad Veracruzana, Xalapa, Veracruz, México.

Glander, K. E., 1975, Habitat and Resource Utilization: An ecological view of social organization in mantled howling monkeys, Dissertation thesis, University of Chicago, Illinois.

Gotelli, N. J., 1998, *A Primer of Ecology*, Second Edition, Sinauer Associates, Inc., Sunderland, MA.

Guevara, S., Laborde, J., Liesenfeld, D., and Barrera, O., 1997, Potreros y ganadería, in: *Historia Natural de Los Tuxtlas*, E. González-Soriano, R. Dirzo, and R. C. Vogt, eds., Universidad Nacional Autónoma de México, México, D.F., pp. 43–58.

Hanski, I., 1991, Single-species metapopulation dynamics: Concepts, models, and observations, *Biol. J. Linn. Soc.* 42:17–38.

Hanski, I., 1999, *Metapopulation Ecology*. Oxford University Press, London.

Harrison, S., 1994, Metapopulations and conservation, in: *Large-Scale Ecology and Conservation Biology*, P. J. Edwards, R. M. May, and N. R. Webb, eds., Blackwell Science, Osney Mead, Oxford, pp. 111–128.

Hoopes, M. F., and Harrison, S., 1998, Metapopulation, source-sink and disturbance dynamics, in: *Conservation Science and Action*, W. J. Sutherland, ed., Blackwell Science, Osney Merd, Oxford, pp. 135–151.

INEGI, 1999, Fotografías aéreas, San Juan Volador. Línea 4 a la 7. Escala 1:20,000. Instituto Nacional de Estadística Geográfica e Informática San Juan Volador. E15 A74.

Jiménez, H. J., 1992, Distribución y abundancia del recurso alimenticio en un fragmento de selva alta perennifolia y su uso por *Ateles* y *Alouatta* en el ejido Magallanes (Municipio de Soteapan, Veracruz), Undergraduate thesis, Universidad Veracruzana, Xalapa, Veracruz, México.

Juan-Solano, S., 2000, A comparative study of resource use by howler monkey groups (*Alouatta palliata*) in isolated rainforest fragments of the region of Los Tuxtlas, Veracruz, Mexico, *ASP Bulletin* 24:8.

Lande, R., and Barrowclough, G. F., 1987, Effective population size, genetic variation, and their use in populations management, in: *Viable Populations for Conservation*, M. E. Soulé, ed., Cambridge University Press, Cambridge, UK, pp. 87–123.

Mas, J., and Correa-Sandoval, J., 2000, Análisis de la fragmentación del paisaje en el área protegida "Los Petenes," Campeche, México, *Boletín del Instituto de Geografía, UNAM* 43:42–59.

McCullough, D. R., ed., 1996, *Metapopulations and Wildlife Conservation*, Island Press, Covelo, CA.

Miranda, F., and Hernández, X. E., 1963, Los tipos de vegetación de México y su clasificación, *Boletín de la Sociedad Botánica de México* 28:29–178.

Rzedowskl, J., 1978, *Vegetación de México*, Editorial Limusa, México, D.F.

Rodríguez-Luna, E., García-Orduña, F., Silva-López, G., and Canales-Espinosa, D., 1987, Primate conservation in Mexico, *Prim. Conserv.* 8:114–118.

Rylands, A. B., Mittermeier, R. A., and Rodríguez-Luna, E., 1995, A species list for the New World primates (Platyrrhini): Distribution by country, endemism, and conservation status according to the Mace-Lande System, *Neotropi. Prima.* 3:113–160.

Saunders, D. A., Hobbs, R. J., and Margules, C. R., 1991, Biological consequences of ecosystem fragmentation: A review, *Conserv. Biol.* 5:18–32.

Silva-López, G., García-Orduña, F., and Rodríguez-Luna, E., 1988, The status of *Ateles geoffroyi* and *Alouatta palliata* in disturbed forst areas of Sierra de Santa Marta, Mexico, *Prim. Conserv.* 9:53–61.

Silva-López, G., Benítez-Rodríguez, J., and Jiménez-Huerta, J., 1993, Uso del hábitat por monos araña (*Ateles geoffroyi*) y aullador (*Alouatta palliata*) en áreas perturbadas, in: *Estudios Primatológicos en México*, E. Rodríguez-Luna, L. Cortés, and J. Martínez-Contreras, eds., Universidad Veracruzana, Xalapa, Veracruz, México, pp. 421–435.

Wiens, J. A., 1996, Wildlife in patchy environments: Metapopulations, mosaics, and management, in: *Metapopulations and Wildlife Conservation*, D. R. McCullough, ed., Island Press, Covelo, CA, pp. 53–84

PRIMATES OF THE BRAZILIAN ATLANTIC FOREST: THE INFLUENCE OF FOREST FRAGMENTATION ON SURVIVAL

Adriano G. Chiarello[*]

1. INTRODUCTION

1.1. Atlantic Forest and Fragmentation

The Atlantic forest has been cleared and disturbed since the arrival of the first Europeans in the early 1500s (Dean, 1995). More than 70% of the Brazilian population is concentrated in this region, and it is estimated that less than 8% of the original forest remains as isolated forest (SOS Mata Atlântica et al., 1998). Nevertheless, this forest still harbors an extremely rich biological diversity, second only to the Amazon region, with about 261 mammal species already recorded, 73 (28%) of which are endemics (da Fonseca et al., 1999; Myers et al., 2000). Until the present, 15 species and 24 subspecies of primates have been recognized for the Atlantic forest (Rylands et al., 1996, 2000), including two endemic genera, *Brachyteles* and *Leontopithecus*. The overall biological richness and the high degree of threat make the Atlantic forest one of the top biodiversity "hotspots" in the world (Myers et al., 2000).

The majority of the remaining Atlantic forest fragments are small in area (<100 ha) and isolated from other fragments by pastures and agriculture (SOS Mata Atlântica and INPE, 1993; Ranta et al., 1998). The effects of fragmentation are not only related to area reduction, but also to an increase in isolation among remaining fragments (Skole and Tucker, 1993). These two factors cause severe alterations in the physical and biological environments of a previously intact forest (Lovejoy et al., 1983; Bierregaard et al., 1992). Another related problem is that forest fragmentation leads to an increase in the amount of forest edge. The modified microenvironment and liana proliferation in the edge result in increased tree mortality (Lovejoy et al., 1986; Kapos, 1989; Ferreira and Laurance, 1997) and prevention of regeneration of several tree species (Janzen, 1986).

---

[*] Programa de Mestrado em Zoologia de Vertebrados, Pontifícia Universidade Católica de Minas Gerais, Belo Horizonte, MG, Brazil. Email: bradypus@terra.com.br.

Primates in Fragments: Ecology and Conservation
Edited by L. K. Marsh, Kluwer Academic/Plenum Publishers, 2003

1.2. Other Main Threats

1.2.1. Biological Imbalances

A wide range of reptilian, avian, and mammalian species can potentially prey on primates of the Atlantic forest (Olmos, 1994; Albuquerque, 1995; Passamani, 1995a; Facure and Giaretta, 1996; Stafford and Ferreira, 1996; Galetti et al., 1997; Garla, 1998; Brito, 2000). The predator-prey relationships can, however, be greatly altered in the fragmented landscape of the Atlantic forest. Jaguars, pumas, and harpy eagles are, for example, absent from most Atlantic forest fragments of small size (Chiarello, 1999; Cullen, 1997), since these predators need vast foraging areas in the range of several hundreds to thousands of hectares. It is likely, therefore, that large-bodied primate genera are less affected by big cats and raptors in these fragments.

On the other hand, smaller cats (*Leopardus* spp. and *Herpailurus yaguarundi*) and other carnivores such as the tayra (*Eira barbara*) do survive in small forest fragments and can potentially prey on primates (Passamani, 1995a; Bianchi, 2001). In fact, these "mesopredators" can be present in higher density in forest fragments than in areas with a full complement of mammalian predators ("mesopredator release;" Palomares et al., 1995). Another related problem is that the colonization and fragmentation of this biome have increased the proximity of urban settlements and farms to forest fragments, and, consequently, domestic dogs and cats can also prey on native fauna, including primates. Local extinctions of primates have been documented for several Atlantic forest fragments (de Oliveira and Oliveira, 1993; Ferrari and Diego, 1995; Chiarello, 1999). With less competitors around, the surviving primate species are theoretically subjected to decreased competition pressure ("density compensation hypothesis"), as has been observed in some Amazonian localities (Peres and Dolman, 2000).

1.2.2. Hunting

That hunting is a powerful force in exterminating the megafauna is well known, even in more extensive and less disturbed forest sites of the Neotropics (Robinson, 1996; Bodmer et al., 1997; de Souza-Mazurek et al., 2000; Peres and Dolman, 2000). Because of this factor, several Neotropical sites have a depauperate fauna, especially in terms of large mammals and birds (cracids and tinamids) (Redford, 1992). If the impact of hunting is severe, even in extensive areas of "pristine" forests, the consequences for isolated forest fragments can be disastrous, as they lack the sources of potential new migrants. Additionally, hunters have greater access to forests in a fragmented landscape (Robinson, 1996).

1.2.3. Logging and Fire Intrusion

As with hunting, logging has a long history of disturbance in the Atlantic forest (da Fonseca, 1985; Dean, 1995). This illegal activity is still widely practiced throughout the biome and has been observed even within biological reserves and national parks (Pardini, 2001; personal observation). Forest remnants are more vulnerable to disturbance caused by fires, which are normally rare events in tropical evergreen forests (Uhl et al., 1988). Degraded pastures are the predominant vegetation cover surrounding Atlantic forest fragments and they are highly flammable (Nepstad et al., 1996). Fire events are even more common in regions subjected to moderate to severe seasonality in rainfall, as is the case of the greater part of the Atlantic forest region. Although the effect of these fires on native fauna and flora has been studied in the Amazon forest (Cochrane and Schulze, 1999) and in the Cerrado (savannah-like vegetation of central Brazil) (Marini and Cavalcanti, 1996; Vieira, 1999), fire effects on the Atlantic forest remain poorly documented (Castellani, 1986).

1.3. Goals of this Chapter

The focus in this chapter is to review data on primate richness and abundance collected in a representative sample of Atlantic forest fragments in Brazil. For this, quantitative data from three different regions and 14 fragments were analyzed or reviewed. Different from the majority of other studies that focus on a particular species or on a particular forest fragment or biological reserve, data from the selected sites were collected for all primate species in a range of fragment sizes using the same methodology, thus allowing a comparison of the possible effects of fragmentation. In two of these regions, lowland and lower-montane of northern Espírito Santo state, data were collected in previous studies (Chiarello, 1999; 2000a; Chiarello and Melo, 2001) or are here published for the first time. In the Plateau Region of western São Paulo state, data were reviewed from Cullen (1997). Three main issues are addressed: 1) the relationship between fragment size, primate density, and species richness; 2) the influence of hunting on Atlantic forest primates, and 3) the size of isolated primate populations and their viability in the long run. Although the main discussion focuses on the role of fragment size and hunting pressure in primate communities, the importance of edge effects, forest disturbance, and predation pressure are also considered.

2. MATERIAL AND METHODS

2.1. Study Sites

Data analyzed are derived from research carried out in six forest fragments located in the lowlands (<200 m asl) of northern Espírito Santo, southeastern Brazil (Chiarello, 1999; 2000a; Chiarello and Melo, 2001), and in three other fragments located in the same state but in the lower-montane region (600 to 900 m asl) (Figure 1). The results of the latter are published here for the first time. I also reviewed results from a similar study in five fragments located in the Plateau Range of western São Paulo state, southeastern Brazil (Cullen, 1997) (Figure 1). General information about the 14 study sites are presented in Table 1. These three sets of fragments/reserves were chosen for analysis because they are representative of the situation found in the Atlantic forest as a whole, and all were sampled using similar protocols. All sites are located in subregions with similar climatic, edaphic, and topographic conditions, thus, I assumed that the fragments within each subregion had similar fauna and flora originally. In all cases, data were collected on the mammal community as a whole, not only on primates, which allowed a much deeper analysis of each fragment. Detailed descriptions and information about the study sites can be found in Chiarello (1997, 1999, 2000a, b) and in Cullen (1997).

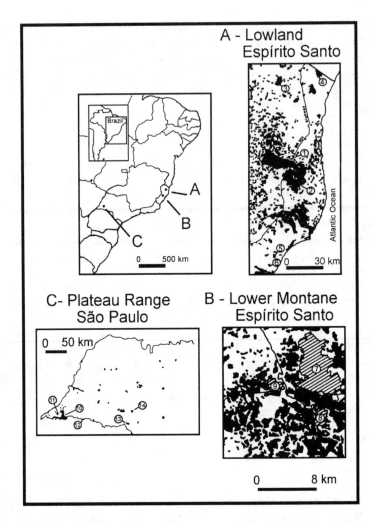

Figure 1. Map of northern Espírito Santo and western São Paulo states, in southeastern Brazil, showing the remaining patches of native forest (adapted from Chiarello, 1999 and Cullen et al., 2000). **Lowland Espírito Santo (A):** 1 - Sooretama Biological Reserve; 2 - Vale do Rio Doce Natural Reserve; 3 - Córrego do Veado Biological Reserve; 4 - Córrego Grande Biological Reserve; 5 - M7; 6 - Putiri. **Lower Montane Espírito Santo (B):** 7 - Augusto Ruschi Biological Reserve; 8 - Santa Lúcia Biological Station; 9 - São Lourenço Municipal Reserve. **Plateu Range São Paulo (C):** 10 - Morro do Diabo State Park; 11 - Tucano Farm; 12 - Mosquito Farm; 13 - Rio Claro Farm and 14 - Caetetus Ecological Station.

2.2.1. Lowland, Northern Espírito Santo

The region of lowland, northern Espírito Santo (LLES) has been disturbed since the early 1500s, but forest destruction increased greatly during the early 20th century (Aguirre, 1951). The principal cause of deforestation was logging (Heinsdijk et al., 1965), which is still taking place in the region, as satellite images indicated that 5.5% of the state's area was cleared of forests between 1990 and 1995 (SOS Mata Atlântica et al.,

Table 1. Location, area, hunting intensity, altitude, and annual precipitation in the study sites.

Site/abbreviation	Latitude/ Longitude	Area (ha)	Hunting intensity	Altitude (m)	mm rain (annual)	Source
Lower-montane, ES (LMES)						
Santa Lúcia Biological Station (SLBS)	19°58′S; 40°32′W	~440	low	550 to 950	1,868	this study
São Lourenço Municipal Reserve (SLMR)	19°56′S; 40°36′W	~500	moderate/ heavy	700 to 950	1,402	this study
Augusto Ruschi Biological Reserve (ARBR)	19°45′S; 40°27′W	3,598	low	800 to 1,100	1,630	this study
Lowland ES (LLES)						
Aracruz Cellulose Forest Fragment (M7)	19°48′S; 40°07′W	260	moderate/ heavy	50 to 90	1,328	Chiarello, 2000b
Aracruz Cellulose Forest Fragment (Putiri)	19°48′S; 40°07′W	210	moderate/ heavy	50 to 90	1,328	Chiarello, 2000b
Córrego Grande Biological Reserve (CGBR)	18°15′S; 39°47′W	1,504	moderate/ heavy	50 to 90	1,419	Chiarello, 2000b
Córrego do Veado Biological Reserve (CVBR)	18°21′S; 40°09′W	2,400	moderate/ heavy	90 to 110	~1,000	Chiarello, 2000b
Vale do Rio Dôce Natural Reserve (VRDNR)	19°12′S; 40°02′W	21,800	low	30 to 70	1,051	Chiarello, 2000b
Sooretama Biological Reserve (SBR)	18°59′S; 40°05′W	24,250	moderate/ heavy	50 to 150	~1,000	Chiarello, 2000b
Plateau Range SP (PRSP)						
Rio Claro Farm (RCF)	22°46′S; 48°55′W	1,700	moderate/ heavy	400 to 500	1,260	Cullen, 1997
Tucano Farm (TF)	22°30′S; 52°30′W	2,000	moderate/ heavy	400 to 500	1,300	Cullen, 1997
Mosquito Farm (MF)	~22°44′S; ~51°00′W	2,100	moderate/ heavy	400 to 500	1,200	Cullen, 1997
Caetetus Ecological Station (CES)	22°44′S; 49°42′W	2,178	low	400 to 500	1,260	Cullen, 1997
Morro do Diabo State Park (MDSP)	22°30′S; 52°16′W	35,000	low	400 to 500	1,200	Cullen, 1997

1998). A recent estimate (1995) revealed that <9% of the state area is still covered with native forests, and the majority of the remnants are small and isolated forest patches of <1,000 ha of area (SOS Mata Atlântica et al., 1998). Three of the study areas are privately owned reserves (VRDNR, M7/317, and Putiri) and the other three (SBR, CVBR, and CGBR) are biological reserves administered by the Brazilian Environmental Agency (IBAMA). The reserves range in size from 210 to 24,250 ha (Table 1), and all exhibit signs of past human disturbance, including illegal hunting, intrusion of fires from adjacent pasturelands, and logging (Chiarello, 2000b). Even so, large tracts of predominantly primary forest are still found in the two larger reserves (SBR: 24,250 ha and LFR: 21,800 ha). The overall terrain is flat with altitudes varying from 30 to 100 m above sea level and never exceeds 200 m.

Four primate species inhabit the reserves and fragments of this region: the brown howler monkey, *Alouatta guariba* (= *A. fusca*), the capuchin monkey, *Cebus nigritus* (= *Cebus apella nigritus* and *C. a. robustus*), the masked titi-monkey, *Callicebus personatus* (= *C. personatus personatus*), and the Geoffroy's tufted-ear marmoset, *Callithrix geoffroyi*. The taxonomy adopted here follows Rylands et al. (2000).

2.1.2. Lower-montane, Northern Espiríto Santo (LMES)

As in the north of the state, colonization started in this region soon after 1500 but until about 1800–1850 forest disturbance was minimum (Wied, 1820; Saint-Hilaire, 1974). After 1850 European immigrants, mainly from Germany and Italy, began clearing forest for firewood and to open space for agriculture, mainly coffee. Coffee, logging, and the cultivation of fruits and vegetables were the main economic activities until about 1960. From 1970 onwards, *Eucalyptus* plantations started to be common in the region, but given the hilly terrain, much of the original forest remains, mostly on inclined slopes and mountain tops where the access of vehicles and machinery is difficult. Apart from forest reserves (the three study sites are the largest forest fragments remaining in the LMES region), the great majority of remaining fragments is ≤50 ha, but connectivity is higher than in the lowland, as fragments are located closer to each other, and they are connected by strips of native forest more frequently.

Overall, the region of LMES has altitudes varying from 500 to 900 m and a milder temperature than that of the lowland sites. All three reserves are located within the municipality of Santa Teresa, a small town with about 25,000 inhabitants. The larger site, the Augusto Ruschi Biological Reserve (3,598 ha) is state owned and administered by IBAMA. The Santa Lúcia Biological Station (≅ 440 ha) is also state owned but administered by Mello Leitão Biology Museum in Santa Teresa. The São Lourenço Municipal Reserve (roughly 500 ha) has both state and private lands, but it has not been officially decreed as a reserve. All three sites are covered by reasonably well preserved forests. Six primate species are found in LMES: two marmosets (*Callithrix geoffroyi* and *C. flaviceps*), the masked titi (*Callicebus personatus*), the capuchin monkey (*Cebus nigritus*), the brown howler monkey (*Alouatta guariba*), and the muriqui (*Brachyteles hypoxanthus*).

2.1.3. Plateau Range, Western São Paulo (PRSP)

Originally, native forests covered about 82% of São Paulo state but today they are restricted to only 8% of the state's area (Kronka et al., 1993). The great majority of the remaining forest is located in the eastern, more humid slopes of 'Serra do Mar' and 'Serra da Mantiqueira.' Little remains of the mesophitic and semi-desciduous forests that once covered the largest part of São Paulo tablelands (Willis and Oniki, 1992). The Plateau forest sites ('Mata de Planalto') are located in the westernmost part of São Paulo state, in a region locally known as 'Pontal do Paranapanema.' The study fragments of this region range in size from 1,700 to 35,000 ha (Table 1). Two of the sites studied by Cullen (1997) are state owned reserves administered by the São Paulo State Secretary for the Environment (Morro do Diabo State Park and Caetetus Ecological Station). The remaining three sites are located on private land (Rio Claro, Mosquito, and Tucano farms). All five sites are located within 300 km from each other (Cullen, 1997). The forests are semi-deciduous and reasonably well preserved in all sites. In the last decades this region has staged much social conflict, since the landless people movement ('Movimento dos Trabalhadores Sem Terra') chose it for the settlement of families of rural people in unproductive private lands. In addition to the social disruption in the area, this movement has proved disastrous for the native fauna, since fragments located close to the settlements are invariably hunted by settlers (Cullen, 1997). Only three primate species are found in PRSP sites: *Alouatta guariba*, *Cebus nigritus,* and *Leontopithecus chrysopygus* (Table 1).

2.2. Data Collection

The sampling protocol consisted of censusing medium- to large-sized species of mammals by line transect method *sensu* Buckland et al. (1993). All species seen during censusing were recorded, except bats, small rodents, and marsupials (<1 kg of body weight), but only encounters with primates are analyzed in this chapter. The censuses were carried out repeatedly between two to four main trails/fragment (1,500 to 2,000 m long) in LLES and LMES sites and from four to eight trails/fragment in PRSP sites (Cullen, 1997). Effort was made to keep the walking speed as constant as possible at 1 km/h. The cumulative distance walked in three sets of fragments is shown in Table 2.

Every time a primate was observed the following data were collected: species, group size, and composition; and distance between the observer and the first animal seen, and between the animal and the trail (perpendicular distance). Distances (e.g., animal to trail) were measured directly with a rangefinder or calculated through basic trigonometry using the angle formed by the first animal seen and the trail. The angles were recorded with a sighting compass. As everywhere in the Atlantic forest, all 14 study sites were subject to illegal hunting, but according to differing degrees of surveillance and protection, sites were ranked in two categories: low hunting pressure and moderate/heavy hunting pressure (Chiarello, 2000b; Cullen, 1997; Cullen et al., 2000) (Table 1). For LLES and LMES, the ranking of sites was done on the basis of the degree to which each site was effectively surveyed and protected against hunters. Low hunting pressure was seen in sites where a team of well trained/efficient forest guards exists and makes daily or almost daily surveillance trips; moderate/heavy hunting pressure was seen in sites with the presence of a small, or poorly trained, or unmotivated team of forest guards, and surveillance trips happen only occasionally [i.e., once a week or less]). Secondarily, the amount of evidence indicating hunting activity in each site (e.g., encounters with hunters, dogs, hunting trails, snares and traps, and shotguns heard) was also used as indicative of hunting pressure. For PRSP sites, the ranking of hunting pressure was determined by "the number of households close to the fragments as a general measure of human activity, the number of gunshots herd by observers walking censuses, the number of active platforms found when opening trails, the number of hunting dogs associated with families, and the number of hunters encountered while preparing or finishing a census" (Cullen et al., 2000).

Table 2. Sampling effort (distance walked) allocated to line-transect sampling in the three study sites. Reserve names as in Table 1.

Study region	Total distance walked (km)	No. of fragments	Average km/ fragment	Source
LMES	185	3	62	this study
LLES	459	6	77	Chiarello, 1999
PRSP	2287	5	381	Cullen, 1997

2.3. Data Analysis

Encounter rates with primate species were used to contrast differences in primate abundance in relation to fragment size and hunting pressure (Chiarello, 1999). The encounter rate was chosen because, apart from being significantly correlated with true density (Spearman rank correlation; $r_s = 0.822$; $N = 51$; $p < 0.001$ for all species and sites analyzed), it allowed the use of data from a larger number of species (due to small sample sizes, it was not possible to calculate densities via transect sampling for some species, see below). Encounter rates were calculated as the total number of encounters per 10 km for all primate species seen during transect sampling. The relationship between primate density (abundance), and fragment size was tested using the Spearman rank correlation (r_s), and differences in primate density between sites of differing hunting intensities (heavy × low) were contrasted, when possible, with the Mann-Whitney test (Zar, 1996).

Density was calculated with DISTANCE software (Laake et al., 1994). See Buckland et al. (1993) for a detailed treatment of the technique. Given that the length of the trail or transect is known, the basic problem of estimating densities through line transect sampling is to know the width of the sampled area. The DISTANCE software uses the perpendicular distances (animal to trail) to estimate this unknown parameter (Buckland et al., 1993). Species population sizes were then estimated for each fragment, multiplying density estimates of each primate species by the corresponding area of the fragment (Table 1). A more detailed description of data analysis is found in Chiarello (2000a), Chiarello and Melo (2001), and Cullen (1997).

3. RESULTS

3.1. Populations in Fragments

3.1.1. Lower-Montane Espírito Santo

This region harbors six species of primates (Table 3). Its topographic diversity, with altitudes varying from 300 to 900 m, allows for species that are typically inhabitants of the lowlands, such as the Geoffroy's marmoset, as well as species that are mostly found in higher altitudes, like the muriqui and the buffy-headed marmoset. Of the three sites studied in this region, only the largest one, the Augusto Ruschi Biological Reserve, still has the full complement of primate species (Table 3). It is very likely that muriquis once roamed the forests in Santa Lucia Biological Station (Ruschi, 1965), but this species

Table 3. Number of primate species and estimates of primate population sizes for reserves located in LMES. Reserve names as in Table 1.

Site	Area (ha)	No. of spp.	Callithrix geoffroyi	Callithrix flaviceps	Callicebus personatus	Cebus nigritus	Alouatta guariba	Brachyteles hypoxanthus
SLBS	~440	5	7 to 10*	13 to 20	14 to 33	12 to 29	23 to 53	EXT
SLMR	~500	5	4 to 6*	16 to 24*	18 to 43	32 to 76	8 to 12	EXT
ARBR	3,598	6	p	133 to 648	40 to 80*	218 to 515	99 to 234	10 to 20**

* Minimum estimate based on calling data. ** No individuals of this species were seen during the present study, but Pinto et al. (1993) report that at least one group survives in this reserve. Abbreviations: "p" (present but not seen); "EXT" (species extinct locally).

seems to be extinct at present (in 2001 one researcher reported having seen muriquis there, but subsequent intensive searches failed to find the species there). The same probably applies to São Lourenço Municipal Reserve.

As shown in Table 3, the primate population sizes estimated for the three reserves of this region are all <650 individuals per species, and for the two reserves of about 500 ha of area (SLBS and SLMR) are <500 individuals for all species. For the larger-sized ARBR (3,596 ha), this population threshold is observed only for *Callithrix flaviceps* and *Cebus nigritus*.

3.1.2. Lowland Espírito Santo

Four primate species inhabit the lowlands of northern Espírito Santo (Table 4). The muriqui is historically absent from the region encompassed by the Rio Doce in the south and southern Bahia in the north (Aguirre, 1971). Therefore, the absence of this species in all six reserves should not be considered a result of recent extinction, at least not extinction caused by colonizers. No data are available on the past impact of hunting by various tribes of indigenous people that inhabited the region until the late 19th and early 20th centuries (Dean, 1995). Recent local extinctions have probably occurred, however, with masked titis in one reserve (CGBR) and with brown howlers in three reserves, all of which are forest fragments <2,500 ha (Table 4). These inferred extinctions should be considered as "probable" because data are not available about the certain past occurrence of these species before the isolation of these fragments (Chiarello and Melo, 2001).

Relatively large population sizes, in the range of thousand of individuals per species, were only found for the two larger sites of this study region (>20,000 ha). Population sizes estimated for the four fragments with 210 to 2,500 ha were <500 individuals per species, but less than 50 individuals per species in the smaller sites (210 to 260 ha) (Table 4).

3.1.3. The Plateau Range, São Paulo

The five study sites located in western São Paulo have only three species of primates, the black-lion tamarin, the brown capuchin monkey, and brown howlers (Table 5). The five sites lie below the southern limit of the natural range of marmosets and masked titis and they are to the west of the western limit of the *Brachyteles arachnoides'* range. Although the population sizes of some species are extremely low, none of the sites

Table 4. Estimates of primate population sizes for reserves/fragments located in LLES. Reserve names as in Table 1.

Site	Area (ha)	No. of spp.	*Callithrix geoffroyi*	*Callicebus personatus*	*Cebus nigritus*	*Alouatta guariba*
Putiri	210	3	6 to 9	10 to 15	16 to 23	EXT**
M7	260	4	7	2 to 4	36 to 53	3 to 4
CGBR	1,504	2	23 to 33	EXT**	52 to 75	EXT**
CVBR	2,400	3	220 to 322	7 to 10*	124 to 181	EXT**
VRDNR	21,800	4	6,125 to 8,971	1,254 to 1,819	3,951 to 5,742	157-227
SBR	24,250	4	5,100 to 7,469	1,521 to 2,212	2,396 to 3,483	3 to 6*

* Minimum estimate based on calling data. **EXT: either the species does not occur there or are extinct locally.

Table 5. Estimates of primate population sizes for five sites located in the PRSP (data from Cullen, 1997). Full names of reserves are given in Table 1.

Site	Area (ha)	No. of spp	*Leontopithecus chrysopygus*	*Cebus nigritus*	*Alouatta guariba*
RCF	1,700	3	55 to 70	112 to 267	160 to 424
TF	2,000	3	20	111 to 266	145 to 324
MF	2,100	3	5	125 to 220	548 to 961
CES	2,178	3	10 to 55	296 to 425	12
MDSP	35,000	3	931	2,604 to 4,610	1,029 to 28,812

studied by Cullen (1997) in this region showed primate extinctions (Table 5). Similar to what was found in the other two systems of forest fragments, primate population sizes >500 individuals per species were not found in the four fragments of about 2,000 ha, and only the Morro do Diabo State Park (35,000 ha) has population sizes around 1,000 individuals per reserve or larger (Table 5).

3.2. Effects of Fragment Size and Hunting

The abundance of primates was weakly correlated with fragment size (Figure 2). When encounter rates are pooled for all primate species in a site (i.e., using a single estimate of primate abundance per site), the resulting abundance, expressed as group sightings per 10 km of transect sampling, was found to be significant only for LMES fragments (Figure 2b). This relationship, however, was not significant when species are examined separately (Figures 3 and 4). Only for the Geoffroy's marmoset was this relationship marginally significant (Figure 3a).

Similarly, the influence of hunting intensity on primate abundance was not significant (Mann-Whitney test, $p>0.05$ in all cases; Figure 5). It comes as no surprise that small-bodied species (marmosets and lion tamarins) and extremely elusive ones like masked titis were not affected, as these are rarely taken by hunters in the Atlantic forest. Larger-bodied species such as capuchins and howlers, which are more likely targets, showed no clear trend of decreasing abundance with increasing hunting intensity.

It is not known if the absence of brown howlers in three fragments of LLES and the muriqui from two fragments of LMES was caused by past hunting activity, but it would be interesting to test if the density of surviving species increased in these fragments as a result of decreased competition pressure. As shown in Figure 6, however, this does not seem to be the case, because, contrary to what was expected, the overall density of primates was actually higher in fragments where howlers were still present, although the results were marginally significant (Figure 6a; $p = 0.05$). The trend was in the opposite direction for the surviving species in LMES fragments, but statistical comparisons were not possible for this set of fragments (Figure 6b).

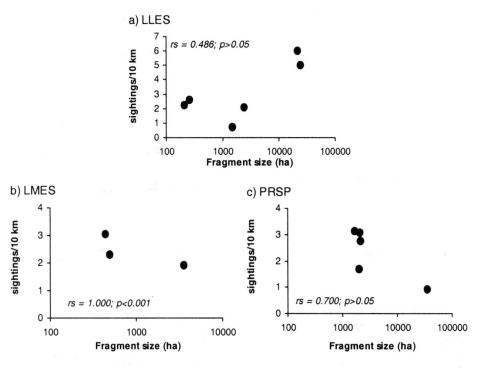

Figure 2. Relationship between fragment size ($\log_{10}$) and relative abundance of primate groups (pooling all species), as measured by the number of primate groups seen per 10 km of line-transect sampling in LLES (a), LMES (b), and PRSP fragments (c). The Spearman rank correlation coefficient (r_s), and the p is also shown. See Table 1 for site names.

4. DISCUSSION

4.1. Forest Fragmentation and Disturbance

The results indicated that the success of Atlantic forest primates in isolated forest remnants, as measured by density, varied considerably both between species and study sites. Even within a given set of forest fragments, which are more comparable to each other due to higher similarities in climatic, floristic, and historical characteristics, the abundance of primates showed no clear relationship with forest area. This contrasts with results from a previous study carried out in the same fragments of LLES in which the densities of medium- to large-size mammals were, in general, lower in small fragments (~200 ha) when compared to the larger reserves (Chiarello, 1999; 2000a). The results of that study showed that large terrestrial mammals such as peccaries, tapirs, and terrestrial frugivores such as agoutis (*Dasyprocta*) are negatively affected by fragmentation, a finding also observed in the fragments of PRSP (Cullen, 1997).

Species with a large size and mainly a frugivorous diet have a higher requirement of space, and, therefore, their populations may be more severely affected by habitat reduction than folivores or diet generalists (Milton and May, 1976; Milton, 1993). Also,

Marmosets/Lion-tamarins   **Masked titis**

a) *C. geoffroyi* (LLES)   b) *C. personatus* (LLES)

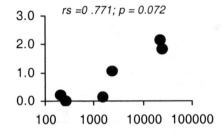

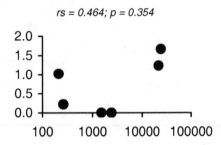

c) *C. flaviceps* (LMES)   d) *C. personatus* (LMES)

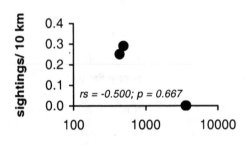

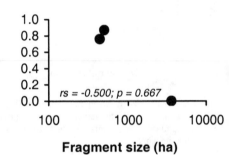

e) *L. chrysopygus* (PRSP)

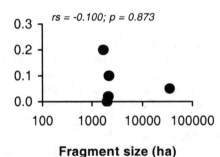

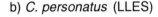

Figure 3. Relationship between fragment size (log$_{10}$) and relative abundance of marmosets/lion-tamarins (left) and masked titis (right) in LLES (a-b), LMES (c-d), and PRSP fragments (e). The Spearman rank correlation coefficient (r_s) and the corresponding *p* is also shown. See Table 1 for site names.

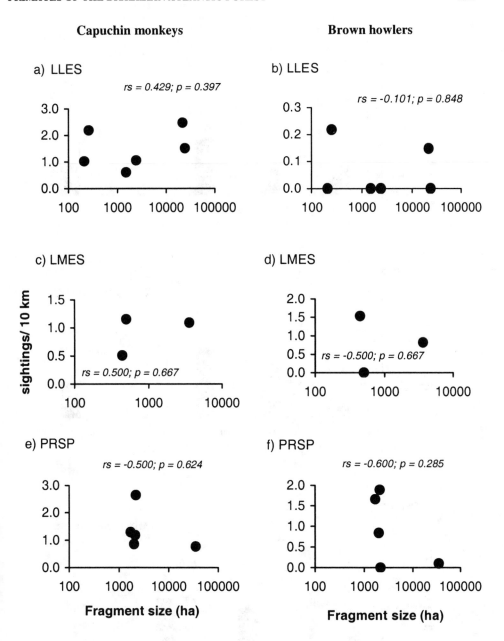

Figure 4. Relationship between fragment size (log₁₀) and relative abundance of capuchin monkeys (left) and brown howler monkeys (right) in LLES (a-b), LMES (c-d), and PRSP fragments (e-f). The Spearman rank correlation coefficient (r_s) and the corresponding p is also shown. See Table 1 for site names.

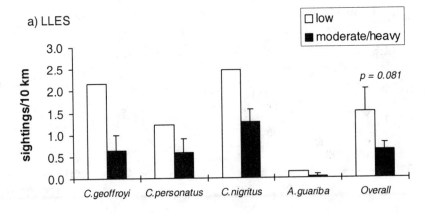

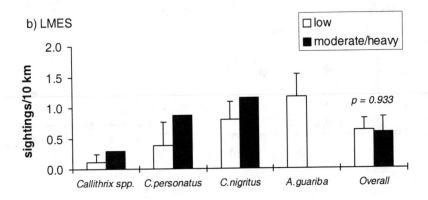

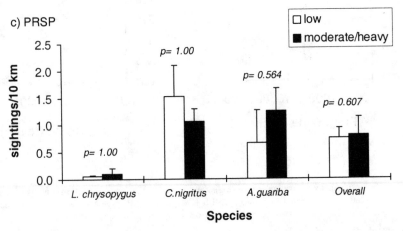

Species

Figure 5. Abundance of primates (sightings/10 km) in low (white bars) and moderate to heavy hunting intensities (black bars) in (a) LLES, (b) LMES, and (c) PRSP fragments. Mann-Whitney test results (*p*) are show above each pair of bars. See Table 1 for site names.

a) LLES

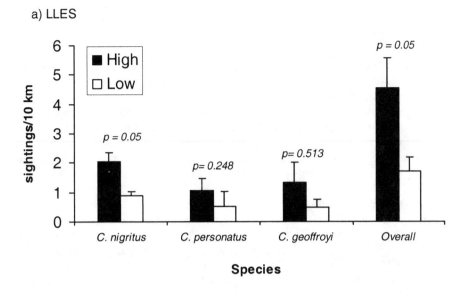

b) LMES

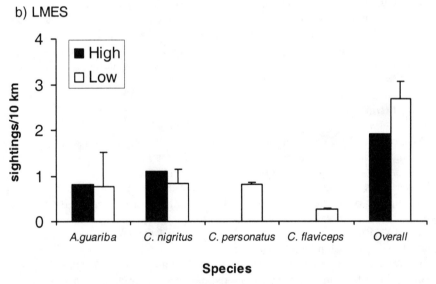

Figure 6. Abundance of primates (sightings/10 km) in fragments with a full complement of primate species (higher competition) (black bars) and in fragments where some primate species were absent (lower competition) (white bars). Mann-Whitney test results (p) are shown above each pair of bars. See Table 1 for site names.

due to the combining effect of forest disturbance and the lack or reduction of vertebrate dispersers, trees that produce large fruits seem to be drastically affected by forest fragmentation. For example, Tabarelli et al. (1999) found a significant increase in relative importance (density) of ruderal tree species in small fragments (5 to 370 ha), with a concomitant decrease in trees of later successional stages, especially fruit-producing trees such as Myrtaceae, Lauraceae, and Sapotaceae, which are the main sources of fruit for several vertebrate species of the Atlantic forest. With less fruit trees available to them, the guilds of frugivorous mammals and birds are the most severely affected in small forest fragments (Chiarello, 1999; Price et al., 1999).

With the possible exception of muriquis, which disappeared in the two smaller fragments of LMES, other primates, on the other hand, appeared to be little affected in terms of abundance in small fragments when compared to larger ones. This indicates that Atlantic forest primates are ecologically flexible, being capable to adapt to the disturbances caused by fragmentation. In fact, it is likely that edge effects, that are more pronounced in small fragments, can provide some species with a more constant or less seasonal food supply. Lianas, for example, are abundant along forest edges and natural gaps of tropical forests, and their phenology is less seasonal than that of trees, especially in terms of production of new leaves (Putz and Windsor, 1987; Morellato, 1996). Folivorous species such as howlers can, therefore, take advantage of the increased supply of new leaves in these fragments (Crockett, 1998). Brown howlers surviving in a small forest reserve of São Paulo (Santa Genebra reserve, 250 ha) are present in very high density (Chiarello, 1993), and lianas made up a significant part of their annual diet (Chiarello, 1994). In terms of food supply, the increased disturbance regime found in these small fragments can also be advantageous for other primates such as marmosets, which eat mainly gums and insects. Gum-producing plants such as *Inga* spp. (Mimosaceae) and several liana species (Leguminosae and Sapindaceae) are abundant along forest edges of Atlantic forest fragments and they are heavily consumed by marmosets (Passamani, 1995b; personal observation). Edible insects, mainly orthopterans and other leaf-eating insects, also proliferate in secondary vegetation, and these are very important for the diet of both marmosets and lion tamarins.

4.2. Illegal Hunting

As shown in Figure 5 results of the 14 sites also failed to reveal a strong influence of illegal hunting in Atlantic forest primates. This is not to say that hunters do not take primates or that this illegal activity does not contribute to deplete primate populations in this region. It is well known that hunting has contributed to local extinction of species and lowering of population sizes even in vast, continuous forest areas like the Amazon (Ayres and Ayres, 1979; Peres, 1996; Bodmer et al., 1997). These "empty forests" (sensu Redford, 1992) are tracts of exuberant forest depleted of game species due to hunting practiced since pre-Columbian times.

Before the European settlement in Brazil in the 1500s, the Atlantic forest was probably subject to a similar scenario, but the indigenous populations that once inhabited these forests have long been exterminated (Dean, 1995). It must be mentioned that different from the Amazon region, hunting in the Atlantic forest is carried out by settlers and colonists and not by indigenous populations. This difference is important as Indians target a much broader range of prey species and have primates as the preferred mammalian order (Redford and Robinson, 1987), while colonists hunt less prey species and have a preference for large terrestrial rodents, such as pacas and agoutis, ungulates (peccaries and deer), tapirs, and armadillos (Cullen, 1997; Cullen et al., 2000; Chiarello, 2000b).

Primates are also a target, but evidence indicates that the larger-bodied *Brachyteles* and *Alouatta* and, to a lesser extent, *Cebus*, are probably the most impacted (Lane, 1990;

Oliver and Santos, 1991; Chiarello, 2000b), not only because of their size, but also due to their lower intrinsic rates of natural increase, when compared to the smaller marmosets and tamarins (Robinson and Redford, 1986). Martuscelli et al. (1994) analyzed the conservation status of the muriqui and state that hunting is the most important factor contributing to the species decline in several locations. Similarly, other studies have reported the widespread incidence of hunting and the absence of larger genera (Ferrari and Diego, 1995), or of howlers (Oliveira and Oliveira, 1993), in the majority of fragments surveyed. Marmosets and lion tamarins are rarely taken as game, but these small and beautiful primates were intensively captured in the past for the pet trade, an illegal activity that, although drastically reduced, still takes place today, especially in Bahia and other states of northeastern Brazil.

A related problem is that due to the proximity of urban zones or rural communities to some forest fragments, domestic dogs and cats can potentially cause increased predation on some primate species. Domestic and feral dogs, alone or in groups of two or three, have been observed in almost all study fragments of northern Espírito Santo, and, in two of them, located in the lower-montane region of Santa Teresa (SLMR and SLBS), domestic dogs have been seen, and photographed by camera traps, hunting deep within the forest almost everyday (unpublished data). Obviously, the most impacted species are large terrestrial mammals, such as agoutis, pacas, armadillos, and brocket deer, but, given that primates sometimes descend to the ground, especially so in a fragmented landscape, they are also potentially preyed upon. Brown howlers and masked titis, for example, have been reported to fall victim to dogs in this region (personal observation).

4.3. Population Sizes

In a previous paper (Chiarello, 2000a) I discussed the problem of small population sizes with data on forest mammals collected in six of the 14 sites considered here. Regarding estimated population sizes, data from that paper were similar to the estimates calculated for the three sites located in lower-montane Espírito Santo and to those estimated by Cullen (1997) for the fragments of the PRSP. These figures are also similar to the estimates reported by Gonzáles-Solís et al. (2001) from a 140,000-ha tract of forest in Paranapiacaba range, southern São Paulo. In brief, large population sizes, in the range of thousands of individuals per species, are only found in reserves and fragments much larger than 3,600 ha, and more likely in the range of 20,000 ha or more, as is the case of VRDNR (21,800 ha), SBR (24,250 ha), and MRSP (35,000 ha). All the remaining 11 sites have estimated population sizes in the range of 500 individuals or smaller, while the two fragments of about 200 ha have <50 individuals per species.

One central question regarding the viability of populations isolated in forest fragments is 'what is the minimum size for long-term survival of a population?' This topic is extremely polemic as authorities disagree about the length of time that should be considered, because different organisms behave in different ways in response to small population sizes and because the incidence of environmental catastrophes adds much uncertainty in such estimates (Shaffer, 1981; Franklin and Frankham, 1998; Lynch and Lande, 1998). Some species show, for example, evidences of genetic deterioration or the deleterious effects caused by prolonged periods of inbreeding, while others do not (see Franklin and Frankham [1998] and Lynch and Lande [1998] for a general discussion). To keep matters simple, if one considers the 'magic' number of 500 individuals, as has been mentioned widely (Redford and Robinson, 1991; Robinson, 1996), as the minimum necessary to avoid inbreeding and genetic deterioration, and to allow for adaptative evolution in the long term (Franklin and Frankham, 1998), the total population size (N) must be even larger than this, that is, in the range of 1,500 to 5,000 individuals. This is because the effective population size (N$_e$) is 3 to 10 times smaller, in general, than total population size (Frankham, 1995). In the set of fragments analyzed here, such large population sizes were only found in reserves ≥20,000 ha. Does this mean that all primate populations surviving in smaller forest remnants are doomed? If that is the case, there is reason to be concerned, because the vast majority of remaining fragments of the Atlantic forest are much smaller then 20,000 ha (Ranta et al., 1998).

As remarked elsewhere (Chiarello, 2000a), the discussion of minimum size of population must, necessarily, take into account the degree to which primates are able to move between 'isolated' fragments. If they can use the matrix of inhospitable pastures or croplands to move from one fragment to another, then the scenario will be that of a metapopulation, with interconnected subpopulations. In this hypothetical scenario, even tiny subpopulations (tens of individuals) could be able to endure in the long run as they will be demographically and genetically in contact with each other. This is the case with black howler monkeys, *Alouatta caraya*, in the Brazilian Pantanal, where groups, and sometimes even solitary individuals, inhabit 'habitat islands,' that are small patches of forest located on higher ground that are not flooded during the wet season (Medri, 2000). These patches of forest are, in general, <5 ha and are separated one from another by grass-dominated vegetation (Medri, 2000). Nevertheless, black howlers are able to cross vast distances through the open grassland. A similar situation is found with red howlers, *Alouatta seniculus*, in the Venezuelan Llanos (Crockett, 1996), as well as other howler species (this volume).

On the other hand, the Atlantic forest is typically a closed forest formation and, therefore, one can reasonably suppose that the native primate species are not well adapted to venture through wide, open places. But is that really the case? Research with primates on this important topic has not been carried out yet, but results with other forest species, notably small marsupials and other small mammals, seem to indicate that Atlantic forest mammals are able to cross reasonably large tracts (300 m) of open grassland (Pires and Fernandez, 1999) or croplands (M. Passamani, personal communication).

5. CONCLUDING REMARKS

It is evident that the problem of fragmentation has become much more complex than when originally conceived. Currently, the concept of fragmentation, helped by an enormous number of publications in recent years, is starting to incorporate a higher degree of complexity (Lovejoy, 1997; Crome, 1997). This makes it more difficult to generalize results from studies carried out in different localities (Laurance et al., 1997). In this way, future research, given the importance of the disturbance history, the characteristics of the landscape matrix, and the biological idiosyncrasies of the organisms under study, will tend to have a more local approach, in which the fragmentation problem will be understood better if framed within a specific set of local circumstances.

For example, the results of the Biological Dynamics of Forest Fragment Project in Manaus 'currently the only integrated, long-term study of forest fragmentation in the tropics' (Laurance, 1997; Gilbert, this volume), should be interpreted in the context of that particular region of the Amazon. Different from the Atlantic forest fragments, the Manaus fragments are not completely isolated from the continuous forest. Parts of the forest cleared in the late 1970s and early 1980s have regenerated around the fragments, linking them with the continuous forest, which is relatively close to the fragments. The regrowth of vegetation has contributed to some surprising results, such as the higher species richness of small mammals (Malcolm, 1997) and frogs (Tocher et al., 1997) in fragments than in the continuous forest. Studies of understory birds have shown that forest regrowth around the fragments has promoted migration between continuous forest and the fragments (Bierregaard and Stouffer, 1997). Currently, areas cleared of forest or covered by disturbed forest, pasturelands, or crops, are the exception rather than the rule in the (still) vast area of Western Brazilian Amazon.

On the other hand, fragments of native forests are the exception in the human-dominated landscape of the Atlantic forest, where the remnants are isolated by long distances and have been disturbed for decades, or, in some cases, for centuries (da Fonseca, 1985; Dean, 1995; Viana et al., 1997). This leads one to conclude that in order to guarantee the long-term conservation of Atlantic forest primates, actions should be take in several directions to diminish the impact of a myriad of disturbing factors. The problem is complex and, accordingly, needs to be treated with a complex array of measures, policies, and strategies. The effective enforcement of the law against hunting and logging, a better surveillance against and control of forest fires, and the creation of wind breaks or buffer zones to lower the disturbance caused by edge effects, will not be enough to guarantee the survival of primates in the long term. Equally important, and perhaps even more important, is the need to improve connectivity between fragments as these will allow current isolated subpopulations to function truly as metapopulations and thus have a better chance to escape the hazards of genetic, demographic, and environmental stochasticity. Without this, fragments where local extinctions take place, as was observed for muriquis in some sites and for brown howlers and masked titis in others, will not be recolonized by dispersing individuals unless they are located close to each other.

6. SUMMARY

In this chapter I discussed the situation of primate populations isolated in small forest fragments of the southeastern Brazilian Atlantic forest, in particular 1) the success of primates, as measured by density and 2) the sizes of their surviving populations in forest fragments of various sizes. There are six main primates in this region: muriquis, howlers, capuchins, masked titi, marmosets, and lion-tamarins. Given the small sizes and isolation of most Atlantic forest fragments, there are correspondingly small population sizes of primates. These populations are subjected to an array of factors both internal (e.g., competitive and predatory interactions, and encroaching edge effects) and external (e.g., edge effects, fire intrusions, selective logging, and illegal hunting) that will greatly diminish their chances of survivorship in the long term. Data were summarized from the last 100 years on forest fragments located in northern Espírito Santo and São Paulo states. Overall abundance showed no clear relationship with fragment area or hunting pressure, a result that indicates a high degree of ecological flexibility and adaptability of Atlantic forest primates. The larger frugivores appeared, however, to be more affected by hunters and by their larger demand of forest area. The abundance of second growth plant species and an increase in insects in forest fragments may allow the remaining species to survive.

7. ACKNOWLEDGMENTS

I thank Laura Marsh for the invitation to participate in the XVIII Congress of the International Primatological Congress, held in Adelaide, Australia, which made possible my contribution to this book. The field study in Espírito Santo was supported by grants from National Geographic Society (5365-94), Fauna and Flora International (94/32/10), World Fund for Nature (CBO 123-94), and Brazilian Science Council (CNPq) (200273/92-2 and 469321/00-8 – Projecto Biodiversidade da Mata Atlântica de Santa Teresa, ES). The Museu de Biologia Mello Leitão helped with logistics and accommodation.

8. REFERENCES

Aguirre, A., 1951, *Sooretama - Estudo sobre o Parque de Reserva, Refúgio e Criação de Animais Silvestres, "Sooretama," no Municipio de Linhares, Estado do Espírito Santo,* Ministério da Agricultura, Serviço de Informação Agrícola, Rio de Janeiro (publicação postuma em 1992), 50 pp.

Aguirre, A. C., 1971, *O Mono* Brachyteles Arachnoides *(E. Geoffroy),* Academia Brasileira de Ciências, Rio de Janeiro.

Albuquerque, J. L. B., 1995, Observations of rare raptors in southern Atlantic rain forest of Brazil, *J. Field Ornithol.* **66(3)**:363–369.

Ayres, J. M., and Ayres, C., 1979, Aspectos da caça no alto rio Aripuanã, *Acta Amazon.* **9**:287–298.

Bianchi, R. de C., 2001, Estudo comparativo da dieta da jaguatirica, *Leopardus pardalis* (Linnaeus, 1758), em Mata Atlântica,. Masters Thesis, Universidade Federal do Espírito Santo, Vitória, ES, Brazil.

Bierregaard, R. O., Jr., and Stouffer, P. C., 1997, Understory birds and dynamic habitat mosaics in Amazonian rainforests, in: *Tropical Forest Remnants, Ecology, Management, and Conservation of Fragmented Communities,* W. F. Laurance and R. O. Bierregaard, eds., The University of Chicago Press, Chicago, pp.138–155.

Bierregaard, R. O., Jr., Lovejoy, T. E., Kapos, V., dos Santos, A. A., and Hutchings, R. W., 1992, The biological dynamics of tropical rainforest fragments, *Bioscience* **42**:859–866.

Bodmer, R. E., Eisenberg, J. F., Redford, K. H., 1997, Hunting and the likelihood of extinction of Amazonian mammals, *Cons. Bio.* **11**:460–466.

Brito, B. F. A., 2000, Ecologia alimentar da onça-parda, *Puma concolor,* na Mata Atlântica de Linhares, Espírito Santo, Brasil, Masters Thesis, Universidade de Brasília.

Buckland, S. T., Anderson, D. R., Burnham, K. P., and Laake, J. L., 1993, *Distance Sampling, Estimating Abundance of Biological Populations,* Chapman and Hall, London.

Castellani, T. T., 1986, Sucessão secundária inicial em mata tropical semi-decídua, após perturbação por fogo, Masters Thesis, Universidade Estadual de Campinas, Campinas, Brazil.

Chiarello, A. G., 1993, Home range of the brown howler monkey, *Alouatta fusca,* in a forest fragment of southeastern Brazil, *Folia Primatol.* **60(3)**:173–175.

Chiarello, A. G., 1994, Diet of the brown howler monkey *Alouatta fusca* in a semi-deciduous forest fragment of southeastern Brazil, *Primates* **35**:25–34.

Chiarello, A. G., 1997, Mammalian community and forest structure of Atlantic forest fragments in southeastern Brazil, PhD. Thesis, University of Cambridge, Cambridge, England.

Chiarello, A. G., 1999, Effects of fragmentation of the Atlantic forest on mammal communities in southeastern Brazil, *Bio. Cons.* **89**:71–82.

Chiarello, A. G., 2000a, Density and population size of mammals in remnants of Brazilian Atlantic forest, *Cons. Biol.* **14**:1,649–1,657.

Chiarello, A. G., 2000b, Influência da caça ilegal sobre mamíferos e aves das matas de tabuleiro do norte do estado do Espírito Santo, *Bol. Mus. Biol. Mello Leitão (N. Ser.)* **11/12**:229–247.

Chiarello, A. G., and Melo, F. R., 2001, Primate population densities and sizes in Atlantic forest remnants of northern Espírito Santo, Brazil, *Int. J. Primatol.* **22**:379–396.

Cochrane, M. A., and Schulze, M. D., 1999, Fire as a recurrent event in tropical forests of the eastern Amazon: Effects on forest structure, biomass, and species composition, *Biotropica* **31**(1):2–16.

Crockett, C. M., 1996, The relation between red howler monkey (*Alouatta seniculus*) troop size and population growth in two habitats, in: *Adaptative Radiations of Neotropical Primates*, M. A. Norconk, A. L. Rosenberger, and P. A. Garber, eds., Plenum Press, New York, pp. 489–510.

Crockett, C. M., 1998, Conservation biology of the genus *Alouatta*, *Int. J. Primatol.* **19**:549–577.

Crome, F. H. J., 1997, Researching tropical forest fragmentation: Shall we keep on doing what we're doing? in: *Tropical Forest Remnants, Ecology, Management, and Conservation of Fragmented Communities*, W. F. Laurance and R. O. Bierregaard, eds., The University of Chicago Press, Chicago, pp. 485–501.

Cullen, L., Jr., 1997, Hunting and biodiversity in Atlantic forest fragments, São Paulo, Brazil, Masters Thesis, University of Florida.

Cullen, L., Jr., Bodmer, R. E., and Pádua, C. V., 2000, Effects of hunting in habitat fragments of the Atlantic forests, Brazil, *Biol. Cons.* **95**:49–56.

da Fonseca, G. A. B., 1985, The vanishing Brazilian Atlantic forest, *Biol. Cons.* **34**:17–34.

da Fonseca, G. A. B., Herrmann, G., and Leite, Y. L. R., 1999, Macrogeography of Brazilian mammals, in: *Mammals of the Neotropics—The Central Neotropics, Vol. 3., Ecuador, Peru, Bolivia, Brazil*, J. F. Eisenberg and K. H. Redford, eds., The University of Chicago Press, Chicago, pp.549–563.

Dean, W., 1995, *With the Broadax and Firebrand, the Destruction of the Brazilian Atlantic Forest*, University of California Press, Berkeley.

de Oliveira, M. M., and Oliveira, J. C. C., 1993, A situação dos cebídeos como indicador do estado de conservação da Mata Atlântica no estado da Paraíba, Brasil, in: *A Primatologia no Brasil, Vol. 4*, M. E. Yamamoto and M. B. C. Souza, eds., Universidade Federal do Rio Grande do Norte, Natal, pp. 155–167.

de Souza-Mazurek, R. R., Pedrinho, T., Feliciano, X., Hilario, W., Geroncio, S., and Marcelo, E., 2000, Subsistence hunting among the Waimiri Atroari Indians in central Amazonia, Brazil, *Biol. Cons.* **9**:579–596.

Facure, K. G., and Giaretta, A. A., 1996, Food habits of carnivores in a coastal Atlantic forest of southeastern Brazil, *Mammalia* **60**(3):499–502.

Ferrari, S. F., and Diego, V. H., 1995, Habitat fragmentation and primate conservation in the Atlantic forest of eastern Minas Gerais, Brazil, *Oryx* **29**:192–196.

Ferreira, L. V., and Laurance, W. F., 1997, Effects of forest fragmentation on mortality and damage of selected trees in central Amazonia, *Cons. Bio.* **11**:797–801.

Frankham, R., 1995, Effective population size/adult population size ratios in wildlife: A review, *Genet. Res.* **66**:95–107.

Franklin, I. R., and Frankham, R., 1998, How large must populations be to retain evolutionary potential? *Animal Cons.* **1**:69–73.

Galetti, M., Martuscelli, P., Piza, M. A., and Simao, I., 1997, Records of harpy and crested eagles in the Brazilian Atlantic forest, *Bull. B. O. C.* **117**(1):27–31.

Garla, R., 1998, Ecologia alimentar da onça pintada (*Panthera onca*) na Mata dos Tabuleiros de Linhares, ES (Carnivora: Felidae), Masters Thesis, Universidade Estadual Paulista, Rio Claro, SP. 63 p.

Gonzáles-Solís, J., Guix, J. C., Mateos, E., and Llorens, L., 2001, Population density of primates in a large fragment of the Brazilian Atlantic rainforest, *Biol. Cons.* **10**:1,267–1,282.

Heinsdijk, D., de Macedo, J. G., Andel, S., and Ascoly, R. B., 1965, *A Floresta do Norte do Espírito Santo*, Departamento de Recursos Naturais Renováveis - Divisão de Silvicultura, Seção de Pesquisas Florestais, Rio de Janeiro.

Janzen, D. H., 1986, The eternal external threat, in: *Conservation Biology, the Science of Scarcity and Diversity*, M. E. Soulé, ed., Sinauer Associates, Inc., Sunderland, MA pp. 286–303.

Kapos, V., 1989, Effects of isolation on the water status of forest patches in Brazilian Amazon, *J. Trop. Ecol.* **5**:173–185.

Kronka, F. J. N., Matsukuma, C. K., Nalon, M. A., del Cali, I. H., Rossi, M., Mattos, I. F. A., Shin-Ike, M. S., and Pontinhas, A. A. S., 1993, *Inventário Florestal do Estado de São Paulo*. Instituto Florestal, São Paulo.

Laake, J. L., Buckland, S. T., Anderson, D. R., and Burnham, K. P., 1994, *Distance User's Guide Version 2.1*, Colorado Cooperative Fish and Wildlife Research Unit, Colorado State University, Fort Collins.

Lane, F., 1990, A hunt for "monos" (*Brachyteles arachnoides*) in the foothills of the Serra da Paranapiacaba, São Paulo, Brazil, *Primate Cons.* **11**:23–25.

Laurance, W. F., 1997, Hyper-disturbed parks: Edge effects and the ecology of isolated rainforest reserves in tropical Australia, in: *Tropical Forest Remnants, Ecology, Management, and Conservation of Fragmented Communities*, W. F. Laurance and R. O. Bierregaard, eds., The University of Chicago Press, Chicago pp. 71–83.

Laurance, W. F., Bierregaard, R. O., Gascon, C., Jr., Didham, R. K., Smith, A. P., Lynam, A. J., Viana, V. M., Lovejoy, T. E., Sieving, K. E., Sites, J. W., Jr., Andersen, M., Tocher, M. D., Kramer, E. A., Restrepo, C., and Moritz, C., 1997, Tropical forest fragmentation: Synthesis of a diverse and dynamic discipline, in: *Tropical Forest Remnants, Ecology, Management, and Conservation of Fragmented Communities*, W. F. Laurance and R. O. Bierregaard, eds., The University of Chicago Press, Chicago pp. 502–514.

Lovejoy, T. E., 1997, Foreword, in: *Tropical Forest Remnants, Ecology, Management, and Conservation of Fragmented Communities*, W. F. Laurance and R. O. Bierregaard, eds., The University of Chicago Press, Chicago pp. ix–x.

Lovejoy, T. E., Bierregaard, R. O., Rankin, J. M., and Schubart, H. O. R., 1983, Ecological dynamics of tropical forest fragments, in: *Tropical Rain Forest: Ecology and Management*, S. L. Sutton, T. C. Whitmore, and A. C. Chadwick, eds., Blackwell Scientific Publication, Oxford pp. 377–384.

Lovejoy, T. E., Bierregaard, R. O., Rylands, A. B., Malcolm, J. R., Quintela, C. E., Harper, L. H., Brown, K. S., Jr., Powell, A. H., Powell, G. V. N., Schubart, H. O. R., and Hays, M. B., 1986, Edge and other effects of isolation on Amazon forest fragments, in: *Conservation Biology, the Science of Scarcity and Diversity. 1st ed.*, M. E. Soulé, ed., Sinauer Associates, Inc., Sunderland, MA pp. 257–285.

Lynch, M., and Lande, R., 1998, The critical effective size for a genetically secure population, *Animal Cons.* 1:70–72.

Malcolm, J. R., 1997, Biomass and diversity of small mammals in Amazonian forest fragments, in: *Tropical Forest Remnants, Ecology, Management, and Conservation of Fragmented Communities*, W. F. Laurance and R. O. Bierregaard, eds., The University of Chicago Press, Chicago pp. 207–221.

Marini, M. A., and Cavalcanti, R. B., 1996, Influência do fogo na avifauna do sub-bosque de uma mata de galeria do Brasil central, *Rev. Brasil. Biol.* 56(4):749–754.

Martuscelli, P., Petroni, L. M., and Olmos, F., 1994, Fourteen new localities for the muriqui *Brachyteles arachnoids*, *Neotrop. Primates* 2:12–15.

Medri, I. M., 2000, Ocorrência de *Alouatta caraya* em capões do Pantanal Sul, in: *Ecologia do Pantanal—Curso de Campo 2000*, J. C. C. dos Santos, J. M. Longo, M. B. Silva, A. G. Chiarello, and E. Fischer, eds., Universidade Federal do Mato Grosso do Sul, Campo Grande pp. 158–163.

Milton, K., 1993, Diet and primate evolution, *Scient. Amer.* 269:70–77.

Milton, K., and May, M., 1976, Body weight, diet, and home range area in primates, *Nature* 259:459–462.

Morellato, P. C., 1996, Reproductive phenology of climbers in a southeastern Brazilian forest, *Biotropica* 28(2):180–191.

Myers, N., Mittermeier, R. A., Mittermeier, C. G., Fonseca, G. A. B., and Kent, J., 2000, Biodiversity hotspots for conservation priorities, *Nature* 403:853–858.

Nepstad, D. C., Moutinho, P. R., Uhl, C., Vieira, I. C., and Cardoso da Silva, J. M., 1996, The ecological importance of forest remnants in an eastern Amazonian frontier landscape, in: *Forest Patches in Tropical Landscapes*, J. Schelhas and R. Greenberg, eds., Island Press, Washington, D.C. pp. 133–150.

Oliver, W. L. R., and Santos, I. B., 1991, Threatened endemic mammals of the Atlantic forest region of southeastern Brazil, *Wildl. Preserv. Trust Special Sci. Rep.* 4:1–126.

Olmos, F., 1994, Jaguar predation on muriqui *Brachyteles arachnoides*, *Neotrop. Primates* 2(2):16.

Palomares, F., Gaona, F., Ferreras, P., and Delibes, M., 1995, Positive effect on game species of top predators by controlling smaller predator populations: An example with lynx, mongooses, and rabbits, *Cons. Biol.* 9:295–305.

Pardini, R., 2001, Pequenos mamíferos e a fragmentação da Mata Atlântica de Una, Sul da Bahia—Processos e Conservação, Ph.D. Dissertation, Universidade de São Paulo, São Paulo. 147 p.

Passamani, M., 1995a, Field observation of a group of Geoffroy's marmosets mobbing a margay cat, *Folia Primatol.* 64:163–166.

Passamani, M., 1995b, Ecologia, comportamento, e manejo do Sagui-da-cara-branca (*Callithrix geoffroyi*) em fragmentos de Mata Atlântica no norte do Espírito Santo, Masters Thesis, Universidade Federal do Espírito Santo, Vitória.

Peres, C. A., 1996, Population status of white-lipped *Tayassu pecari* and collared peccaries *T. tajacu* in hunted and unhunted Amazonian forests, *Biol. Cons.* 77:115–123.

Peres, C. A., and Dolman, P. M., 2000, Density compensation in Neotropical primate communities: Evidence from 56 hunted and nonhunted Amazonian forests of varying productivity, *Oecologia* 122:175–189.

Pinto, L. P. S., Costa, C. M. R, Strier, K. B., and da Fonseca, G. A. B., 1993, Habitat, density, and group size of primates in a Brazilian tropical forest, *Folia Primatol.* 61:135–143.

Pires, A. S., and Fernandez, F. A. S., 1999, Use of space by the marsupial *Micoureus demerarae* in small Atlantic forest fragments in southeastern Brazil, *J. Trop. Ecol.* 15:279–290.

Price, O. F., Woinarski, J. C. Z., and Robinson, D., 1999, Very large area requirements for frugivorous birds in monsoon rainforest of the Northern Territory, Australia, *Biol. Cons.* 91:169–180.

Putz, F. E., and Windsor, D. M., 1987, Liana phenology on Barro Colorado Island, *Biotropica* 19(4):334–341.

Ranta, P., Blom, T., Niemel, J., Joensuu, E., and Siitonen, M., 1998, The fragmented Atlantic rain forest of Brazil: Size, shape, and distribution of forest fragments, *Biol. Cons.* 7:385–403.

Redford, K. H., 1992, The empty forest, *Bioscience* 42:412–422.

Redford, K. H., and Robinson, J. G., 1987, The game of choice: Patterns of Indian and colonist hunting in the Neotropics, *Am. Anthropol.* 89:650–667.

Redford, K. H., and Robinson, J. G., 1991, Park size and the conservation of forest mammals in Latin America, in: *Latin American Mammalogy, History, Biodiversity, and Conservation,* M. A. Mares and D. J. Schmidly, eds., University of Oklahoma Press, Norman and London pp. 227–234.

Robinson, J. G., 1996, Hunting wildlife in forest patches: An ephemeral resource, in: *Forest Patches in Tropical Landscapes,* J. Schelhas and R. Greenberg, eds., Island Press, Washington D.C. pp. 111–130.

Robinson, J. G., and Redford, K. H., 1986, Intrinsic rate of natural increase in Neotropical forest mammals: Relationship to phylogeny and diet, *Oecologia* 65:516–520.

Ruschi, A., 1965, Lista dos Mamíferos do Estado do Espírito Santo, *Bol. Mus. Biol. Mello Leitão, Ser. Zool.* 24A:1–40.

Rylands, A. B., da Fonseca, G. A. B., Leite, Y. L. R., and Mittermeier, R. A., 1996, Primates of the Atlantic forest: Origin, distributions, endemism, and communities, in: *Adaptative Radiations of Neotropical Primates,* M. A. Norconk, A. L. Rosenberger, and P. A. Garber, eds., Plenum Press, New York pp 21–51.

Rylands, A. B., Schneider, H., Langguth, A., Mittermeier, R. A., Groves, C., and Rodriguez-Luna, E., 2000, An assessment of the diversity of New World primates, *Neotrop. Primates* 8:61–93.

Saint-Hilaire, A., 1974, Viagem ao Espírito Santo e Rio Doce, (Translated by Milton Amado). Coleção Reconquista do Brasil, 4. Editora da Universidade de São Paulo e Livraria Itatiaia, Belo Horizonte, MG. 121 pp.

Shaffer, M. L., 1981, Minimum population sizes for species conservation, *Bioscience* 31:131–134.

Skole, D. L., and Tucker, C., 1993, Tropical deforestation and habitat fragmentation in the Amazon: Satellite data from 1978 to 1988, *Science* 260:1,905–1,910.

SOS Mata Atlântica and INPE, 1993, *Atlas da Evolução dos Remanescentes Florestais e Ecossistemas Associados do Domínio da Mata Atlântica no Período 1985–1990,* Fundaçăcao SOS Mata Atlântica & Instituto Nacional de Pesquisas Espaciais, São Paulo.

SOS Mata Atlântica; INPE, and ISA, eds., 1998, *Atlas da Evolução dos Remanescentes Florestais e Ecossistemas Associados no Domínio da Mata Atlântica no Período 1990-1995,* Fundação SOS Mata Atlântica, Instituto Nacional de Pesquisas Espaciais e Instituto Socioambiental, São Paulo.

Stafford, B. J., and Ferreira, F. M., 1996, Predation attempts on Callitrichids in the Atlantic coastal rain forest of Brazil, *Folia Primatol.* 65:229–233.

Tabarelli, M., Mantovani, W., and Peres, C. A., 1999, Effects of habitat fragmentation on plant guild structure in the montane Atlantic forest of southeastern Brazil, *Biol. Cons.* 91:119–127.

Tocher, M. D., Gascon, C., and Zimmerman, B. L., 1997, Fragmentation effects on a central Amazonian frog community: A ten-year study, in: *Tropical Forest Remnants, Ecology, Management, and Conservation of Fragmented Communities,* W. F. Laurance and R. O. Bierregaard, eds., The University of Chicago Press, Chicago pp. 124–137.

Uhl, C., Buschbacher, R., and Serrão, E. A. S., 1988, Abandoned pastures in eastern Amazonia, I: Patterns of plant succession, *J. Ecol.* 76:663–668.

Viana, V. M., Tabanez, A. A. J., and Batista, J. L. F., 1997, Dynamics of restoration of forest fragments in the Brazilian Atlantic moist forest, in: *Tropical Forest Remnants, Ecology, Management, and Conservation of Fragmented Communities.* W. F. Laurance and R. O. Bierregaard, eds., The University of Chicago Press, Chicago pp.351–365.

Vieira, E. M., 1999, Small mammal communities and fire in the Brazilian Cerrado, *J. Zool. Lond.* 249:75–81.

Wied, P. M., 1820, *Travels in Brazil in 1815, 1816, and 1817,* Translated from the German, Sir Richard Phillips and Co., London.

Willis, E. O., and Oniki, Y., 1992, Losses of São Paulo birds are worse in the interior than in the Atlantic forests, *Ciência e Cultura* 44:326–328.

Zar, J. H., 1996, *Biostatistical Analysis,* 3rd edition. Prentice-Hall International, Inc., Upper Saddle River, NJ.

DYNAMICS OF PRIMATE COMMUNITIES ALONG THE SANTARÉM-CUIABÁ HIGHWAY IN SOUTH-CENTRAL BRAZILIAN AMAZONIA

Stephen F. Ferrari, Simone Iwanaga, André L. Ravetta, Francisco C. Freitas, Belmira A. R. Sousa, Luciane L. Souza, Claudia G. Costa, and Paulo E. G. Coutinho[*]

1. INTRODUCTION

All Neotropical primates (Platyrrhini) are highly specialized for an arboreal way of life (Hershkovitz, 1977) and rarely, if ever, come to the ground under natural conditions, despite the fact that savannas and open woodlands cover more than a third of tropical South America. It remains unclear why there are no New World ecological equivalents of the terrestrial or semi-terrestrial Old World baboons, macaques, vervets, and patas monkeys. Whatever the reasons, what is clear is that all platyrrhines are particularly vulnerable to the effects of habitat fragmentation that result from modern-day human occupation of the New World's tropical forests. Tracts of the original vegetation are isolated from one another by open fields of pasture or crops. Distances between fragments are rarely less than a few hundred meters and are usually a number of kilometers. These conditions create effective barriers to dispersal, even for those forms, such as howlers (*Alouatta*) or marmosets (*Callithrix*) that are relatively tolerant of habitat disturbance.

In Brazil, habitat fragmentation has reached critical levels in the Atlantic Forest biome (Chiarello, this volume), threatening all endemic species with extinction (Rylands et al., 1997). Long-term conservation of these species will depend on the management of remnant metapopulations formed by isolated subunits. Overall, deforestation is still limited in Brazilian Amazonia, at least in relative terms, although there are "hotspots" located mainly in the southern half of the basin where a large proportion of the original forest cover has been removed. Deforestation in these hotspots has already become

[*] Stephen F. Ferrari, Simone Iwanaga, Francisco C. Freitas, Belmira A.R. Souza, Luciane L. Souza, Claudia G. Costa, and Paulo E.G. Coutinho, Department of Genetics, Universidade Federal do Pará, Caixa Postal 8607, 66.075-900 Belém – PA, Brasil, André L. Ravetta, Department of Zoology, Museu Paraense Emílio Goeldi, 66.075-000 Belém – PA, Brasil. Correspondence to S. F. Ferrari (email: ferrari@ufpa.br).

Primates in Fragments: Ecology and Conservation
Edited by L. K. Marsh, Kluwer Academic/Plenum Publishers, 2003

problematic for a handful of primates with relatively small and localized geographic ranges.

The most marked example of this kind of problem is the piebald tamarin (*Saguinus bicolor bicolor*). This monkey has a diminutive range–by Amazonian standards–that coincides with the metropolitan area of Manaus, the largest urban center in central Amazonia (Egler, 1992). The long-term prospects for this tamarin, which has been reduced in the wild to a series of small, isolated subpopulations, are probably little better than those of some Atlantic Forest callitrichids, such as *Leontopithecus chrysomelas* (Rylands, 1989) or *Callithrix flaviceps* (Ferrari and Diego, 1995).

Most other species have much larger ranges, but many are endemic to deforestation hotspots in the southern Amazon basin. Smaller-bodied forms tend to have relatively small geographic ranges (Ayres and Clutton-Brock, 1992; Ferrari, in press), but medium- and large-bodied species are included here. A prominent example is the Ka'apor capuchin (*Cebus kaapori*), a medium-sized platyrrhine that is endemic to the Hylea, east of the Tocantins River, the most densely populated and deforested region of Brazilian Amazonia. In addition to its restricted geographic range, *C. kaapori* appears to be naturally rare, and the loss of more than half of the region's original forest cover apears to have significantly reduced surviving numbers in the wild. As a result, this species may be one of the most endangered of Amazonian mammals (Ferrari and Queiroz, 1994; Lopes and Ferrari, 1996).

Further west, the white-fronted spider monkey (*Ateles marginatus*; Figure 1) is endemic to the Xingu-Tapajós interfluvium, in which the present study area is located. Deforestation is less widespread in this region in comparison with the eastern extreme of the biome. However, *A. marginatus*, like other atelids, is far more vulnerable to hunting pressure than *C. kaapori*, given that the body weight of an adult spider monkey is two to three times that of a capuchin. Atelid populations can be decimated by hunting even before the advance of habitat fragmentation (Peres, 1991), and the ongoing colonization of a large part of the interfluvium has led to the classification of *A. marginatus* as an endangered species (Rylands et al., 1997).

Even in hotspots such as eastern Amazonia and the Brazilian state of Rondônia, something like half of the original forest cover still remains, which is in stark contrast with less than ten percent in the Atlantic Forest. Thus, there is still a certain amount of leeway for the planning of management strategies, supported by empirical evidence from studies of the fragmentation process, such as that of Bierregaard et al. (1992). Some consequences of the process are predictable on the basis of the ecological characteristics of different species, although recent studies at a number of sites in southeastern Amazonia (e.g., Bobadilla, 1998; Lopes and Ferrari, 2000) have revealed unexpected complexity. The results of the present study in south-central Amazonia further reinforce the complexity of this process, varying between predictable effects of factors such as fragment size and unexpected variation in the diversity and abundance of different species.

2. STUDY AREA

The study area is located to the south of the city of Santarém in the western extreme of the Brazilian state of Pará. The area is just east of the lower Tapajós River (Figure 2). The northern limit of the area coincides with km 63 of the Santarém-Cuiabá Highway,

Figure 1. Adult female white-fronted spider monkey, *Ateles marginatus*. Photo by Stephen F. Ferrari.

the BR-163, and the southern limit coincides with the town of Rurópolis, located on the intersection with the Transamazon Highway.

This study area was chosen for a number of reasons, in particular, the proximity of the Tapajós National Forest (Flona Tapajós). This forest is a protected area encompassing over half a million hectares of continuous forest. This area is representative of the original ecosystem and provides a control for comparisons between sites representing different degrees of habitat fragmentation. Other characteristics of the study area include its relatively low human population density and the linear pattern of human colonization along the BR-163. Finally, the location of the study area facilitated logistical support from the Federal University of Pará in Santarém and the Brazilian Institute for the Environment and Natural Resources (IBAMA), that administrates Flona Tapajós.

An additional and nonetheless important factor was the occurrence in the area of the white-fronted spider monkey, *Ateles marginatus*. As the species' endangered status is based primarily on its vulnerability to the effects of human colonization (Rylands et al., 1997), the data collected in the present study will provide an initial, important step in the development of conservation strategies.

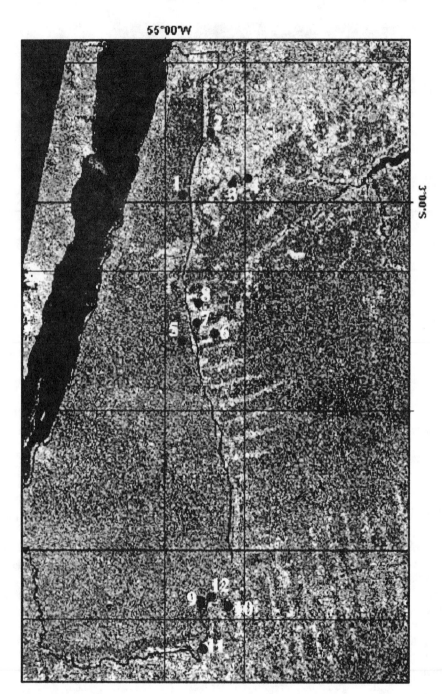

Figure 2. Location of the study sites in central Amazonia. On this Landsat image, forest is represented by the darker stippled shading, and open areas in lighter shades. The black area (water) to the west (left) is the Tapajós River. The Tapajós National Forest is outlined. The "herring-bone" pattern of colonization is clearly visible along the Santarém-Cuiabá Highway (north-south) and the Transamazon Highway (east-west).

3. THE PRIMATE FAUNA

Seven diurnal and one nocturnal species occur in the study area (Table 1), representing all four platyrrhine subfamilies, as defined by Schneider and Rosenberger (1996). There are a number of ecological differences among members of these subfamilies, which can be characterized in broad terms as distinct ecological guilds (Rosenberger, 1992; Ferrari, in press). Atelines are large-bodied, specialized frugivore-folivores, and callitrichines are small-bodied, gummivore-insectivores. The medium-sized cebines and pitheciines are generalists and seed predators, respectively.

Vulnerability to the effects of habitat fragmentation varies both among and within guilds, according to the ecological specializations of different genera. Marmosets, for example, may actually benefit from habitat disturbance, in particular, because of their morphological specializations for the systematic exploitation of plant gums as an alternative source of carbohydrates (Ferrari, 1993). Howlers are similarly resistant to reduced habitat quality because of their ability to switch to a relatively folivorous diet where or when fruit is scarce (Milton, 1980; Mendes, 1989; Marsh, 1999). However, *Ateles*, the second ateline genus studied here, appears to be an obligate frugivore depending on relatively large areas of continuous forest to guarantee access to fruiting trees throughout the year (van Roosmalen, 1985; Symington, 1988). Even selective logging can have a significant effect on spider monkey populations, and almost any kind of habitat fragmentation is potentially disastrous.

A similar contrast in vulnerability to habitat fragmentation distinguishes the two pitheciine species. Like spider monkeys, bearded sakis are highly specialized frugivores and would thus seem likely to be sensitive to habitat disturbance (Johns and Ayres,

Table 1. Primate species occurring in the study area along BR-163 in western Pará. Formal nomenclature follows Schneider and Rosenberger (1996) and Rylands et al. (2000) and common names follow Rowe (1998).

Taxon	Common name	Geographic range
Cebidae		
Cebinae		
Aotus infulatus	Night monkey	Southern Amazon basin.
Cebus apella	Tufted capuchin	Tropical South America.
Saimiri sciureus	Common squirrel monkey	Amazon basin.
Callitrichinae		
Mico argentatus	Silvery marmoset	Southern Amazon basin, between the Tapajós and lower Tocantins.
Atelidae		
Atelinae		
Alouatta belzebul	Red-handed howler	Southeastern Amazonia to northern Atlantic Forest.
Ateles marginatus	White-fronted spider monkey	Xingu-Tapajós interfluvium.
Pitheciinae		
Callicebus moloch	Dusky titi	Southern Amazon basin, west of the Tocantins and east of the Madeira.
Chiropotes albinasus	White-nosed bearded saki	Between the Xingu and Madeira/Jiparaná rivers.

1987). However, recent studies (e.g., Bobadilla and Ferrari, 2000; Ferrari et al., 2002a; Port-Carvalho and Ferrari, 2002) have indicated that these monkeys may be relatively tolerant of fragmentation, at least under certain circumstances. On the other hand, titi monkeys of the *Callicebus moloch* species group appear to be much better adapted ecologically for more marginal habitats (Peres, 1993; Ferrari et al., 2000).

4. PROCEDURES

4.1. Selection of Study Sites

The experimental design of the present study was inspired by that of the Biological Dynamics of Forest Fragments (BDFF) Project in northeastern Amazonas state (Rylands and Keuroghlian, 1988; Bierregaard et al., 1992). In the BDFF Project, blocks of forest of different sizes were isolated from the original forest cover to study the effects of habitat fragmentation. There are two principal differences here. First, as the present study was concerned specifically and exclusively with mammals, the smallest fragments studied were of between 50 and 100 ha, in contrast with the one- and 10-ha blocks used in the BDFF Project. Even this size of fragment is relatively small in comparison with the known home ranges of many platyrrhine genera at sites in the Amazon basin (e.g., Terborgh, 1983; Spironelo, 1987; Peres, 1996). Two other categories of fragment size–hundreds and thousands of ha–were also used here.

A second difference is that, rather than being established "artificially" in accordance with the objectives of the project, the fragments studied here were formed during the colonization of the study area over the past thirty years by settlers, the vast majority of whom are immigrants from other parts of southeastern Amazonia, principally the state of Maranhão. The fragments were thus highly irregular in size and shape, with varying degrees of isolation from the surrounding continuous forest and human populations. Despite these complications, these sites are clearly more representative of the typical process of human colonization, both in the region and throughout most of Brazilian Amazonia.

The first step in the selection of study sites was the examination of recent satellite images of the study area for the identification of appropriate fragments in the three different size categories. Sites were arranged in three distinct nuclei, centered on the IBAMA guard stations at Bases 2, 3, and 5 (Table 2; Figure 2). Each nucleus consisted of a control (continuous forest within Flona Tapajós in the proximity of each guard station) and three sites on neighboring private properties to the east of the highway. The three sites represented fragments of the order of dozens, hundreds, and thousands of ha (small, medium, and large, respectively). This arrangement was designed to ensure the comparability of sites, at least within each nucleus, as well as sampling possible differences in relation to the history of colonization and proximity to urban centers.

Potential sites were visited to confirm the characteristics of each fragment *in situ* and to request permission from the landowners to carry out fieldwork on their properties. The objectives of the research and the activities involved were explained carefully to each landowner. In only one case was permission refused, even though members of the research team were allowed to visit fragments on the property. Most other landowners not only authorized fieldwork on their land, but were also overtly supportive and provided invaluable information, manpower, and infrastructure.

Table 2. Characteristics of the study sites selected along the Santarém-Cuiabá Highway in western Pará (see Figure 2).

	Study site	Geographic coordinates	Fragment category and size
1	Base 2 Flona Tapajós	03°02'46"S, 54°57'18"W	Control (continuous forest)
2	Massafra	02°50'06"S, 54°53'14"W	Large (4500 ha)
3	São Benedito	02°57'05"S, 54°47'41"W	Medium (900 ha)
4	São Benedito	02°56'31"S, 54°47'44"W	Small (90 ha)
5	Base 3 Flona Tapajós	03°21'19"S, 54°56'57"W	Control (continuous forest)
6	Sítio do Oswaldo	03°21'06"S, 54°52'40"W	Large (3000 ha)
7	Fazenda do Goiano	03°19'54"S, 54°55'24"W	Medium (450 ha)
8	Fazenda Tocantins	03°14'18"S, 54°55'25"W	Small (70 ha)
9	Base 5 Flona Tapajós	03°56'33"S, 54°52'39"W	Control (continuous forest)
10	Sr. Ceará	03°56'58"S, 54°51'11"W	Large (2500 ha)
11	Antônio Medeiros	04°03'56"S, 54°54'20"W	Medium (300 ha)
12	Sr. Ceará	03°57'00"S, 54°52'05"W	Small (90 ha)

4.2. Data Collection

The primate populations at the study sites were surveyed using standard line transect methods, following procedures established in previous studies in Brazilian Amazonia (Ferrari et al., 1999; Lopes and Ferrari, 2000). Preparatory work at each site included selection of a representative area of forest (in the case of the controls and larger-sized fragments), and the establishment of a system of straight-line trails. Where fragment dimensions permitted, a standard 2 km by 1 km layout (6 km circuit) was used, as in Ferrari et al. (1999), but this was adapted in smaller fragments. Trails were cut with minimal removal of undergrowth, swept clean of debris, and marked at 100-m intervals.

During surveys, trails were walked carefully at a mean velocity of approximately 1.5 km per hour. At each encounter with a primate, a standard series of data was collected, including species, group composition, and sighting and perpendicular distances, measured with a surveyor's tape. Surveys began just after dawn, as soon as visibility was adequate, and continued until either the programmed distance had been walked or it was necessary to interrupt data collection because of adverse weather conditions (rain or high winds) or reduced visibility. Data collection was also suspended temporarily whenever visibility was reduced significantly by changing weather conditions.

Line transect surveying has been used successfully at a number of sites in Brazilian Amazonia (e.g., Emmons, 1984; Peres, 1997a; Lopes and Ferrari, 2000), although sighting rates of primate species rarely exceed one per 10 km surveyed, and in many cases, it is very much less than this. As most surveys have been based on a total transect length of 100 km or less, such low sighting rates create problems not only for the estimation of parameters such as population density–whether using quantitative (e.g., Buckland et al., 1993) or qualitative (National Research Council, 1981) methods–but also the confirmation of the occurrence of relatively rare species.

With this in mind, we originally hoped to conduct transects with a total length of 1,000 km at each study site in an attempt to guarantee sample sizes (numbers of sightings) adequate for the calculation of reliable estimates of population density. This decision was based primarily on the results of previous surveys in southeastern Amazonia (e.g., Bobadilla, 1998; Lopes and Ferrari, 2000). However, the results of the first three surveys conducted during the present study indicated that this increase in sampling effort was still insufficient to provide adequate samples for rare species (Ferrari et al., 2002b), which led to the modification of procedures.

The results of the first three surveys indicated that a transect of 400 km would be the optimal sample size for the study area. A transect of this length should provide an adequate number of records of the more common species, such as *Alouatta belzebul*, as well as confirmation of the occurrence of the rarer species, such as *Ateles marginatus* and *Chiropotes albinasus*. Reducing transect length permitted an increase in the number of study sites from the original eight to the 12 presented here, providing three replicates of each size of fragment. This replication, it was hoped, would support a more reliable analysis of the effects of fragment size on primate communities. Preliminary observations in fragments of the smallest size class indicated that transect length could be further reduced without decreasing the effectiveness of sampling. In this case, standard transect length was 200 km.

Despite variations in effort between sites (Table 3), samples were considered to be adequate for the systematic evaluation of the effects of habitat fragmentation on primate communities. An additional consideration for data collection was controlling for possible seasonal effects. At all study sites, sampling effort was divided equally between wet (December to May) and dry (June to November) season periods. An analysis of the results of the surveys at sites 1, 2, and 3 did in fact indicate that climatic conditions influenced the sighting rates of some species (Ferrari et al., 2002c).

To confirm the composition of the primate community at each site, survey records were complemented with any relevant additional observations made by researchers at the site, as well as information collected in informal interviews with experienced local residents. As in previous studies (Ferrari et al., 1999; Lopes and Ferrari, 2000), direct questions on the occurrence of given species were avoided until indicated by the interviewee.

Table 3. Sampling effort at the different study sites along the Santarém-Cuiabá Highway in western Pará.

Category of fragment size	Study site	Transect length (km)
Control (continuous forest)	1	300[a]
	5	1000
	9	400
Large	2	400
	6	1000
	10	404
Medium	7	1000
	3, 11	400
Small	4, 8, 12	200

[a]Additional sampling at this site occurred following selective logging in the proximity of the transect.

5. RESULTS

5.1. Communities

As expected, all the region's seven diurnal primate species are found in the Flona Tapajós (Table 4), although spider monkeys and bearded sakis were not recorded in surveys at Base 2 (site 1) nor during additional observations at the site. Ravetta (2001) also found no direct evidence of the occurrence of *Ateles* in adjoining areas of the national forest surrounding Base 2, although a number of residents indicated that it can be found in this area. While contiguous with the rest of the reserve, Base 2 is located in the most heavily impacted portion of the Flona Tapajós, and, among other factors, the study area is adjacent to plots that have been selectively logged. The lack of direct observations indicates that the species is, at the very least, extremely rare in this area. However, it may mean that this species has become locally extinct within the past few years. This conclusion is supported by its confirmed absence from all the neighboring sites (2, 3, and 4).

At the opposite extreme, the smallest communities contained only three species, although they were distributed equally among the different categories of fragment size (sites 2, 3, and 8). As might be expected, these communities were composed of the ecologically most resilient platyrrhines, in particular, the howlers and capuchins. These two species are the most widely distributed of New World primates (Freese and Oppenheimer, 1981; Neville et al., 1988). The ecological flexibility of the marmosets, and their tolerance of disturbed habitats, is also well documented (Ferrari, 1993; Tavares and Ferrari, in press), and *M. argentatus* is relatively abundant in both natural (Albernaz and Magnusson, 1999) and man-made (present study) habitat fragments in the region. It is interesting to note that the remnant communities at sites 2 and 3 are equivalent to that typically found in the gallery forests of the *cerrado* biome of the central Brazilian plateau.

Table 4. Diurnal primate species recorded at the different study sites along the Santarém-Cuiabá Highway in western Pará. See Table 2 for site names and size.

Taxon	Presence (*P* = observed during fieldwork[a], *R* = reported in interviews) or absence (*A*) of species at site											
	1	2	3	4	5	6	7	8	9	10	11	12
Alouatta belzebul	P	P	P	P	P	P	P	P	P	P	P	P
Ateles marginatus	R	A	A	A	P	R	P	A	P	P	P	A
Callicebus moloch	P	A	A	P	P	P	P	P	P	P	P	P
Cebus apella	P	P	P	P	P	P	P	P	P	P	P	P
Chiropotes albinasus	R	A	A	A	P	P	P	A	P	P	P	P
Mico argentatus	P	P	P	P	P	P	P	A	P	P	P	P
Saimiri sciureus	R	A	A	P	P	P	P	A	P	A	P	P
Total species	6[b]	3	3	5	7	6[b]	7	3	7	6	7	6

[a]In a few cases, the species was observed at a site, but not recorded in surveys, in which case no estimate of abundance is available.

[b]Most likely number, based on overall observations at the site. In both cases, the fauna excludes *Ateles marginatus*.

The absence of marmosets from the small fragment at site 8 is thus somewhat enigmatic. That *M. argentatus* should have been "substituted" by *Callicebus moloch* at this site is less curious, given that titis of this group are also relatively tolerant of habitat disturbance (Peres, 1993; Ferrari et al., 2000).

As observations in the continuous forest indicated that marmosets have a patchy distribution in this environment, a simple explanation might be that the fragment originally coincided with a local lacuna in the distribution of *M. argentatus*. While the present-day fragment may offer more ideal conditions for marmosets, it is now separated from neighboring areas of continuous forest by at least one km of open pasture. This pasture formed a considerable barrier to the dispersal of these diminutive monkeys, the adult body weight of which is approximately 350 g. However, an adult titi monkey, *Callicebus moloch* (body weight around 1 kg), was once encountered crossing an access road almost exactly halfway between the fragment and the continuous forest. The monkey was surprised by the approaching vehicle and leapt up onto a fence post before descending into the dense grass of the surrounding pasture.

Little is known of the dispersal of platyrrhines between forest fragments. However, this observation, together with the composition of the fauna in the fragment, suggests that body size may be an important limiting factor. Predation pressure tends to increase with decreasing body size, and a solitary marmoset crossing long distances over open ground may be exceptionally vulnerable.

A complementary factor may be the differences in the social organization of the respective species. In contrast with the other three genera, in particular *Callicebus*, marmosets are highly tolerant of isosexual adult group members, even when unrelated. Whereas maturing titi monkeys are expelled from their natal groups by the parent of the same sex (Kinzey, 1981), marmosets often remain as nonreproductive helpers for many years (Ferrari and Digby, 1996). It would seem reasonable to assume that this tolerance, reinforced by the dangers of dispersal over open ground, would be a major determinant of marmoset dispersal patterns in a fragmented landscape, such as that observed at site 8.

Between these two extremes of community size, the influence of fragment size on primate diversity is far from clear (Figure 3), except for the fact that *Ateles* was absent from all three of the smallest fragments. *Chiropotes* has a similar distribution, although it was present in the small fragment at site 12. The absence of both species from fragments of less than 100 ha is predictable on the basis of known home range sizes for the two genera, although recent observations in southeastern Amazonia (Ferrari et al., 2002a; Port-Carvalho and Ferrari, 2002) have shown that bearded sakis are able to survive over the long term in much smaller fragments.

The remaining species, *Saimiri sciureus*, was present at just over half the sites, but exhibited no clear tendency in relation to fragment size. Squirrel monkeys are relatively abundant in small forest fragments in southeastern Amazonia (Lima, 2000; Lopes and Ferrari, 2000), and the evidence indicates that the presence of riverine habitat (Terborgh, 1983) is a more important determinant of the presence of the species than fragment size.

Overall, the only clear pattern is the variation among nuclei and the complexity of this variation. When comparing fragments of equivalent size, species diversity is almost invariably lower in nucleus 1 in comparison with the remaining two areas (Table 4). However, with the exception of site 8, there is little difference between nuclei 2 and 3, although the occurrence of the less-tolerant species–*Ateles* and *Chiropotes*–was more consistent in nucleus 3 (see below).

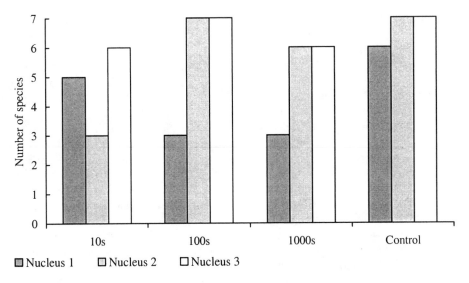

Figure 3. Number of primate species recorded at 12 sites along the Santarém-Cuiabá Highway. Sites are grouped according to fragment size: 10s = fragments between 50 and 100 ha; 100s = fragments of 100 to 1000 ha; 1000s = fragments > 1000 ha; Control = continuous forest (Flona Tapajós).

The determinants of this variation remain unclear. The area in which nucleus 1 is located is characterized by more extensive habitat fragmentation (Figure 2) and a higher population density than the other nuclei, which suggests that its reduced diversity may be related primarily to hunting pressure. However, a number of details contradict this hypothesis. Site 11, for example, has a full complement of species, despite being a medium-sized fragment located within three km of the town of Rurópolis. Local extinction may not necessarily be related to population density or proximity to urban centers. In addition, while hunting was a common activity within the area of nucleus 1, brocket deer (*Mazama* spp.) were more abundant at sites 1 and 2 than anywhere else within the study area. As deer are among the most preferred game species (Redford and Robinson, 1987), it seems unlikely that *Ateles* or *Chiropotes* would have suffered exceptionally intense hunting pressure at all sites within this nucleus.

Given this, it seems possible that both species were rare within the area before human colonization, and thus exceptionally vulnerable to the combined effects of habitat fragmentation and hunting pressure. Ravetta (2001) found good evidence that *A. marginatus* is naturally absent from the floodplain of the Tapajós, west of nucleus 1, but is (or at least was until recently) present on the neighboring plateau, indicating the influence of habitat-related factors. Local rarity may also explain the apparent absence of the species from site 6, given that it is present in the smaller, neighboring fragment at site 7. In addition, there is a marked correlation between the abundance of *Ateles* and *Chiropotes* (see below), which suggests that they are influenced by similar environmental factors.

5.2. Abundance

To facilitate among-site comparisons (Lopes and Ferrari, 2000), sighting rates (number of groups recorded per 10 km of transect walked) were used to estimate the abundance of each species at the different study sites. Sighting rates varied considerably among species and study sites, but rarely exceeded one sighting per 10 km, except in the case of *Alouatta belzebul*. Sighting rates for howlers varied between 0.78 and 8.65 groups per 10 km. The latter value, recorded at site 8, is exceptional, and is in fact greater than the total for all primate species recorded at the remaining 11 study sites (Figure 3). Despite its reduced diversity (Table 4), primates appeared to be far more abundant at this site than at any other (total sighting rate = 10.95 groups per 10 km). The next highest sighting rate was recorded at site 11 (8.15/10 km).

As for diversity, there is little clear relationship between abundance and fragment size (Figure 4), although overall abundance is invariably lowest in the continuous forest in comparison with the fragments within the same nucleus. In fact, in only one case (site 2, 2.9 groups per 10 km) is the sighting rate in a fragment lower than that recorded at a control site (site 9, 3.1 groups per 10 km).

What is clear is that while habitat fragmentation commonly results in a loss of species diversity (Figure 3), it is accompanied by increasing abundance (Figure 4). Although only in the case of nucleus 1 is there a systematic tendency of increasing abundance with decreasing fragment size. In nucleus 2, there is a dip in abundance at the medium-sized fragment (site 7), which coincides with a potentially important increase in diversity in comparison with site 6, that is, the presence of *Ateles marginatus* (Table 4). In addition, sighting rates were lower at site 7 in comparison with site 6 for all species

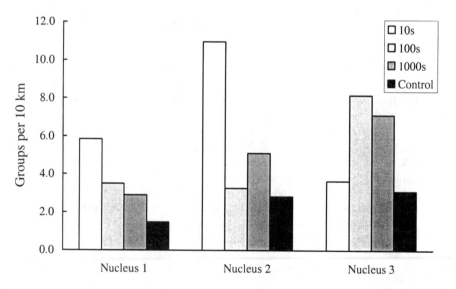

Figure 4. Overall abundance (sightings per 10 km surveyed) of primates at the 12 study sites along the Santarém-Cuiabá Highway. The legend refers to fragment size: 10s = fragments between 50 and 100 ha; 100s = fragments of 100 to 1000 ha; 1000s = fragments >1000 ha; and Control = continuous forest (Flona Tapajós).

except *Chiropotes albinasus*, for which there was an increase from 0.06 to 0.25 groups per 10 km walked.

This would seem to suggest that, while fragment 7 was much smaller than fragment 6, it offered more favorable conditions for the ecologically less tolerant species (*Ateles marginatus* and *Chiropotes albinasus*). Such conditions may have been either natural (e.g., habitat quality, presence of key plant species) or the result of human activities, in particular, selective logging, which appeared to have been more extensive at site 6 (data under analysis). These data also suggest that interspecific competition may be an important determinant of abundance in fragments.

In nucleus 3, abundance increases progressively from the control through the medium-sized fragment, but falls off in the smallest fragment, where diversity is relatively high, which appears to have a negative effect on abundance in small fragments (Figure 5). Diversity varies little within nucleus 3, however, and in nucleus 1 the smallest fragment (site 4) has both the highest diversity and abundance. Once again, the specific local factors underlying the observed pattern remain unclear.

Overall, howlers accounted for 54.8% of all sightings, and thus a large part of the variation in abundance observed among study sites (Figure 6). With the exception of site 3, the relative abundance of howlers within a given nucleus was very similar to that recorded for the primate communities as a whole (Figure 4). The exclusion of howlers (Figure 7) does bring a certain degree of equilibrium to the data, however. In this case, abundance in the smallest fragments appears to be the most consistent, despite the variation in diversity (Figure 5). Abundance in the medium-sized fragments was the most variable.

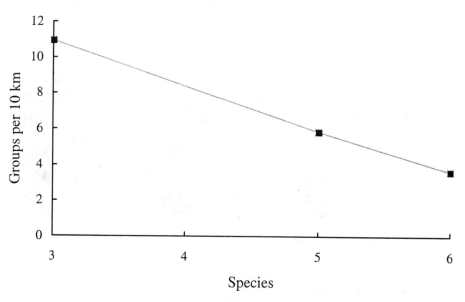

Figure 5. Relationship between abundance (survey records) and species diversity in the three small (<100 ha) fragments surveyed along the Santarém-Cuiabá Highway in western Pará.

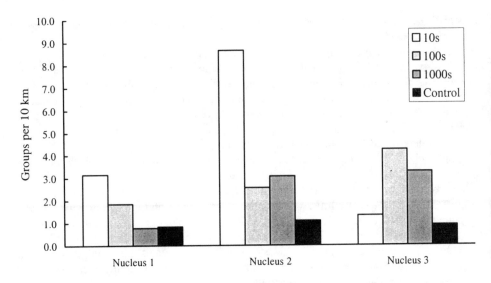

Figure 6. Abundance (sightings per 10 km surveyed) of red-handed howlers (*Alouatta belzebul*) at the 12 study sites along the Santarém-Cuiabá Highway. Legend as in Figure 4.

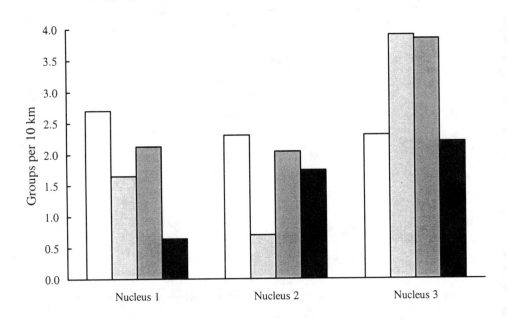

Figure 7. Abundance (sightings per 10 km surveyed) of all primates except *Alouatta belzebul* at the 12 study sites along the Santarém-Cuiabá Highway. Legend as in Figure 4.

The overall differences between nuclei 1 and 2 are also reduced with the exclusion of *Alouatta*. In fact, the mean sighting rate at the four sites in nucleus 1 (1.78±0.9 groups per 10 km) becomes slightly higher than that in nucleus 2 (1.70±0.7), although in nucleus 3 (3.06±0.9) it remains much higher.

The only other species recorded at all 12 study sites was the tufted capuchin (*Cebus apella*), although it presented abundance patterns similar to those seen in Figure 7, which might be expected, given that this species provided almost half (46.74%) of these records. A third of the study area's most widespread species, the silvery marmoset (*Mico argentatus*), presented quite a distinct pattern (Figure 8). For all categories of fragment size, the species was more abundant in nucleus 1 in comparison with the other two nuclei, in direct contrast with overall abundance patterns (Figure 4).

As marmosets tend to be more abundant in disturbed habitats, this pattern may reflect the more extensive habitat fragmentation observed within the area of nucleus 1, which is at least partially related to its longer history of colonization. However, the species was also more than three times more abundant in the continuous forest at site 1 in comparison with other controls (sites 5 and 9), and the lowest sighting rates were recorded at sites 7, 8, and 12, the three smallest fragments in nuclei 2 and 3. So, once again, while there is a general tendency for marmosets to be relatively abundant in fragments, there is no systematic relationship with fragment size.

An alternative interpretation of the relative abundance of marmosets in nucleus 1 would be that this species was naturally common in this area before colonization. This natural abundance could be related to the same complex of factors that determined the scarcity of spider monkeys and bearded sakis in the same area. The variation in

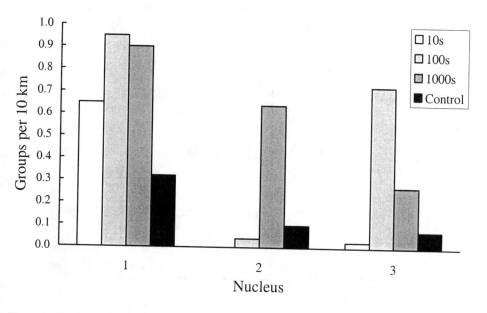

Figure 8. Abundance (sightings per 10 km surveyed) of silvery marmosets (*Mico argentatus*) at the 12 study sites along the Santarém-Cuiabá Highway. Legend as in Figure 4.

abundance recorded in the remaining nuclei is less easily accounted for, but may reflect the local factors determining the species's patchy distribution, as mentioned above.

Despite also being relatively tolerant of habitat disturbance, titi monkeys exhibited an almost exactly opposite pattern of abundance to that of *Mico argentatus* (Figure 9). However, once again, abundance is generally greater in fragments in comparison with the continuous forest, at least within the same nucleus and where the species was present. It is interesting to note that the highest sighting rate for *Callicebus* was recorded at site 8, the only site from which *Mico* was absent. The highest rates for marmosets were recorded at the two sites (2 and 3) from which titis were absent (Figure 8). Nevertheless, both marmosets and titis were extremely rare in the medium fragment (site 7) in nucleus 2, for example, and relatively abundant in the medium fragment in nucleus 3 (site 11).

Once again, local variation (both between and within nuclei) was more pronounced than any clear tendency in relation to factors such as fragment size, although the results do suggest that the species may have been naturally more abundant in the area of nucleus 3 before human colonization. Titis were extremely rare in continuous forest in both nuclei 1 and 2 (site 1, 0.03 sightings per 10 km surveyed; site 5, 0.01), but relatively abundant in nucleus 3 (0.55 sightings per 10 km). They were also relatively abundant at all three fragments within this nucleus.

The remaining three species were recorded in surveys at five to seven sites. Squirrel monkeys were rare at all sites and were sighted at a rate exceeding 0.10 surveyed only in the continuous forest at site 9 (0.17/10 km). As mentioned above, the local distribution of this species tends to be influenced by that of riverine habitats, which almost certainly affected sighting rates at most, if not all sites, according to the location of transects in relation to watercourses. Given this, the present study provides little evidence for the evaluation of the effects of habitat fragmentation on the abundance of this species.

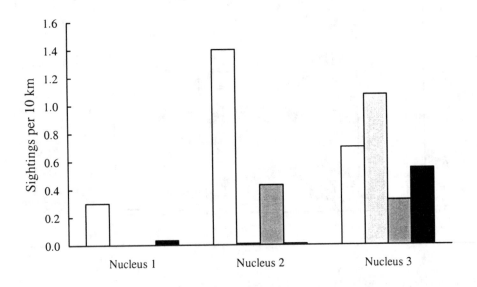

Figure 9. Abundance (sightings per 10 km surveyed) of dusky titi monkeys (*Callicebus moloch*) at the 12 study sites along the Santarém-Cuiabá Highway. Legend as in Figure 4.

As mentioned above, the remaining species, *Ateles marginatus* and *Chiropotes albinasus*, were not recorded during the surveys carried out at sites in nucleus 1. At other sites, sighting rates varied between 0.05 and 0.89 per 10 km surveyed for *Ateles* and 0.06 and 0.52 for *Chiropotes* (Figures 10 and 11). As expected, both species were more abundant in continuous forest than in fragments in nucleus 2 (Figure 10). However, the exact opposite was true in the case of nucleus 3 (Figure 11), with the exception of the smallest fragment, from which *Ateles* was absent.

The relatively low rates recorded in the continuous forest at site 9 were particularly unexpected, especially as the large fragment at site 10, located only a few kilometers away, provided the study's highest sighting rates for both species. In marked contrast, *Ateles* was absent from the large fragment in nucleus 2, where the lowest sighting rate for *Chiropotes* was recorded, even though both species were relatively abundant in the neighboring continuous forest (site 5). This appears to reinforce the idea that the distribution and abundance of these two species are influenced by local variations in factors such as habitat quality and anthropic interference. In support of this, there is a strong correlation (Spearman's $r_s = 0.90$, $p = 0.037$, $n = 5$) between the sighting rates of the two species at the same study sites, suggesting that their abundance is influenced by similar factors.

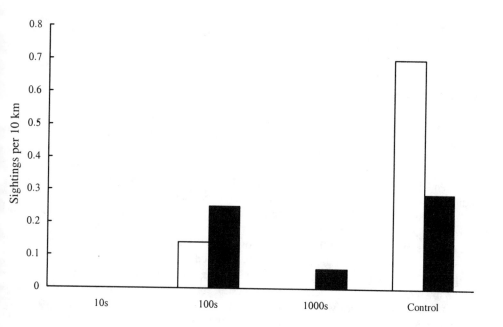

Figure 10. Abundance (sightings per 10 km surveyed) of white-fronted spider monkeys, *Ateles marginatus* (unshaded), and white-nosed bearded sakis, *Chiropotes albinasus* (shaded), at the four sites in study nucleus 2 on the Santarém-Cuiabá Highway.

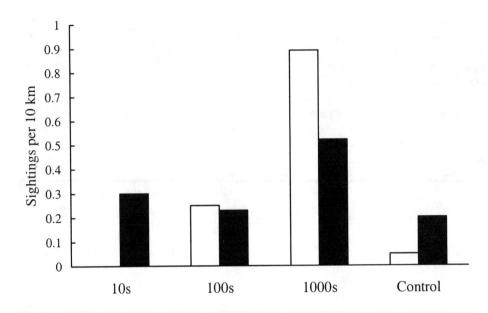

Figure 11. Abundance (sightings per 10 km surveyed) of white-fronted spider monkeys, *Ateles marginatus* (unshaded), and white-nosed bearded sakis, *Chiropotes albinasus* (shaded), at the four sites in study nucleus 3 on the Santarém-Cuiabá Highway.

6. DISCUSSION

The results of the present study provide disappointingly few systematic insights into the effects of habitat fragmentation on primate populations in southern Amazonia. Decreasing diversity, for example, is a predictable consequence of habitat fragmentation, but, while most fragments have fewer species than continuous forest, this is not always true, and in fact no clear relationship was found between fragment size and species number, even when comparing sites within the same nucleus.

The only universally consistent pattern of diversity was the presence of howlers and capuchins at all sites and the absence of spider monkeys from the smallest fragments. Despite the predictability of the latter phenomenon, the sum of the evidence indicates that the number of species at a given site was determined by a complex of factors other than fragment size alone, including natural variations in the distribution of different species and habitat quality, possibly reinforced, at least in some cases, by anthropic interference.

An additional pattern was the apparent local extinction of both *Ateles* and *Chiropotes* from the colonized area of nucleus 1. Despite the fact that this coincides with the most densely populated part of the study area, it remains unclear whether hunting pressure was the primary determining factor. Ravetta (2001) found some evidence of hunting of *Ateles marginatus* within the study area, but this species clearly did not suffer the kind of pressure recorded for ateline populations by Peres (1991) in western Amazonia.

The circumstantial evidence is conflicting, although the data from site 1 (continuous forest) suggest that primates were originally rare in this area before human colonization.

In this case, both *Ateles* and *Chiropotes* may have been excluded from at least some of the fragments through random "sampling" factors, as in the case of marmosets at site 8, for example. If habitat quality was a primary determinant of species rarity, it would seem reasonable to conclude that habitat fragmentation would have greatly reduced the potential for the survival of remnant populations in most fragments. Hunting pressure would obviously reinforce this, but it remains unclear whether it would be necessary for extinction.

In contrast with species diversity, increased abundance in fragments, in comparison with continuous forest, was a universal pattern, and especially evident when comparing sites within the same nucleus. Nevertheless, each nucleus presents a different pattern of abundance in relation to fragment size, once again suggesting the interplay of a complex of factors. In addition to natural variation in the distribution and abundance of species, presumably determined by fluctuations in habitat quality, interspecific competition may be an important factor (e.g., Figures 5, 8, and 9).

Increased abundance in fragments was not restricted, as expected, to the more tolerant species, but included *Ateles* and *Chiropotes* in nucleus 3 (Figure 11), implying once again the influence of specific local factors. At least two factors may be involved here in addition to those mentioned above. The first is methodological. As mentioned previously, line transect surveying is potentially vulnerable to sampling biases, which may be further reinforced by the location of the transect in relation to the characteristics of the study area (especially important in the case of *Saimiri*). However, such bias would be random, in contrast with the relatively consistent pattern shown in Figure 11. The second factor is a short-term "refugee" effect, resulting from ongoing fragmentation, but even if this had been occurring, it is difficult to imagine, once again, how such an effect could be so consistent across sites and species.

Overall, rather than defining pattern and process, the results of the present study tend to emphasize the complexity of the effects of anthropic habitat fragmentation. While a number of factors are involved, a predominant component of this complexity would appear to be the natural heterogeneity of the original populations, related, in turn, to variation in habitat quality (e.g., Peres, 1997b; Ferrari et al., 1999; Iwanaga and Ferrari, 2002a, 2002b). Anthropic habitat fragmentation clearly adds a new layer to this complexity, with only a few predictable consequences (loss of diversity, increasing abundance).

Given this, a more definitive analysis of the process would appear to depend on a detailed knowledge of the original populations within the study area, although this is obviously difficult to achieve, even where the process is planned, as in the Biological Dynamics of Forest Fragments Project (Bierregaard et al., 1992). A complementary approach would be the identification of the determining factors in relation to habitat quality.

Despite the fact that much of the colonization in Brazilian Amazonia, like that along the BR-163, is government-sponsored, most is very poorly planned in relation to either the viability of agriculture or its effects on the environment. While colonists are required by law to maintain reserves of natural vegetation on their plots, such reserves tend to be randomly located in relation to the characteristics of the local flora and fauna and to neighboring reserves. More careful planning, based on detailed biological and geomorphological surveys, would benefit both colonist and the environment, but would obviously require investments that most governments are reluctant or even unable to make.

7. SUMMARY

Recent human colonization of the Brazilian Amazon has been bolstered by the construction of highways, which provide access to previously isolated areas of *terra firme* forest. The Santarém-Cuiabá highway has been the backbone for colonization in southern central Amazonia for almost thirty years, creating a landscape of forest fragments isolated predominantly by cattle pasture. The effects of this process on local primate communities was evaluated through population surveys at twelve sites, representing different fragment sizes and proximity to urban centers. Standard line transect surveys were conducted, with a total length of 200 to 1,000 km at a given site. In general, larger fragments contained more species than smaller ones, but additional factors appeared to influence diversity at a given site. By contrast, abundance tended to increase in smaller fragments. This tendency was especially marked in the case of the howler monkey, *Alouatta belzebul*, which reached exceptional densities in some of the smallest fragments. As expected, the specialized frugivores *Ateles marginatus* and *Chiropotes albinasus* were most often absent from sites, although this was only partly related to fragment size. While a predictable reduction in diversity was found in the smallest fragments (<100 ha), the characteristics of communities in larger fragments appeared to be influenced by a complex of factors, including hunting pressure and natural variations in the distribution of different species.

8. ACKNOWLEDGMENTS

This study was financed by the National Program for Biodiversity of the Brazilian Environment Ministry (PROBIO/MMA/BIRD/GEF/CNPq). Additional support was received from IBAMA/Santarém and the Kapok Foundation. We are grateful to Robson Mendes and the IBAMA personnel under his supervision, Daniela Oliveira and the PROBIO staff, and the very many local people from the study area who contributed to the collection of data in their many different ways. We also thank Laura Marsh for her comments on the drafts of this manuscript.

9. REFERENCES

Albernaz, A. L., and Magnusson, W., 1999, Home-range size of the bare-ear marmoset (*Callithrix argentata*) at Alter do Chão, Central Amazonia, Brazil, *Int. J. Primatol.* **20**:665–677.
Ayres, J. M., and Clutton-Brock, T. H., 1992, River boundaries and species range size in Amazonian primates, *Am. Nat.* **140**:531–537.
Bobadilla, U. L., 1998, Abundância, Tamanho de Agrupamento e Uso do Hábitat por Cuxiús de Uta Hick *Chiropotes satanas utahicki* Hershkovitz, 1985 em Dois Sítios na Amazônia Oriental: Implicações para a Conservação, Masters dissertation, Universidade Federal do Pará, Belém.
Bobadilla, U. L., and Ferrari, S. F., 2000, Habitat use by *Chiropotes satanas utahicki* and syntopic platyrrhines in eastern Amazonia, *Am. J. Primatol.* **50**:215–224.
Bierregaard, R. O., Jr., Lovejoy, T. E., Kapos, V., dos Santos, A. A., and Hutchings, R. W., 1992, The biological dynamics of tropical rainforest fragments, *Bioscience* **42**:859-866.
Buckland, S. T., Anderson, D. R., Burnham, K. P., and Laake, J. L., 1993, *Distance Sampling: Estimating Abundance of Biological Populations*, Chapman & Hall, London.
Egler, S. G., 1992, Feeding ecology of *Saguins bicolor bicolor* (Callitrichidae: Primates) in a relict forest in Manaus, Brazilian Amazonia, *Folia Primatol.* **46**:61–76.

Emmons, L. H., 1984, Geographic variation in densities and diversities of non-flying mammals in Amazonia, *Biotropica* **16**:210–222.

Ferrari, S. F., 1993, Ecological differentiation in the Callitrichidae, in: *Marmosets and Tamarins: Systematics, Ecology, and Behaviour*, A. B. Rylands, ed., Oxford University Press, Oxford, pp. 314–328.

Ferrari, S. F., in press, Biogeography of Amazonian primates, in: *A Primatologia on Brasil–8*, S. L. Mendes and A. G. Chiarello, eds., Sociedade Brasileira de Primatologia, Santa Teresa.

Ferrari, S. F., and Diego, V. H., 1995, Habitat fragmentation and primate conservation in the Atlantic Forest of eastern Minas Gerais, Brazil, *Oryx* **29**:192–196.

Ferrari, S. F., and Digby, L. J., 1996, Wild *Callithrix* groups: Stable extended families? *Am. J. Primatol.* **38**:19–27.

Ferrari, S. F., Ghilardi, R., Jr., Lima, E. M., Pina, A. L. C. B., and Martins, S. S., 2002a, Mudanças a longo prazo nas populações de mamíferos da área de influência da Usina Hidrelétrica de Tucuruí, Pará, in: *Resumos do XXIVº Congresso Brasileiro de Zoologia, Itajaí, SC*, pp. 540–541.

Ferrari, S. F., Iwanaga, S., Coutinho, P. E. G., Messias, M. R., Cruz Neto, E. H., Ramos, E. M., and Ramos, P. C. S., 1999, Zoogeography of *Chiropotes albinasus* (Platyrrhini, Atelidae) in southwestern Amazonia, *Int. J. Primatol.* **20**:995–1,004.

Ferrari, S. F., Iwanaga, S., Messias, M. R., Ramos, E. M., Ramos, P. C. S., Cruz Neto, E. H., and Coutinho, P. E. G., 2000, Titi monkeys (*Callicebus* spp., Atelidae: Platyrrhini) in the Brazilian state of Rondônia, *Primates* **41**:191–196.

Ferrari, S. F., Iwanaga, S., Souza, L. L., Costa, C. G., Ravetta, A. L., Freitas, F. C., and Coutinho, P. E. G., 2002b, A problemática do tamanho de amostra em levantamentos de transecção linear de populações de mamíferos em ambiente de floresta, in: *Resumos do XXIVº Congresso Brasileiro de Zoologia, Itajaí, SC*, p. 540.

Ferrari, S. F., Iwanaga, S., Souza, L. L., Costa, C. G., Ravetta, A. L., Freitas, F. C., and Coutinho, P. E. G., 2002c, Influência da sazonalidade sobre os resultados de levantamentos de transecção linear de populações de mamíferos na Amazônia central, in: *Resumos do XXIVº Congresso Brasileiro de Zoologia, Itajaí, SC*, p. 540.

Ferrari, S. F., and Queiroz, H. L., 1994, Two new Brazilian primates discovered endangered, *Oryx* **28**:31–36.

Freese, C., and Oppenheimer, J. R., 1981, The capuchin monkeys, genus *Cebus*, in: *Ecology and Behavior of Neotrpical Primates*, R. A. Mittermeier and A. F. Coimbra-Filho, eds., Academia Brasileira de Ciências, Rio de Janeiro, pp. 331–390.

Hershkovitz, P., 1977, *Living New World Monkeys, Volume 1, with an Introduction to Primates*, Chicago University Press, Chicago.

Iwanaga, S., and Ferrari, S. F., 2002a, Geographic distribution and abundance of woolly (*Lagothrix cana*) and spider (*Ateles chamek*) monkeys in southwestern Brazilian Amazonia, *Amer. J. Primatol.* **56**: 57-64.

Iwanaga, S., and Ferrari, S. F., 2002b, Geographic distribution of red howlers, *Alouatta seniculus* (Platyrrhini, Alouattini) in southwestern Brazilian Amazonia, with notes on *Alouatta caraya*, *Int. J. Primatol.* **23**:in press.

Johns, A. D., and Ayres, J. M., 1987, Bearded sakis beyond the brink? *Oryx*, **21**:164–167.

Kinzey, W. G., 1981, The titi monkeys, genus *Callicebus*, in: *Ecology and Behavior of Neotropical Primates*, R. A. Mittermeier and A. F. Coimbra-Filho, eds., Academia Brasileira de Ciências, Rio de Janeiro, pp. 241–276.

Lima, E. M., 2000, Ecologia comportamental de um grupo Silvestre de Macaco-de-Cheiro (*Saimiri sciureus*) no Parque Ecológico de Gunma, Santa Bárbara do Pará-PA, Masters dissertation, Universidade Federal do Pará, Belém.

Lopes, M. A., and Ferrari, S. F., 1996, Preliminary observations on the Ka'apor capuchin, *Cebus kaapori*, from eastern Brazilian Amazonia, *Biol. Conserv.* **76**:321–324.

Lopes, M. A., and Ferrari, S. F., 2000, Effects of human colonization on the abundance and diversity of mammals in eastern Brazilian Amazonia, *Cons. Biol.* **14**:1,658–1,665.

Marsh, L. K., 1999, Ecological effect of the black howler monkey (*Alouatta pigra*) on fragmented forests in the Community Baboon Sanctuary, Belize, Ph.D. thesis, Washington University, St. Louis.

Mendes, S. L., 1989, Estudo ecológico de *Alouatta fusca* (Primates: Cebidae) na Estação biológica de Caratinga, MG, *Rev. Nordest. Biol.* **6**:71–104.

Milton, K., 1980, *The Foraging Strategy of Howler Monkeys*, Columbia University Press, New York.

National Research Council, 1981, *Techniques for the Study of Primate Population Ecology*, National Academy Press, Washington D.C.

Neville, M. K., Glander, K. E., Braza, F., and Rylands, A. B., 1988, The howling monkeys, genus *Alouatta*, in: *Ecology and Behavior of Neotropical Primates, Vol. 2*, R. A. Mittermeier and A. B. Rylands, eds., World Wildlife Fund, Washington D.C., pp. 349–453.

Peres, C. A., 1991, Humboldt's woolly monkey decimated by hunting in Amazonia, *Oryx* **25**:89–95.

Peres, C. A., 1993, Structure and spatial organization of an Amazonian terra firme forest primate community, *J. Trop. Ecol.* **9**:259–276.

Peres, C. A., 1996, Use of space, spatial group structure, and foraging group size of gray woolly monkeys (*Lagothrix lagotricha cana*) at Urucu, Brazil, in: *Adaptive Radiations of Neotropical Primates,* M. A. Norconk, A. L. Rosenberger, and P. A. Garber, eds., Plenum Press, New York, pp. 467–488.

Peres, C. A., 1997a, Primate community structure at twenty western Amazonian flooded and unflooded forests, *J. Trop Ecol.* **12**:1–25.

Peres, C. A., 1997b, Effects of habitat quality and hunting pressure on arboreal folivore densities in Neotropical forests: A case study of howler monkeys (*Alouatta* spp.), *Folia Primatol.* **68**:199–222.

Port-Carvalho, M., and Ferrari, S. F., 2002, Estimativas da abundância de cuxiú-preto (*Chiropotes satanas satanas*) e outros mamíferos não-voadores em fragmentos antrópicos de floresta da região Tocantina, Amazônia oriental, in: *Resumos do XXIV° Congresso Brasileiro de Zoologia, Itajaí, SC,* pp. 531–532.

Ravetta, A. L., 2001, O Coatá-de-Testa-Branca (*Ateles marginatus*) do baixo Rio Tapajós, Pará: Ecologia e Status de Conservação, Masters dissertation, Museu Goeldi, Belém.

Redford, K. H., and Robinson, J. G., 1987, The game of choice: Patterns of Indian and colonist hunting in the Neotropics, *Amer. Anthropol.* **89**:650–667.

van Roosmalen, M. G. M., 1985, Habitat preferences, diet, feeding strategy, and social organization of the black spider monkey (*Ateles paniscus paniscus* Linnaeus 1758) in Surinam, *Acta Amazonica* **15(3 suppl.)**:1–238.

Rosenberger, A. L., 1992, Evolution of New World monkeys, in: *The Cambridge Encyclopedia of Human Evolution,* S. Jones, R. Martin, and D. Pilbeam, eds., Cambridge University Press, Camridge, pp. 209–216.

Rowe, N., 1998, *A Pictorial Guide to the Primates,* Pongonias Press, New Hampton.

Rylands, A. B., 1989, Sympatric callitrichids: The black tufted-ear marmoset, *Callithrix kuhli,* and the golden-headed lion tamarin, *Leontopithecus chrysomelas, J. Hum. Evol.* **18**:679–695.

Rylands, A. B., and Keuroghlian, A., 1988, Primate populations in continuous forest and forest fragments in central Amazonia: Preliminary results, *Acta Amaz.* **18**:291–307.

Rylands, A. B., Mittermeier, R. A., and Rodríguez-Luna, E., 1997, Conservation of Neotropical primates: Threatened species and an analysis of primate diversity by country and region, *Folia Primatol.* **68**:134–160.

Rylands, A. B., Schneider, H., et al., 2000, An assessment of the diversity of New World primates, *Neotrop. Primates,* **8**:61–93.

Schneider, H., and Rosenberger, A. L., 1996, Molecules, morphology, and platyrrhine systematics, in: *Adaptive Radiations of Neotropical Primates,* M. A. Norconk, A. L. Rosenberger, and P. A. Garber, eds., Plenum Press, New York, pp. 3–19.

Spironelo, W. R., 1987, Range size of a group of *Cebus apella* in central Amazonia, *Int. J. Primatol.* **8**:522.

Symington, M. M., 1988, Demography, ranging patterns, and activity budgets of black spider monkeys (*Ateles paniscus chamek*) in the Manu National Park, Peru, *Amer. J. Primatol.* **15**:45–67.

Tavares, L. I., and Ferrari, S. F., In press, Diet of the silvery marmoset (*Callithrix argentata*) at ECFPn: Seasonal and longitudinal variation, in: *Caxiuanã, Biodiversidade e Sustentabilidade,* P. L. B. Lisboa, ed., CNPq/MCT, Belém.

Terborgh, J., 1983, *Five New World Primates: A Study in Comparative Ecology,* Princeton University Press, Princeton.

PRIMATES AND FRAGMENTATION OF THE AMAZON FOREST

Kellen A. Gilbert[*]

1. INTRODUCTION

Many areas of what was once primary Neotropical forest are now areas of deforestation, areas of habitat fragmentation, and areas of isolated forest fragments. This change in the forest is mainly attributed to increased human activity. In the case of the Brazilian Amazon, deforestation is more widespread in the states of Pará and Rondônia than in Amazonas. With increasing human population and associated socioeconomic pressures, however, significant deforestation in this central region of the Amazon basin may soon match the other states (INPE, 1998). What happens to the flora and fauna in this new mosaic of isolated forest remnants and secondary growth is only beginning to be understood. Examining the effects of fragmentation on the medium- and large-sized rain forest mammals is often the first and simplest place to begin, given their relative ease of observation and measurement (Gilbert and Setz, 2001; Malcolm, 1988, 1990; Spironelo, 1987).

Early studies of primate populations and habitat fragmentation focused on the effects of selective logging on primate populations (Wilson and Wilson, 1975; Wilson and Johns, 1982) and the effects on primate species diversity in areas disturbed by agricultural activities (Bernstein et al., 1976). More recently, habitat loss from widespread deforestation has been taken into account to explain decreases in primate diversity and declining populations in forests throughout the Paleo- and Neotropics (White, 1994; Ferrari and Diego, 1995; Hsu and Agoramoorthy, 1996; O'Brien and Kinnaird, 1996; Garcia and Mba, 1997; Thoisy and Richard-Hansen, 1996). A consistent finding in these primate studies is that, if human degradation of forest environments either directly or indirectly continues unchecked, the survival of most of the endemic primates is threatened.

Primates in the isolated and continuous forest reserves of the Biological Dynamics of Forest Fragments Project (BDFFP) have been studied for over 18 years. Some of the same groups have been observed in reserves both before and after forest areas were

---

[*] Department of Sociology and Criminal Justice, Southeastern Louisiana University, Hammond, Louisiana 70402, USA. Email: kgilbert@selu.edu.

experimentally isolated (Rylands and Keuroghlian, 1988). The results from BDFFP studies of primate ecology and behavior in isolated reserves and a continuous forest reserve serving as a control provide unique and valuable data about what happens to primate populations in a fragmented landscape over time.

1.1. Background

The BDFFP was established 20 years ago to examine the micro- and macro-effects of landscape fragmentation on continuous forest in the central Amazonian basin. The project is a joint effort of Brazil's Instituto Nacional de Pesquisas da Amazônia (INPA) and the Smithsonian's Tropical Research Institute. While the Brazilian government began to encourage agricultural development, mainly cattle ranching, and development of the zona franca, or the trade free zone, in the state of Amazonas in the 1970s, Brazilian law mandated that landowners developing rain forest property had to keep half of their property forested. Thomas Lovejoy, then Vice President for Science at the World Wildlife Fund (WWF), saw an excellent opportunity being created in the central Amazon basin to take advantage of the Brazilian law while at the same time initiating a long-term and large-scale experiment to examine the effects of habitat fragmentation. Thus, studies could be conducted before and after fragmentation to determine the minimal sizes of the fragments for species survival. It was to be a real test of island biogeography theory put forth by MacArthur and Wilson in 1967. Lovejoy then brought INPA aboard and in 1979, the Minimal Critical Size of Ecosystems (MCSE) project (BDFFP's original name) was launched, as joint WWF/INPA effort. In Amazonas, the region north of Manaus, the owners of three cattle ranches (Dimona, Porto Alegre, and Esteio) signed a long-term agreement allowing a series of reserves to be created and maintained within their property boundaries (Bierregaard and Gascon, 2001).

The MCSE project staff initially isolated areas of continuous forest into 1-, 10- and 100-ha plots (Figure 1). Other areas were delineated in the same manner, but remain part of the continuous forest and serve as controls. The fragments were isolated by clearing and burning, in some cases to form cattle pastures. This maintenance has contained them, with the need for re-isolating the fragments about every four years. After the first five years of the project it became clear that the original goal of isolating 10,000-ha fragments would not be met. Rather the mosaic of the smaller fragments surrounded by secondary growth, pasture, or abandoned pasture reflected the real-world situation in central Amazonia. The philosophy of the project shifted from determining the minimal size for the preservation of species to the effects of fragmentation on species, hence the name change to the Biological Dynamics of Forest Fragments Project (Bierregaard and Gascon, 2001).

1.2. Study Site

The study area is located 80 km north of Manaus (2°30'S, 60°W) within a larger area of about 20 km by 50 km. The central Amazon basin is upland *terra firme* evergreen tropical moist forest. Average annual rainfall is about 2,300 mm. The terrain is moderately hilly and dissected by small creeks that form the headwaters of tributaries of small rivers. The soils are nutrient-poor, yellow latosols with high clay content (Bierregaard et al., 1992; Fearnside and Leal Fihlo, 2001). This upland area is far from

Figure 1. Satellite photo of the portion of the central Amazonian basin with the BDFFP reserves. Dark gray indicates primary forest, white, secondary forest, and light gray, pasture (Photo courtesy of BDFFP, INPA/SI, 1995).

the Rio Negro and Rio Solimões and their associated riverine habitats of igapó and várzea forests.

Over 1,000 tree species have been identified in the reserves. Some of the most common families include the Sapotaceae, Lecythidaceae, Burseraceae, and Leguminosae (Rankin-de Merona et al., 1992). While the species diversity of trees is high, the region is considered fruit-poor. This characterization is in contrast to other moist forest sites such as Cochu Cashu, Peru, or Barro Colorado Island, Panamá, where there is a greater abundance of trees that produce "monkey fruits," such as *Ficus* spp. (Milton, 1980; Terborgh, 1983). The diversity of tree species in the BDFFP area is high, particularly of Sapotaceae (de Oliveira, 1997). All of the primates eat these fruits; the sakis eat the immature fruits and seeds. However, the year-round abundance of fleshy fruits, preferred by spider monkeys, appears to be low and sporadic (Gilbert, 1994a,b).

Six primate species inhabit the reserves of the BDFFP: red howling monkey (*Alouatta seniculus*), black spider monkey (*Ateles paniscus*), tufted or brown, capuchin (*Cebus apella*), bearded saki (*Chiropotes satanas*), white-faced saki (*Pithecia pithecia*), and golden-handed tamarin (*Saguinus midas*). The night monkey, *Aotus* sp. may occur in the region but has only been observed once in a reserve (M. van Roosmalen, personal communication). With the exception of the golden-handed tamarin, all six species are found in moist forest habitats from French Guiana to eastern Peru, where they have been recorded and studied at Cocha Cashu in Manu National Park (e.g., Terborgh, 1983). Among these six species, the white-faced saki is the rarest (Rylands and Keuroghlian, 1988).

There are five felids that use the fragments, including the jaguar (*Panthera onca*). Potential avian predators common to rare in the region are the harpy eagle (*Harpyia harpja*), and the crested hawk eagle (*Morphnus guianensis*).

1.3. History of Primate Studies in the BDFFP Reserves

The initial primate studies in the BDFFP reserves occurred before the small areas were isolated. In 1981, Rylands and Keuroghlian (1988) began a series of surveys in four areas designated for experimental fragmentation. This provided a unique opportunity to collect baseline data in continuous forest, then to do repeated surveys in the same locations after 10-ha and 100-ha areas had been isolated from what was continuous forest. They worked for over a year and observed immediate changes with the larger primates, spider monkeys, and black-bearded sakis, abandoning the fragments. These initial surveys provided valuable pre- and post-isolation data.

The next study of BDFFP reserve primates focused on the relationship between forest characteristics and primate distribution in four isolated 10-ha fragments (Schwarzkopf and Rylands, 1989). They found more primate species such as red howling monkeys, white-faced sakis, and golden-handed tamarins, in the more structurally complex isolated reserves, defined as those with a high mean number of trees and lianas, a low mean percentage of trees greater than 10 cm DBH (diameter at breast height), and streams within the reserve boundaries. They found no significant relationship between the distance from an isolated reserve to primary forest and the number of primate species present in a reserve (Schwarzkopf and Rylands, 1989).

In the late 1980s and early 1990s individual primate species in the continuous forest reserves and isolated fragments were studied. In the continuous forest, Frazão (1992) studied the foraging strategy of a group of black-bearded sakis. In the same reserve, Spironelo (1987) examined the ranging patterns of brown capuchins. He found the capuchin group ranged over 800 ha, a much larger area than that used by capuchins at other sites. The larger home range is related to poor soils and low density and abundance of fruiting trees (Spironelo, 2001). Primate studies in the isolated fragments include a study of the diet of white-faced sakis (Setz, 1994) and parasitic infection in red howling monkeys (Gilbert 1994b; 1997). The prevalence of endoparasitic infection was higher in howler groups in the smaller (10 ha) isolated fragments than in the continuous forest.

Recently, I have worked in three isolated 10-ha fragments, two 100-ha fragments, and a continuous forest reserve conducting annual primate surveys. I have also surveyed four 1-ha reserves. The fragments have been re-isolated over the past 15 years but at varying intervals (Figure 2). Thus, they do not serve as true replicates. I only included in my surveys the fragments surrounded by less than 4 m of vegetation of secondary growth.

2. SURVEY METHODS

Using the existing trail system in the reserves, I used the same line transect method as used in all previous surveys (Rylands and Keuroghlian 1988). I walked all the trails in each reserve, stopping every 50 m to look and listen for primates (National Research Council 1981). The surveys began at 0600 and again at 1400 until approximately 1800 to

Figure 2. Edge of a 10-ha fragment surrounded by burned second growth. Photo by Kellen Gilbert.

take advantage of the general peaks in primate activity (personal observation). When a primate individual or group was detected (visually or by vocalization), I recorded the species, the number of individuals seen, location, animal to observer distance, height of the individual(s), activity, and mode of detection. I also tried to identify individuals by age and sex. Since most primates in the reserves are not well habituated and can be wary, I repeated surveys in each reserve to increase the probability of detection.

In each fragment I measured the height of the secondary growth on the edges and identified the vegetation. The most common plant species along the edges were identified in the field by BDFFP botanists or collected and identified by INPA and BDFFP herbarium staff in Manaus.

3. PRIMATES IN FRAGMENTS

3.1. 1-ha Reserves

No baseline data exist on the primates in the 1-ha reserves before isolation. Of the four 1-ha fragments, only one (Colosso) has primates. This particular reserve was isolated in 1980 and has been re-isolated four times. It is surrounded by pasture but is less than 150 m from the continuous forest. At the time of this study, this reserve had a group of howling monkeys and a single adult male bearded saki. The howling monkey group had three members in 1997 and six in 2001, including a new infant. A group of golden-handed tamarins were also observed on the periphery of the fragment and in the secondary growth adjacent to the fragment. It is unknown if the group is a permanent resident of the fragment. These are the same species sighted from 1997 to 2001 (Table 1).

3.2. 10-ha Reserves

The resident howling monkey groups remained after the initial isolation of four 10-ha fragments in the early 1980s (Table 2) (Rylands and Keuroghlian, 1988). One of the fragments (Cidade Powell) initially isolated has never been reisolated or maintained as a 10-ha fragment. To date, it is indistinguishable as a fragment from the surrounding forest. The three 10-ha fragments that have been maintained as isolates still have resident howler groups (Figure 3). Group size ranges from four to nine individuals with a mean of 6.2 individuals.

Table 1. Primate observed in the BDFFP isolated 1-ha forest fragments by survey period.

Fragment Name	Primates Present			
	1991 to 92	1992 to 93	1995 to 97	2000 to 01
Colosso	*Alouatta*	*Alouatta*	*Alouatta* *Chiropotes*	*Alouatta* *Chiropotes*
Dimona A	0	0	0	0
Dimona B	0	0	0	0
Porto ·Alegre	0	0	0	0

Table 2. Primates observed in the BDFFP isolated 10-ha forest reserves by survey period (Cidade Powell has not been maintained as an isolate).

Fragment Name	Primates Present				
	1983 to 84	1991 to 92	1992 to 93	1995 to 97	2000 to 01
Cidade Powell	*Alouatta* *Saguinus*	-	-	-	-
Colosso	*Alouatta* *Saguinus*	*Alouatta* *Saguinus* *Pithecia*	*Alouatta* *Saguinus* *Pithecia*	*Alouatta* *Saguinus* *Pithecia* *Chiropotes* *Ateles*	*Alouatta* *Saguinus* *Pithecia* *Chiropotes* *Ateles*
Dimona	*Alouatta* *Pithecia*	*Alouatta* *Pithecia* *Saguinus* *Chiropotes*	*Alouatta* *Saguinus*	*Alouatta*	*Alouatta* *Pithecia* *Saguinus*
Porto Alegre	*Alouatta* *Saguinus*	*Alouatta* *Pithecia*	*Alouatta*	*Alouatta*	*Alouatta*

Figure 3. Adult male red howling monkey. Photo by Andrew Whitaker.

White-faced sakis were absent from three of the four 10-ha fragments after initial isolation (Rylands and Keuroghlian 1988). In 1991, a saki group, ranging from three to six individuals, was observed in three fragments, although a group left the Porto Alegre 10-ha fragment in 1997. No sakis have been seen in this reserve for seven years.

Immediately after isolation, golden-handed tamarin groups left the 10-ha fragments but have since recolonized all but Porto Alegre. It is common to see groups on the fragment edges.

The Colosso 10-ha fragment has had an adult male bearded saki and an adult female black spider monkey as well as a howling monkey group, a white-faced saki group and a tamarin group. The saki and the spider monkey have been present since 1997. An adult female bearded saki was seen with the male saki in 2001. Many times the sakis and spider monkey are close to or travel with the howling monkey group.

3.3. 100-ha Reserves

Groups of howling monkeys, white-faced sakis, and golden-handed tamarins remained in the two 100-ha fragments after isolation (Table 3). A group of brown capuchins has also remained in the Porto Alegre 100-ha fragment. In 2000, a group of 14 bearded sakis and a capuchin group were observed in the Dimona 100-ha fragment, and in the Porto Alegre 100-ha fragment a single adult male bearded saki was observed foraging with a group of white-faced sakis.

3.4. Continuous Forest Reserve

A 100-ha area within a continuous forest reserve has served as a control area for primate surveys since 1991 (Figure 4). All six primate species inhabit the area, with five groups of howling monkeys, one group of white-faced sakis, and one group of tamarins, though these are the most rarely seen. The 100-ha area is also part of the range for two groups of capuchins, one group of bearded sakis, and itinerant spider monkeys.

3.5. Primates in Fragments

Does fragment size affect the number of species present over time? To answer this, I examined the coefficient of variance for three 10-ha fragments and the two 100-ha fragments for number of species present by time (Table 4). There was significant difference in the number of species present between the two size classes (Kruskal Wallis T.S. = 3, df = 1, p = 0.08). Within fragment size classes there was variance in number of species turnover over time. The 10-ha fragments had more variation in the number of turnovers than the two 100-ha fragments using coefficient of variance (Kruskal Wallis T.S. = 8.4, df = 1, p = 0.005). I also examined changes in primate biomass, determined by the average adult weight of males and females by species, per reserve area. There was no significant difference in changes of biomass over time by fragment size class.

3.6. Fragments and Secondary Growth

Secondary vegetation around the reserves is primarily *Cecropia sciadophylla* and *Vismia* sp (Figure 5). *Cecropia* is one of the most common colonizing genera in the

Table 3. Primates observed in the BDFFP isolated 100-ha forest fragments by survey period.

Fragment Name	Primates Present				
	1983 to 84	1991 to 92	1992 to 93	1995 to 97	2000 to 01
Dimona	-	*Alouatta*	*Alouatta*	*Alouatta*	*Alouatta*
		Pithecia	*Pithecia*	*Pithecia*	*Pithecia*
		Saguinus	*Saguinus*		*Saguinus*
					Chiropotes
Porto Alegre	*Alouatta*	*Alouatta*	*Alouatta*	*Alouatta*	*Alouatta*
	Pithecia	*Pithecia*	*Pithecia*	*Pithecia*	*Pithecia*
	Saguinus	*Saguinus*	*Saguinus*	*Saguinus*	*Saguinus*
	Cebus		*Cebus*	*Cebus*	*Chiropotes*

Figure 4. Canopy of upland terra firme wet forest, north of Manaus, Brazil. Photo by Philip C. Stouffer.

Table 4. Number of primate species in the BDFFP isolated forest reserves and the continuous forest by survey period.

Fragment number	Fragment Size (ha)	Number of Species Present				
		1983 to 84	1991 to 92	1992 to 93	1995 to 97	2000 to 01
1	10	2	3	3	5	5
2	10	2	4	2	1	3
3	10	2	2	1	1	1
4	100	4	3	4	4	4
5	100	-	3	3	2	4
6	10,000+	-	6	6	6	6

Neotropics. *Vismia* is more common when pasture has been burned (Nee, 1995). Correlations with time since re-isolation (which corresponds to the height of secondary growth) and the number of primate species present per reserve were calculated for the two 100-ha reserves (100-ha (4) r = 0.61, and 100-ha (5), r = 0.909; Table 4).

Figure 5. Edge of a 10-ha isolated fragment surrounded by abandoned pasture. Photo by Kellen Gilbert.

4. DISCUSSION

Despite their relatively large size, diurnal activity, and charisma, primates in the BDFFP reserves have not been studied consistently. We do not have a large sample size or adequate replicates but we can still reach some conclusions about what happens to arboreal primates after the creation of isolated forest fragments. The immediate effect of isolation was the loss of the large frugivores. These species–the black spider monkeys, bearded sakis, and brown capuchins–require an area larger than 100-ha. The primates that are able to survive and reproduce in the small fragments have small range requirements and, in the case of howling monkeys, can subsist on a large portion of leaves in the diet and are not constrained by the availability of fruit (Rylands and Keuroghlian, 1988; Gilbert, 1994a).

Low turnovers of the number of species present suggests recolonization of the isolated forest fragments has not occurred rapidly. Individual monkeys certainly move in and out of the six isolated reserves. The groups of howling monkeys, however, appear to be the most stable and least likely to move, particularly in terms of the 10-ha reserves. Groups of howling monkeys, white-faced sakis, and golden-handed tamarins have resided in the 100-ha reserves now for nearly 20 years. Movements in and out of the 100-ha reserves by groups of capuchins and bearded sakis occurred after five years of secondary growth succession surrounded the fragmented reserves. However, it is clear that a 100 ha isolated forest fragment is too small to sustain permanent groups of three of the six local primate species.

Estimates on the density of primates in the reserves remain incomplete. With the small number of reserve replicates and unequal sampling efforts it is difficult to make statistical comparisons. It appears though, that the density of primates in the BDFFP reserves is markedly lower than in intact Neotropical moist forests (Mittermeier and van Roosmalen, 1981; Terborgh, 1983; Rylands and Keuroghlian, 1988; Gilbert, 1994a).

The primates in central Amazonia, with the exceptions of the tufted capuchin, howling monkeys, and golden-handed tamarin, which do come to the ground occasionally, are nearly completely arboreal, restricting their movement to forested areas. As a result, forest fragments even separated by pasture 100 m from continuous forest may be temporary islands for some species. With rapid vegetation regrowth around the fragments, they do not remain islands long (Bierregaard et al., 2001). The presence of individual monkeys in 1-, 10-, and 100-ha fragments shows they do move through tall secondary growth. Connectivity, vegetation that connects fragments to tall second growth to continuous forest may be most important for these primates. Nearly 20 years of sporadic primate studies have shown the initial effects of forest fragmentation on six Amazonian primate species.

5. SUMMARY

Nearly two decades of primate surveys in the upland Central Amazonian basin forest fragment reserves of the BDFFP have been conducted. The immediate result of forest fragmentation is local species extinction. Fragmented areas have fewer primate species and fewer individuals. The results show that the larger frugivorous species such as the black spider monkey (*Ateles paniscus*) and the black-bearded sakis (*Chiropotes satanas*) abandon areas fragmented into 1-ha and 10-ha. A 1-ha fragment may sustain a group of highly folivorous red howling monkeys, and a 10-ha area appears to be large enough to sustain one group each of red howling monkeys (*Alouatta seniculus*), white-faced saki (*Pithecia pithecia*), and golden-handed tamarins (*Saguinus midas*). These three species inhabit 100-ha fragments as well, with groups of capuchin monkeys (*Cebus apella*) and black-bearded sakis recolonizing the area when secondary growth surrounding the fragments is over 2 m in height. Individuals may use tall secondary growth as arboreal corridors and small fragments as temporary stopping points as they move within the larger areas of continuous and fragmented Amazonian forest.

6. ACKNOWLEDGMENTS

I am grateful for the assistance of Heraldo Vanconcelos, Phil Stouffer, Ana Subira, and many BDFFP mateiros. Field work was supported by the BDFPP, INPA, Primate Conservation, Inc., and Southeastern Louisiana University Faculty Development Grants. I thank BDFFP and UNPA for permission to conduct field work in Brazil. This is publication number 375 in the BDFFP Technical Series.

7. REFERENCES

Bernstein, I. S., Balcaen, P., Dresdale, L., Gouzoules, H., Kavanagh, M., Patterson, T., and Neyman-Warner, P., 1976, Differential effects of forest degradation on primate populations, *Primates* 17:401-411.

Bierregaard, R. O., Jr., and Gascon, C., 2001, The Biological Dynamics of Forest Fragments Project, in *Lessons from Amazonia*, R. O. Bierregaard, C. Gascon, T. E. Lovejoy, and R. Mesquita, eds., Yale University Press, New Haven, pp. 5–12.

Bierrregaard, R. O., Jr., Lovejoy, T. E., Kapos, V., dos Santos, A. A., and Hutchings, R. W., 1992, The biological dynamics of tropical rainforest fragments, *Bioscience* 42:859–866.

Bierregaard, R. O., Jr., Laurance, W. F., Gascon, C., Benitez-Malvido, J., Fearnside, P. M., Fonseca, C. R. Ganade, G., Malcolm, J. R., Martins, M. B., Mori, S., Oliveira, M., Rankin-de Mérona, J., Scariot, A. Spironello, W., and Williamson, B., 2001, Principles of forest fragmentation and conservation in the Amazon, in: *Lessons from Amazonia*, R. O. Bierregaard, C. Gascon, T. E. Lovejoy, and R. Mesquita, eds. Yale University Press, New Haven, pp. 371–385.

Fearnside, P. M., and Leal Filho, N., 2001, Soil and development in Amazonia, in: *Lessons from Amazonia*, R. O. Bierregaard, C. Gascon, T. E. Lovejoy, and R. Mesquita, eds., Yale University Press, New Haven, pp 291–312.

Ferrari, S. F, and Diego, V. H., 1995, Habitat fragmentation and primate conservation in the Atlantic forest of eastern Minas Gerais, Brazil, *Oryx* 29:192–196.

Frazão, E. da R., 1992, Dieta e estrategia de forragear de *Chiropotes satanas chiropotes* (Cebidae: Primates) na Amazônia central Brasileira, M.Sc. thesis, Instituto Nacional de Pesquisas da Amazônia, Manaus, Brazil.

Garcia, J. E., and Mba, J., 1997, Distribution, status, and conservation of primates in Monte Alen National Park Equatorial Guinea, *Oryx* 31:67–76.

Gilbert, K. A., 1994a, Endoparasitic infection in red howling monkeys (*Alouatta seniculus*) in the central Amazonian basin: A cost of sociality? Ph.D. diss., Rutgers University, New Brunswick.

Gilbert, K. A., 1994b, Parasitic infection in red howling monkeys in forest fragments, *Neotrop. Prim.* 2:10–12.

Gilbert, K. A., 1997, Use of defecation sites as a parasite avoidance behavior by red howling monkeys, *Anim Behav.* 54: 451–455.

Gilbert, K. A., and Setz, E. Z. F., 2001, Primates in a fragmented landscape: Six species in central Amazonia in: *Lessons from Amazonia*, R. O. Bierregaard, C. Gascon, T. E. Lovejoy, and R. Mesquita, eds., Yale University Press, New Haven, pp. 262–270.

Hsu, M. J., and Agoramoorthy, G., 1996, Conservation status of primates in Trinidad, West Indies, *Oryx* 30:285–291.

INPE, 1998, Deforestation 1995-97, Amazônia. Instituto Nacional de Pesquisas Espaciais and Ministério d Ciência e Tecnologia, Brasília DF, Brazil.

MacArthur, R. H., and Wilson, E. O., 1967, *The Theory of Island Biogeography*, Princeton University Press Princeton, NJ.

Malcolm, J. R., 1988, Small mammal abundances in isolated and non-isolated primary forest reserves near Manaus, Brazil, *Acta Amaz.* 18:67–83.

Malcolm, J. R., 1990, Estimation of mammalian densities in continuous forest north of Manaus, in: *Four Neotropical Rainforests*, A. H. Gentry, ed., Yale University Press, New Haven, pp.339–357.

Milton, K., 1980, *The Foraging Strategy of Howler Monkeys: A Study in Primate Economics*, Columbia University Press, New York.

Mittermeier, R. A., and van Roosmalen, M. G. M., 1981, Preliminary observations on habitat utilization and diet in eight Suriname monkeys, *Folia Primatol.* 36:1–39.

National Research Council, 1981, *Techniques for the Study of Primate Population Ecology*, National Academy Press, Washington, D.C.

Nee, M., 1995, *Flora preliminar do Projeto Dinâmica Biológica de Fragmentos Florestais (PDBFF)*, New York Botanical Garden and INPA/PDBFF, Manaus, Brazil.

O'Brien, T. G., and Kinnaird, M. F., 1996, Changing populations of birds and mammals in North Sulawesi *Oryx* 30:150–156.

Oliveira, A. A. de, 1997, Diversidade, estrutura e dinâmica do componente arbóreo de uma floresta de terra firme de Manaus, Amazonas, Ph.D. diss., Universidade de São Paulo, São Paulo.

Rankin-de Mérona, J. M., Prance, G. T., Hutchings, R. W., da Silva, M. F., Rodrigues, W. A., and Uehling, M E., 1992, Preliminary results of large-scale tree inventory of upland rain forest in the Central Amazon *Acta Amaz.* 22:493–534.

Rylands, A. B., and Keuroghlian, A., 1988, Primate populations in continuous forest fragments in central Amazonia, *Acta Amazónica* 18:291–307.

Schwarzkopf, L., and Rylands, A. B., 1989, Primate species richness in relation to habitat structure in Amazonian rainforest fragments, *Biol. Conserv.* **48**:1–12.

Setz, E. Z. F., 1994, Feeding ecology of golden-faced sakis, *Neotrop. Prim.* **2**:13–14.

Spironelo, W. R., 1987, Range size of a group of *Cebus apella* in central Amazonia, *Am. J. Primatol.* **8**:522.

Spironelo, W. R., 2001, The brown capuchin monkey (*Cebus apella*): Ecology and home range requirements in central Amazonia, in: *Lessons from Amazonia*, R. O. Bierregaard, C. Gascon, T. E. Lovejoy, and R. Mesquita, eds., Yale University Press, New Haven, pp. 271–283.

Terborgh, J., 1983, *Five New World Primates: A Study in Comparative Ecology*, Princeton University Press, Princeton.

Thoisy, B., de and Richard-Hansen, C., 1997, Diet and social behaviour changes in a red howler monkey (*Alouattta seniculus*) troop in a highly degraded rain forest, *Folia Primatol.* **68**:357–361.

White, L. J. T., 1994, Biomass of rain forest mammals in the Lope Reserve, Gabon, *J. Anim. Ecol.* **63**:499–512.

Wilson, C. C., and Wilson, W. L., 1975, The influence of selective logging on primates and some other animals in East Kalimantan, *Folia Primatol.* **23**:245–274.

Wilson, W. L., and Johns, A. D., 1982, Diversity and abundance of selected animal species in undisturbed forest, selectively logged forest, and plantations in East Kalimantan, Indonesia, *Biol. Conserv.* **24**:205–218.

SECTION II: BEHAVIORAL ECOLOGY

Laura K. Marsh

1. INTRODUCTION

Until recently, very few scientists have directly or indirectly focused on the consequences of habitat fragmentation on sociality and behavior of vertebrates. Yahner and Mahan (1997) state that there are at least two challenges facing researchers in this area: 1) it is difficult to extend interpretations of behavioral phenomena obtained from individual animals or groups in a localized area to the broader landscape level; and 2) it will be problematic to invoke causal relationships between behavioral phenomena. Understanding the effects of fragmentation on behavior is especially difficult when no other study species remain in similar intact landscapes for comparison (Yahner and Mahan, 1997). Authors in this section discuss one to several groups or species of primates interacting within fragments.

The floristic composition of a given forest fragment determines primate survival. The composition itself may be determined by a chance loss of species due to low density, by increased mortality of trees following change in environmental conditions, or by disappearance of seed dispersers or pollinators that reduces the reproduction and recruitment of food species important to primates (Lovejoy et al., 1986). These effects would occur even though they would not be felt in the canopy levels for at least 50 to 75 years, although if pollination is disrupted, then fruit production effects could occur almost immediately (Howe, 1984; Redford, 1992). Established forest fragments (>10 years), in general, tend to have a reduction in trees or tree saplings with large fruits, in part due to the lack of seed dispersers (Putz et al., 1990). This may or may not affect the species within them over time depending on other available resources.

2. LANDSCAPE AND DEMOGRAPHY

Umapathy and Kumar in Chapter 10 demonstrate consequences of population fragmentation in the lion-tailed macaque (*Macaca silenus*) and Nilgiri langur (*Trachypithecus johnii*) in the Indira Gandhi Wildlife Sanctuary, India. Small fragments were more likely to be privately owned and have lower tree density, basal area, canopy height, and canopy cover and were more disturbed. Large fragments were more likely

Primates in Fragments: Ecology and Conservation
Edited by L. K. Marsh, Kluwer Academic/Plenum Publishers, 2003

owned by the Forestry Department, had greater tree density, basal area, canopy height and canopy cover, and were less disturbed. Demographic changes for lion-tailed macaque were reduction in birth rate and proportion of juveniles, and an increase in group size, number of adults, and variability in adult sex ratios with decreasing area and increasing disturbance levels. In the Nilgiri langur, there was also a decrease in birth rates and proportion of juveniles; however, there was no change in group size and adult sex ratio. A primary difference between the two species was the langurs' ability to disperse between fragments.

3. FEEDING ECOLOGY AND BEHAVIORAL PLASTICITY

Ramos-Fernández and Ayala-Orozco discuss in Chapter 11 the population size and habitat use by spider monkeys (*Ateles geofroyii*) at Punta Laguna, Mexico. Population size, habitat characteristics, ranging, and foraging behavior were compared. The highest density of spider monkeys (89 individuals per km^2) were found in the largest fragment (0.6 km^2). The monkeys fed on 50 species of trees with 10 comprising three-quarters of their diets. High per capita birth rates and a net increase in the size of groups suggest the population over all fragment sizes is expanding.

Norconk and Grafton compared in Chapter 12 the forest structure and plant species composition on a medium-sized island in Lago Guri, Venezuela, in terms of white-faced saki (*Pithecia pithecia*) survival. They compared two plots, one interior and one exposed. Between 1989 and 2002 the exposed plot showed a drastic decline in the number of surviving small stems in the range of 5 cm dbh, as well as heavy losses for trees of all sizes, and a low level of recruitment of very small stems (<5 cm). The loss was primarily due to edge effects and windthrow; however, feeding tree species diversity is complex and changed slowly compared to structural changes. The foremost pressure on the monkeys was not dietary. The primates at this site are at a higher risk of local extinction due to stochastic events and lack of gene flow than they are from starvation from changing forest conditions.

Rodriguez-Luna et al. in Chapter 13 discuss foraging strategy changes in *Alouatta palliata mexicana* after they were released on an island in Lake Catemaco, Veracruz State, Mexico. An experiment of this kind reveals the mechanisms of ecological adaptation on a scale of a few years. It also acts as a comparison to sites in Los Tuxtlas fragments on land. A founder population of 10 individuals was released. After 7 years on the island, the group grew to 57 individuals. No significant differences were found in the number of food species used between the founders and the group seven years later. Lianas, vines, and leaves in general were added to the diet over time as fruit resources became more limited due to the increase in population. Food choice and species choice changed over time even though source species were retained throughout. This study contributes to understanding the plasticity of the howler monkey.

Silver and Marsh in Chapter 14 discuss the dietary flexibility, behavioral plasticity, and survival of *Alouatta pigra*. Howlers in Belize from the fragmented forests of the Community Baboon Sanctuary were translocated to the continuous forest of the Cockscomb Basin Wildlife Sanctuary. Two main lessons were learned from translocated howlers that may reveal their ability to colonize and thrive in forest fragments when other species cannot: 1) howlers appear able to exploit novel dietary items immediately upon introduction to a new area, which demonstrates a high degree of dietary flexibility; and 2) by making behavioral adjustments that presumably minimize voluntary energy expenditures, howler monkeys can survive in habitats with large differences in food abundance. Behavioral adaptations in adjusting time budgets for resting and foraging allow for additional flexibility for negotiating fragments.

4. SEED DISPERSAL

Serio-Silva and Rico-Gray in Chapter 15 studied the role of *Alouatta palliata mexicana* as seed dispersal agents of strangler figs in disturbed and preserved habitat in southern Veracruz, Mexico. Primates are important seed dispersers of many tree species, but *Ficus* is particularly important in their diet. An experiment was conducted using methods of fig seed germination that included seeds from fruits (control) and howler feces (experiment). Controlled germination trials were conducted in the lab and planter boxes in the field. The results showed howlers had a positive and significant effect on the germination of fig trees. The authors feel fragmentation may alter the effectiveness of howlers as seed dispersers of some fig species.

5. COMPILATION

Bicca-Marques in Chapter 16 summarizes the literature available on howler monkeys (*Alouatta* sp.) in fragments. The author looked at six species of howlers in 27 sites throughout Central and South America. He addressed the question of howler survival strategies through use of space, diet, and activity budgets. In general most aspects of *Alouatta* ecology and behavior analyzed did not vary among species and were not predicted by fragment size. Fragment size predicted home range size, the number of plant species used for food, and the diversity of leaf and fruit resources. It did not predict day range, contribution of food items to diet, the number of plant species consumed daily, main food genera selected, and the activity budget. Thus, howlers manage fragmentation without showing directional changes in most aspects of their feeding ecology and behavior, relying on behavioral and dietary flexibility to cope with fragmentation.

6. SUMMARY

Behavioral studies on primates in fragments begin to get at the heart of understanding natural history parameters that effect the survival of species in a disturbed habitat. Whenever possible, studies that compare intact forest to fragments demonstrate variation or the lack of in primate responses to fragmentation. These chapters show the depth of primates experiencing disturbed landscapes. Howlers seem to be the most flexible of those species reported here, in terms of overall adaptability to fragment conditions, while lion-tailed macaques may be having a more difficult time. Spider monkeys are found in fragments, even relatively small ones, but manage with a more mature matrix. Isolation may make the difference to some populations, such as sakis in Lago Guri or howlers in Lake Catemaco, where dispersal is not an option. More studies on primate behavioral ecology are needed to facilitate sound management practices.

7. REFERENCES

Howe, H. F., 1984, Implications of seed dispersal by animals for tropical reserve management, *Biol. Cons.* **30**:261–281.

Lovejoy, T. E., Bierregaard, R. O., Jr., Rylands, A. B., Malcolm, J. R., Quintela, C. E., Harper, L. H., Brown, K. S., Jr., Powell, A. H., Powell, G. V. N., Schubart, H. O. R., and Hays, M. B., 1986, Edge and other effects on isolation on Amazon forest fragments, in: *Conservation Biology: The Science of Scarcity and Diversity*, M. E. Soulé, ed., Sinauer Assoc., Sunderland, MA.

Putz, F. E., Leigh, E. G., Jr., and Wright, S. J., 1990, Solitary confinement in Panama, *Garden* **2**:18–23.

Redford, K. H., 1992, The empty forest, *BioScience* **42(6)**:412–422.

Yahner, R. H., and Mahan, C. G., 1997, Behavioral considerations in fragmented landscapes, *Cons. Biol.* **11(2)**:569–570.

IMPACTS OF FOREST FRAGMENTATION ON LION-TAILED MACAQUE AND NILGIRI LANGUR IN WESTERN GHATS, SOUTH INDIA

Govindasamy Umapathy and Ajith Kumar[*]

1. INTRODUCTION

1.1. The Western Ghats

India has been identified as one of the 12 megadiversity countries (Myers, 1992) with about 126,000 species. India has 8% of the global biodiversity, even though it covers only 2.4% of the land area of the world (Khoshoo, 1995). The Eastern Himalayas and Western Ghats are also among the 18 biodiversity hotspots in the world (Khoshoo, 1995). The Western Ghats covers only 5% of the land area of India, but has 30% of India's species. It has about 5,000 flowering plant species of which 1,500 are endemic (Nair, 1991). There are 58 endemic plant genera, of which 42 are monotypic. About 490 species of trees occur of which 308 (62.5%) species in 58 families are endemic. About 267 species of orchids occur belonging to 72 genera, of which 130 species are endemic.

There are about 65 species of non-volant mammals in the Western Ghats, of which 11 are endemic. Endemism is not high among mammals compared to reptiles (80 species out of 170) and amphibians (90 species out of 120). Among the mammal endemics, the Malabar civet (*Viverra civettina*) is the most endangered species. The lion-tailed macaque (*Macaca silenus*), Nilgiri langur (*Trachypithecus johnii*), Nilgiri marten (*Martes gwatkinsi*), brown palm civet (*Paradoxurus jerdoni*), Travancore flying squirrel (*Petinomys fuscocapillus*), jungle striped squirrel (*Funambulus tristriatus*), spiny dormouse (*Platacanthomys lasiurus*), and Nilgiri tahr (*Hemitragus hylocrius*) are endemic at species level. The brown mongoose (*Herpestes fuscus*), stripe-necked mongoose (*H. vitticollis*), grizzled giant squirrel (*Ratufa macroura*), Layardi's striped squirrel (*F. layardi*), and dusky striped squirrel (*F. sublineatus*) are endemic at sub-species level, and are also found in Sri Lanka. The Western Ghats contains the largest population (15,000 individuals) of Asian elephant (*Elephas maximus*) and also contains

[*] Govindasamy Umapathy: Laboratory for the Conservation of Endangered Species, CCMB, Uppal Road Hyderabad, India, 500 007. Ajith Kumar: Salim Ali Centre for Orthinology and Natural History, Coimbatore, India, 641108. Correspondence to G. Umapathy (email: gupathy@yahoo.com).

Primates in Fragments: Ecology and Conservation
Edited by L. K. Marsh, Kluwer Academic/Plenum Publishers, 2003

healthy populations of other large mammals that include gaur (*Bos gaurus*), sambar (*Cervus unicolor*), spotted deer (*Axis axis*), tiger (*Panthera tigris*), leopard (*Panthera pardus*), and Indian wild dog or dhole (*Cuon alpinus*). There are 508 species of birds, of which 15 are endemic (Ali and Ripley, 1987; Daniels, 1997).

1.2. Loss and Fragmentation of Forests

The existing forests of the Western Ghats, especially wet evergreen forests, are highly fragmented. Between 1920 and 1990, when forest loss was nearly 40%, the number of forest fragments was estimated to increase nearly fourfold, from 179 to 769, with an 83% reduction in average fragment size (Menon and Bawa, 1997). Moreover, there has been an increase in perimeter-to-area ratio of the fragments. Only a few areas have more than 200 km^2 of continuous wet evergreen forests. These are Agasthyamalai Hills, Cardamom Hills, Silent Valley-New Amarambalam Forests, and southern parts of the South Kannada District in Karnataka State. There are several proposed hydroelectric projects that could submerge some of the best remaining lowland wet evergreen forests (e.g., Pooyamkutty Project). New roads and railways could cut across the few remaining large rain forest areas (e.g., the proposed road from Papanasam in Tamil Nadu to Thiruvananthapuram in Kerala across the Agasthyamalai Hills and the railway to Sabarimala Temple in Kerala).

The objectives of the project were to assess the variation in the occurrence and abundance of primates in forest fragments in relation to several landscape and habitat parameters; assess the changes in the demographic parameters of lion-tailed macaque and Nilgiri langur due to habitat fragmentation; and use the findings from the study to suggest appropriate measures to enhance the survival of these primates in forest fragments.

2. STUDY AREA

The Indira Gandhi (formerly Anamalai) Wildlife Sanctuary (IGWS) in Tamil Nadu is one of the largest sanctuaries in south India (Figure 1). Created in 1976, it covers an area of about 987 km^2, extending 45 km north to south and 25 km east to west (10°12' and 10°54' N and 76°44' and 77°48' E). It is located about 90 km from Coimbatore City, mainly in the Valparai Taluk, but extends to Pollachi and Udumalpet Taluks of Coimbatore District and Kodaikanal Taluk of Dindugal District. Three major public roads from Pollachi Town pass through IGWS—the Pollachi to Chalakudi Road through Valparai, the Pollachi to Parambikulam Road through Topslip, and the Pollachi to Munnar Road through Udumalpet Range. A network of roads connects Valparai Town to various estate settlements (Figure 2).

Almost in the center of IGWS is 180 km^2 of tea and coffee estates that are under private ownership, and in its center is Valparai Town. IGWS is bordered on the southwest by Parambikulam Wildlife Sanctuary (287 km^2), on the south by the Reserve Forest of Chalakudi Forest Division and Eravikulam National Park (97 km^2), and on the southeast by Chinnar Wildlife Sanctuary (90 km^2), which are all in Kerala State. In the east it is mostly surrounded by cultivated plains. These sanctuaries along with the Reserve Forest of Nelliyampathi Hills form a large conservation area for large and wide-ranging species such as elephant, gaur, and tiger.

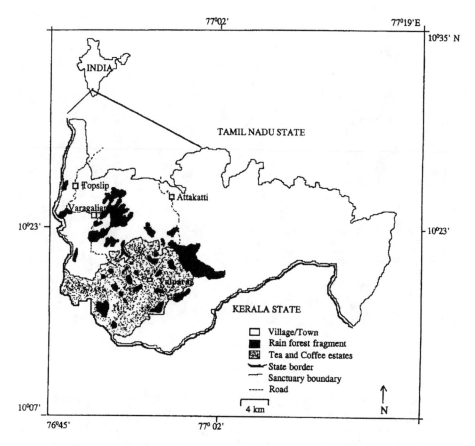

Figure 1. Indira Gandhi Wildlife Sanctuary showing the rain forest fragments.

The altitude of IGWS ranges from 220 m in the plains at the foothills in the east to 2,513 m atop Thanakkanmalai in the Grass Hills. Rainfall varies considerably, ranging from 500 mm in the eastern slopes of IGWS to 5,000 mm in the western slopes. IGWS receives both southwest (June to September) and northeast (October and November) monsoons, with about 80% of the rainfall during the former. The daytime temperature varies considerably from 23°C to 40°C at the foothills (200 to 350 m) to 20°C to 30°C at higher elevations (1,800 to 2,300 m). In the night, temperatures range from 15°C to 25°C at the foothills and from 10°C to 20°C at mid-elevation of 900 m to 1,200 m. The temperature is lower at higher elevations, going down to 0°C in December and January at about 2,000 m. March to May are the hottest months.

The natural vegetation in this area includes wet evergreen forest, montane shola-grassland, moist deciduous, dry deciduous, and thorn forests (Figure 3). Tropical wet evergreen forest is found in an altitude of 600 to 1,600 m.

Figure 2. An example of a tea estate settlement within the Indira Gandhi Wildlife Sanctuary. Photo by G. Umapathy.

Figure 3. One of the degraded rain forest fragments (Korangumudi) within Indira Gandhi Wildlife Sanctuary. Photo by G. Umapathy.

2.1. Forest Fragments

Fragments of wet evergreen forest were identified during a survey of the study area in January and February 1994 (Table 1). A fragment was defined as a patch of natural vegetation originally of wet evergreen type, isolated from other such forests except for narrow corridors, if any. It is often surrounded by human-made vegetation such as plantations and human settlements. The fragments were initially identified based on personal knowledge of the area (Umapathy, personal observation) and inquiry with local forest officials, estate managers, and Taluk administrative officers. These fragments were marked on a Survey of India Map and later verified by field visits.

We identified 25 total forest fragments in the study area within a radius of about 35 km of Valparai Town (Figure 1). A total of about 400 km were surveyed on foot, ranging from 500 m up to 10 km per fragment depending on its area. Habitat parameters were recorded from a total of 350 circular plots with 5-m radius. An analysis of the data revealed that several of the habitat and landscape parameters were closely interrelated. These interrelationships are presented first, before examining the occurrence and abundance of the primates in the fragments.

Even though all the 25 fragments that we identified came within geographical limits of IGWS, 14 of them were privately owned and 11 were within the administrative control of IGWS (Table 1). All privately owned fragments were formed when the surrounding forests were cleared for planting tea, particularly by Tata Tea, Kothari, Parry Agro, and Non-conventional Energy Power Corporation (NEPC), as well as cardamon, coffee, or eucalyptus. The fragments were retained for soil and water conservation (on hilltops and slopes), for cardamom and coffee cultivation, or to meet firewood needs of estate workers. The first of these fragments was probably created in the 1880s, and the last one in the 1930s (Congreve, 1938). Some of the privately owned fragments have been fully or partly underplanted with cardamom (5 out of 14). In most of the privately owned fragments (10 out of 14), trees have been lopped and logged repeatedly, lianas removed, and undergrowth replaced with cardamom and coffee or have been invaded by weeds such as *Lantana camera* and *Eupatorium odoratum.* All the cardamom and coffee planted fragments have labor settlements nearby, the neighboring people often remove firewood, small timber, and forest produce from the fragments. A road passes through or near most of the private fragments. Within privately owned fragments there was a clear difference between those owned by large tea estates and those owned by smaller coffee and cardamom estates. The former were better protected from firewood and timber removal compared to the latter. In contrast, only a few of the fragments owned by the Forest Department have been underplanted with cardamom or coffee (e.g., Akkamalai and Sankarankudi).

Most of the fragments owned by the Forest Department were created after the 1930s and up to the 1970s, when surrounding vegetation was cleared for teak plantation (e.g., Varagaliar) or when a chain of reservoirs was built under the Parambikulam-Aliyar Project. Some of the Forest Department-owned fragments were formed when the surrounding forests were leased to private companies and individuals for tea plantations (e.g., Andiparai and Sankarankudi) between 1880s and 1930s. Most of the Forest Department-owned fragments are away from human settlements and thus are not under intensive human pressure. Many of these fragments have been selectively logged in the past (20 to 50 years ago). Presently, privately owned tea and coffee estates cover more than 180 km^2 in the center of IGWS (Sundararaju, 1987).

Table 1. The presence/absence of lion-tailed macaque and Nilgiri langur and the status of vegetation in 25 rain forest fragments. For definitions of habitat parameters, see Table 2.

Name of the fragment	Area ca (ha)	Size@ class	Ownership#	Presence of* LTM	Presence of* NL	Tree density (/ha)	Canopy height (m)	Canopy cover (%)	Basal area (m²/ha)
Akkamalai shoal	2,500	VL	F	1	1	492	19	67	97.54
Varagaliar	2,000	VL	F	1	1	455	31	57	49.81
Kuruvampalli	500	VL	F	1	1	410	30	55	58.73
Vellamalai top	200	L	F	0	1	323	13	33	33.71
Monampoly	200	L	F	1	1	347	23	52	55.46
Sankarankudi	180	L	F	1	1	184	21	33	37.48
Surilimalai	75	L	P	1	1	244	36	52	75.18
Iyerpadi	100	L	P	0	1	441	16	60	85.86
Andiparai shoal	185	L	F	1	1	357	22	77	29.41
Iyerpadi church	50	M	F	0	1	403	17	59	43.90
Korangumudi Est.	35	M	P	1	1	161	17	23	37.82
Puthuthottam Est.	50	M	P	1	0	128	19	62	67.23
Tata Estate	24	M	P	1	1	93	25	56	11.01
Chinnakallar	40	M	F	0	1	336	18	56	39.53
Sholaiyar P.House	10	S	F	0	1	244	13	27	31.41
Varattuparai-II	2	S	P	0	1	233	18	42	51.59
Varattuparai-III	4	S	P	0	1	182	10	21	20.29
Varattuparai-IV	1	S	P	0	0	127	9	18	8.01
Nirar dam	8	S	F	0	1	164	8	75	6.54
Sholaiyar dam	6	S	P	0	1	370	17	55	85.96
Monica Estate	2	S	P	0	0	127	20	5	18.48
Pannimedu-I	5	S	P	0	0	216	15	44	43.70
Pannimedu-II	10	S	P	0	0	306	14	48	42.82
Urulikkal I	5	S	P	0	0	178	16	36	61.87
Urulikkal II	2	S	P	0	0	111	9	35	19.80

@ = size of fragments: S=small (<10 ha), M=medium (11-50 ha), L=large (51-200 ha) and VL=very large (>200 ha).
# = Ownership of fragment: F=Forest Department, P=private estate.
* = Presence of primates: 1=present, 0=absent.
LTM= lion-tailed macaque; NL- Nilgiri langur

2.2. Matrix Around the Fragments

The matrix around the fragments showed considerable variation depending on ownership of the fragments. Most of the Forest Department-owned fragments were surrounded by teak or eucalyptus (e.g., Varagaliar Shola) or tea plantations (e.g., Andiparai Shola), and none had large human settlements on the edge. The fragments owned by large tea estates were mostly surrounded by eucalyptus and coffee plantations (and reservoirs in some cases) and had large human settlements, but not close to the fragments. The fragments owned by small cardamom and coffee estates were also mostly surrounded by tea estates. All of these, however, had labor settlements either within or at the edge of the fragment (e.g., Puthuthottam and Korangumudi Estates). Orchards raised around these settlements included bananas (*Musa* spp.), guava (*Psidium* spp.), jack fruit (*Artocarpus heterophyllus*), and mango (*Mangifera indica*) and formed an integral part of the fragment.

3. METHODS

3.1. Study Animals

The lion-tailed macaque is an endangered primate (IUCN, 1996) and is listed in Schedule I of the Indian Wildlife (Protection) Act, 1972 (Figure 4). It is endemic to the wet evergreen forest of the Western Ghats in south India. Its distribution ranges from

Figure 4. Lion-tailed macaque (male) in a *Maesopsis eminii* tree. Photo by G. Umapathy.

Agasthyamalai Hills in the south to a few kilometers north of Sharavathi River in the north (Kumar, 1995a) in an elevation ranging from 150 to 1,500 m.

In the recent past, its distribution was contiguous from southernmost Western Ghats up to the state of Maharashtra (Kumar, 1995b). The severe loss and fragmentation of wet evergreen forests resulted in the local extinction of the lion-tailed macaque populations by the 1950s in Goa and Maharashtra. The lowland wet evergreen forest in Kerala has also been destroyed, confining the population to higher elevations (Kumar, 1995a). At present the species is restricted to three states, in which Kerala has about 2,000 animals, and Tamil Nadu and Karnataka have about 1,000 animals each (Kumar, 1995a). The current population is fragmented into many subpopulations, only five are large with more than 10 groups each (Mookambika-Someshwara, Kodagu, Silent Valley-Amarambalam, Cardamom Hills, and Agasthyamalai) and the remaining are highly fragmented with one to five groups forming isolated populations (Kumar, 1995a).

Lion-tailed macaques live in groups of 8 to 40, with an average of about 18 individuals (Kumar, 1995a). The number of adult males (>8 years of age) in a group varies from 1 to 4. Most of the groups have only one adult male, one subadult male, and five to seven adult females, the remaining being juveniles and infants. Larger groups often have more than one adult male. The average adult sex ratio is about five females to one adult male. Births occur throughout the year with a peak between December and February (Kumar, 1987; Kumar and Kurup, 1993). The lion-tailed macaque feeds on a variety of food items that include seeds, fruits, flowers, nectar, invertebrates, bird eggs, and, rarely, small vertebrates such as lizards and infants of giant and flying squirrels. The lion-tailed macaque gets its food mainly from the upper canopy, but it also eats flowers and fruits of vines, small trees, and shrubs.

Nilgiri langur is a vulnerable primate (IUCN, 1996) endemic to the Western Ghats (Figure 5). Unlike the lion-tailed macaque, it occurs in a variety of forest types such as moist deciduous, riverine, and montane wet evergreen. Nilgiri langur is restricted to the Western Ghats between 8° to 12° N from Agasthyamalai in Kerala in the south to Kodagu in Karnataka in the north. Due to habitat destruction and hunting for medicinal purposes, it had been destroyed in some areas by the late 1960s (Kurup, 1973). Following the Wildlife (Protection) Act, 1972, there has been a recovery of populations in most areas. At present Nilgiri langur is confined to the relatively undisturbed areas of the Western Ghats, extending over a range of elevation from 150 to 2,500 m. In the Mundanthurai Plateau in Tirunelveli District, Tamil Nadu, it is found at the foothills of the Western Ghats along the riverine forest (Oates, 1979).

Nilgiri langur lives in groups of 3 to 25 animals (Srivastava et al., 1996), with an average of eight to nine animals (Prater, 1980). Most groups consist of one adult male, four to five adult females, and the rest are infants and juveniles. Adult females form one-third of a group (Srivastava et al., 1996). Langurs feed on young and mature leaves, flowers, fruits, seeds, petiole, resin, and bark. In Agasthyamalai they used 115 plant species in a year, with fiber and condensed tannin content characterizing foliage selection (Oates et al., 1980). Social interactions are infrequent, often without physical contact and for short durations (Poirier, 1970).

Figure 5. A male Nilgiri langur on a *Mesua ferrea* tree. Photo by G. Umapathy.

3.2. Estimation of Primate Abundance

After the fragments were located, data on their ownership, area, time since isolation, management history, and present land use pattern around them were collected from forest and revenue department officials, concerned estate managers, and literature, especially Congreve (1938). The fragments were surveyed from January to May 1994 to assess the occurrence and abundance of primates and habitat quality. Each forest fragment was surveyed three or four times along existing trails to assess the occurrence and abundance of primates. The survey was carried out between 0700 and 1000 hours. The survey team was comprised of three or four persons, walking at 1.5 km to 2 km per hour. The distance surveyed in a fragment varied from 500 m to 10 km depending on its area. When a primate was sighted, the species, number of individuals, age/sex composition, sighting distance, and sighting angle from the transect were recorded. The sighting distance was visually estimated and the sighting angle was recorded using a compass. Since the number of sightings was insufficient to estimate density, the abundance of lion-tailed macaque and Nilgiri langur was estimated as the number of groups sighted per kilometer.

3.3. Estimation of Habitat Parameters

Several habitat parameters (Table 2) were recorded from a 5-m-radius circular plot laid at an interval of 75 m on either side of the transect. Since the length of the transect in a fragment was proportional to its area, the number of sample points increased with area of the fragment. Several landscape parameters were also noted, such as presence of

Table 2. Methods used for estimating habitat parameters.

Variable	Definition	Method of Measurement	Unit
Tree density	Number of trees/ha (>30 cm DBH)	5-m-radius circular plot	Trees/ha
Basal area	Area covered by tree stems per ha (>30 cm DBH)	Same as above	m^2/ha
Canopy cover	Percentage of area covered by canopy of trees in a plot	Gridded motorcycle rear view mirror	Percentage of cover
Canopy height	Canopy height at the plot	Visual estimation	Meters
Shrub cover	Percentage cover of undergrowth	Visual estimation	Percentage cover/plot
Litter cover	Percentage of ground covered by leaf litter	Visual estimation	Percentage cover/plot
Soil type	Nature of the soil such as rocky, sandy, etc.	Visual estimation	Percentage cover
Cattle grazing	Intensity of cattle grazing	Visual estimation	in rank

corridor or connectivity, location of the fragment on the landscape (such as valley, hilltop, and slope), surrounding vegetation types, and land use pattern. Distance to the nearest village or settlement, to the nearest water source, and to the main forest area was also recorded.

3.4. Demographic Data

The demographic parameters estimated for the lion-tailed macaque and Nilgiri langur were group size, age/sex composition, birth rate, and growth rate of groups. The group size and age/sex composition were estimated during surveys conducted in eight fragments between March and May 1996. We attempted to get demographic data from as many groups and fragments as possible. Accurate data could be obtained only from 11 groups in eight fragments, including the four main study groups. The age/sex classification was based on a comparison with individuals of estimated age in the main study groups. The birth rate was estimated for each group as the number of infants (<1 year old) as a proportion of the adult females. For the main study groups the age/sex composition in May 1996 was used for analysis.

Unlike the lion-tailed macaque, the Nilgiri langur was relatively easy to locate given its higher density. However, obtaining accurate data on age/sex composition was much more difficult because most groups were very shy and fled or hid among foliage. Moreover, in the absence of striking sexual dimorphism, sex identification of even adults was difficult. Demographic data on the Nilgiri langur are therefore limited and come from the main study groups.

The exponential growth rate of groups was calculated for the four main study groups of lion-tailed macaque and Nilgiri langur for one year (from March 1995 to February 1996), as $r = \ln (N_1/N_0)$, where N_0 is the group size in 1995 and N_1 is the group size in 1996.

3.5. Analysis of Data

Data on habitat parameters, animal abundance, and demography were analyzed using nonparametric statistics (Siegel and Castellan, 1988). Differences between two independent samples were tested using Mann-Whitney (M-W) U. The differences between more than two samples were tested using Kruskal-Wallis (K-W). The association between two variables was estimated using Spearman rank correlation (r_s); the frequencies were tested using the Chi-square (χ^2) test or the Fisher exact probability test when sample size was low. The significance level decided *a priori* was 0.05 two tailed. Most of the analyses were done using the SPSS package (Norusis, 1990).

Given the correlation between landscape and habitat parameters, and the apparent effect of these on the occurrence of the two species, we explored the independent effect of these parameters using logistic regression (Hosmer and Lemeshow, 1989). The independent variables included were area, ownership (categorical variable), tree density, basal area, canopy height, and canopy cover. Since the sample size is small, the regression results can be considered only indicative.

4. RESULTS

4.1. Fragment Dynamics

Fragment area data was collected from official records maintained by the Forest and Revenue Departments, estate owners, and our visual assessment in the field. These often did not tally, especially for the privately owned fragments, since the records were created nearly a century ago and forest loss increased. The Forest Department also did not have records for each forest fragment they controlled. For these reasons, analysis of area and its relationship to landscape and habitat parameters was made at two levels. To examine correlations with other parameters, our visually estimated area was used. Spearman rank correlation coefficient was used to overcome biases in the estimation of area. While testing for significance, "area" was grouped into four size classes (Table 2); small (<10 ha), medium (11 to 50 ha), large (51 to 200 ha), and very large (>200 ha). These were further grouped into small (<50 ha) and large (>50 ha) whenever sample size requirements for analysis (Chi-square and M-W U tests) had to be met.

4.2. Landscape Parameters

The wet evergreen forest fragments varied considerably in area, from less than 2 ha (Varattuparai-IV) to more than 2,000 ha (Table 1). There was a definite relationship between area and ownership. Larger fragments were owned by the Forest Department and the smaller fragments were under private ownership (Table 1). All but two of the 11 small fragments were privately owned, all three very large and four of the six large fragments were owned by the Forest Department. When the fragments were regrouped into small (<50 ha) and large fragments (>50 ha) a significantly greater percentage (63.6%) of the Forest Department-owned fragments were large ($\chi^2 = 6.5$ df = 1, $p<0.05$) compared to those privately owned (14.3%).

The lowest altitude at which wet evergreen forest occurred in the study area was 650 m at Varagaliar, and the highest altitude that this study considered was 1,500 m. Most of the study area was within the 900- to 1,200-m range (Table 3). The lack of wet evergreen forests in lower altitudes is likely a result of clear felling. All the naturally occurring montane forest fragments in the shola-grassland ecosystem occur in the Grass Hills area in the southern part of IGWS above 1,500 m. Most of the privately owned small fragments (78.6%) were in the medium altitudes where most of the estates were confined. The Forest Department-owned fragments occurred over a wider range of altitude.

A fragment had connectivity if it had canopy continuity with another forest fragment through a narrow belt of either natural forest or human-made vegetation, such as teak or eucalyptus plantations. Out of 25 fragments only 11 had connectivity or canopy contiguity with another wet evergreen forest. The length of these corridors ranged from less than 200 m of natural forest (e.g., between Varagaliar Shola and Kuruvampalli Shola) to a few kilometers of teak plantations (between Kuruvampalli Shola and Anaikunthi Shola). Nearly 90% of the large (>50 ha) fragments had connectivity, compared to only 18% of the smaller fragments.

Table 3. A comparison of habitat parameters with reference to landscape variables.

Landscape variables	Number of fragments	Tree density /ha	Basal area m^2/ha	Canopy height (m)	Canopy cover (%)
Area (ha)					
<10	11	205.27	35.45	13.55	36.91
11 to 50	5	231.80	48.29	19.00	50.80
51 to 200	6	309.67	45.86	22.06	51.00
>200	3	452.33	68.69	26.67	59.67
K-W [a]		n.s	n.s	n.s	n.s
Ownership					
Forest Dept	11	329.09	46.67	19.09	54.55
Private	14	215.14	42.81	17.57	38.93
M-W		<0.05	n.s	n.s	<0.05
Connectivity					
Yes	11	342.45	56.78	2.27	52.55
No	14	204.64	34.86	15.07	40.50
M-W		<0.01	<0.05	<0.05	n.s
Altitude (m)					
600 to 900	5	309.80	41.28	24.40	49.40
901 to 1,200	15	204.45	41.05	16.47	40.13
1,201 to 1,500	5	403.02	58.08	17.40	59.20
K-W		<0.01	n.s	n.s	n.s
Location					
Top	7	242.10	43.79	15.57	46.00
Slope	10	272.10	46.32	21.30	46.50
Valley	8	290.50	42.86	16.75	44.75
K-W		n.s.	n.s	n.s	n.s

[a] K-W = Kruskal-Wallis test; M-W = Mann-Whitney U test; n.s. = not significant
See Table 2 for category definitions.

4.3. Habitat Parameters

Even though we measured several habitat parameters in forest fragments (Table 2), only those that were indicative of the status of the habitat with reference to the primates have been analyzed here. These are tree density, basal area, canopy cover, and canopy height.

All four habitat parameters varied considerably among the fragments (Table 1). Tree density varied from 93 trees per ha to 492 trees per ha. The lowest tree density was recorded in Tata Estate (24 ha) and highest in Akkamalai Shola (2,500 ha). Tree density decreased with decreasing area of the fragment ($r_s = 0.81$, $p = 0.001$). The large fragments (>50 ha) had a significantly higher tree density on average (357/ha) than small fragments (213/ha; M-W U = 20.5 p = 0.0035). The ownership of fragments also influenced tree density. The fragments owned by the Forest Department had a significantly higher tree density (329.09/ha) than those privately owned (215.14/ha; M-W U = 37 p = 0.028; Table 3). Some privately owned fragments had higher tree densities than other privately owned fragments (Iyerpadi, Pannimedu, and Sholaiyar). Among fragments that were between 50 and 200 ha in area, those owned by the Forest Department had a higher tree density (331/ha) than those privately owned (205/ha, Table 4).
Location of the fragments in the landscape (valley, slope, and hilltop) did not influence tree density. Fragments in lower (<900 m) and higher (>1,200 m) altitudes had greater tree densities (310/ha and 403/ha, respectively) than the medium-altitude fragments (204/ha). Fragments with connectivity had significantly higher tree densities (342/ha) than those without (205/ha) (M-W U = 24.5 $p<0.01$).

Since tree density was related both to area and ownership, we compared tree density holding ownership constant and *vice versa* (Table 4). Within privately owned fragments, there was no significant difference between small (<10 ha) and large fragments (>10 ha) even though the latter had greater values. The same was the case with fragments owned by the Forest Department. Thus, there seems to be an area effect, the absence of significance being due to small sample size (N = 5).

Table 4. A comparison of habitat parameters with reference to area within ownership and ownership within area.

Ownership	Area (ha)	No. of fragments	Tree density/ha	Basal area m²/ha	Canopy height (m)	Canopy cover (%)
Private	<10	9	205.05	40.60	14.67	32.89
	>10	5	213.40	47.03	22.80	49.80
M-W [a]			n.s.	n.s.	n.s.	n.s.
Forest Dept.	50 to 200	4	302.75	39.01	19.75	48.75
	>200	3	452.33	68.69	26.67	59.67
M-W			n.s.	n.s.	n.s.	n.s.
Forest Dept./ Private	50 to 200	6	331.33	46.90	18.83	51.83
		5	205.80	47.03	22.80	49.80
M-W			n.s.	n.s.	n.s.	n.s.

[a] M-W = Mann-Whitney U test; n.s. = not significant

The variations in basal area, canopy cover, and canopy height were similar to tree density. All increased with increasing fragment area (Table 3) and were greater in fragments with connectivity and in fragments owned by the Forest Department (Table 4). The absence of statistical significance was mostly due to small sample sizes. All basal areas were lowest in medium-altitude fragments, most of which were privately owned. Within both privately owned and Forest Department-owned forests, basal area, canopy cover, and canopy height were greater in the larger fragments, thus showing the effect of area (Table 4).

4.4. Occurrence of Primates

4.4.1. Area

Of the 25 fragments that were surveyed, 10 (40%) fragments had lion-tailed macaque, 19 (76%) had Nilgiri langur, and five had none (Table 1). The lion-tailed macaques are found in only 67% of the large fragments. Lion-tails were found in 60% of the medium-sized fragments and none in the small fragments. The Nilgiri langur were found in the medium-sized fragments with 80% remaining and 55% remaining in small fragments (<10 ha).

When fragments were lumped into large (>50 ha) and small (<50 ha), a significantly greater proportion (77.8%) of the large fragments had lion-tailed macaque than the small fragments (18.8%; Fisher exact test $p = 0.006$). Nilgiri langurs also differed significantly between large and small fragments (88% and 63%, respectively; Fisher exact test $p = 0.04$).

4.4.2. Ownership

Lion-tailed macaques occurred in 54.5% of the Forest Department-owned fragments and only in 28.6% of privately owned fragments. Nilgiri langurs were present in all Forest Department-owned and 57% of the privately owned fragments (Table 5). The occurrence of Nilgiri langur differed significantly with ownership (Fisher exact test $p = 0.02$); whereas, the difference was not significant in the case of lion-tailed macaque (Fisher exact test $p = 0.18$).

4.4.3. Location of Fragments

Both primates were less likely to be present on hilltop fragments, compared to either valley or hill slope. The lion-tailed macaque seemed to be particularly likely to disappear from hilltop fragments, with occurrence in only 17.3%, compared to 71.4% for Nilgiri langur. Both primates were more frequently absent from fragments in the medium-altitude range (900 to 1200 m, Table 5). When the high and low altitudes were pooled and tested against medium elevation, the difference was not significant (Fisher test $p = 0.11$ for lion-tailed macaque and $p = 0.28$ for Nilgiri langur). Most of the small fragments were in medium elevation. Therefore, it was not possible to examine the effect of altitude controlling for area.

Table 5. Percentage of fragments with the lion-tailed macaque and Nilgiri langur with reference to landscape variables.

Landscape parameters	Number of fragments	Lion-tailed macaque	Nilgiri langur
Area			
Small (<50 ha)	16	18.70	62.50
Large (>50 ha)	9	77.70	100.00
Fisher's test (p)		*0.01*	*0.04*
Ownership			
Forest Dept	11	54.50	100.00
Private	14	28.60	57.14
Fisher's test (p)		*0.18*	*0.02*
Location			
Hilltop	7	14.30	71.40
Others	18	50.00	77.80
Fisher's test (p)		*0.12*	*0.55*
Altitude			
Medium (900 to 1200 m)	15	26.40	60.00
Others	10	60.00	100.00
Fisher's test (p)		*0.11*	*0.03*
Connectivity			
Yes	11	63.60	90.90
No	14	21.40	64.30
Fisher's test (p)		*0.04*	*0.14*

4.4.4. Connectivity

The occurrence of both primates was high in fragments with connectivity to another wet evergreen forest (Table 5). Lion-tailed macaques occurred in a higher percentage of fragments (63.6%) with connectivity than those without (21.4%, Fisher exact test $p = 0.04$). Nilgiri langurs occurred 90.9% in fragments with connectivity when compared to 64.3% in fragments without connectivity. Thus, the lion-tailed macaque was more affected if there was no connectivity.

4.4.5. Habitat Parameters

Canopy height was the best independent predictor of the occurrence of the lion-tailed macaque (slope = 0.9254, $p = 0.04$). Inclusion of other independent variables, including area, did not significantly improve predictability. Using the derived equation ($p = 1/(1 + e^{-y})$, where y = -17.381 + 0.9254 × canopy height), correct prediction of occurrence could be achieved for 23 of the 25 fragments. (It should be noted, however, that this was done on the same data set that was used to derive the equation.) For Nilgiri langur, tree density was the best and only independent predictor (slope = 0.0112, $p = 0.0662$). Using the derived equation (as above but with y = -1.3930 + 0.0112 × tree density) correct prediction of occurrence could be made for only 76% of the 25 fragments. Nilgiri langur was absent from five fragments where it was predicted to be present.

4.4.6. Diet

The major changes in feeding ecology for the lion-tails and Nilgiri langurs in fragments were: 1) a reduction in the number of species in small fragments, 2) the greater dependence on shrubs compared to lianas in the larger and less disturbed fragments, and 3) in the lion-tailed macaque there was a reduction in the relative proportion of invertebrates in the diet, which might affect immature growth and survival (for greater details see Umapathy and Kumar, 2000).

4.5. Population Density

4.5.1. Lion-tailed Macaque

In fragments with this species, the lowest density was in a large fragment, Akkamalai (0.28 groups/km^2), and the highest in the smallest fragment, Tata Estate (1.42 groups/km^2). Abundance seemed to increase with decreasing fragment area (r_s = -0.72, p = 0.01). In fragments larger than 200 ha, abundance did not vary as in small fragments (from 0.28 to 0.42 groups/km^2). Except for canopy cover, all habitat parameters were negatively correlated with density of the lion-tailed macaque (Table 6). Tree density and basal area were highly negatively correlated, when only fragments with the primates were considered. When fragments with lion-tailed macaques absent were included in the analysis, the relationship disappeared. The variance in the abundance of lion-tailed macaque seemed to increase with decreasing area abruptly in fragments less than 200 ha in area. Tree density also showed a similar tendency in fragments with less than 400 trees per ha, most of which were small fragments.

4.5.2. Nilgiri Langur

The abundance of Nilgiri langur ranged from 0.34 to 6.6 groups per km^2 over all fragments where they were present. The lowest abundance was in a large fragment, Sankarankudi (180 ha), and the highest in the smallest fragment, Varattuparai-II (2 ha). Abundance varied considerably in the small fragments (<20 ha) from 0.54 to 6.6 groups per km^2 than in larger fragments (0.34 to 1.03 groups/km^2). Except area, none of the habitat parameters were significantly correlated with abundance of Nilgiri langur (Table 6). As in the case of lion-tailed macaque, the variance in abundance decreased with habitat parameters and increasing area (especially above 20 ha).

Table 6. Spearman rank correlation coefficients between the abundance of arboreal mammals and habitat parameters with probability value in parenthesis. Only fragments with primates have been included.

Primate Species	No. of fragments	Area	Tree density	Basal area	Canopy cover	Canopy height
Lion-tailed macaque	10	-0.72 *(0.01)*	-0.75 *(0.01)*	-0.67 *(0.02)*	0.07 *(0.42)*	-0.30 *(0.19)*
Nilgiri langur	19	-0.43 *(0.03)*	-0.14 *(0.28)*	-0.20 *(0.20)*	-0.14 *(0.28)*	-0.215 *(0.46)*

4.6. Demography

4.6.1. Lion-tailed Macaque

Group size and composition were recorded for 11 groups from eight forest fragments. Both the largest (36 animals) and the smallest (8 animals) groups were from medium-sized fragments (Table 7) in Puthuthottam-I and Pannimedu, respectively. There was no correlation between group size and fragment area (Table 8).

Table 7. Demographic parameters in 11 groups of lion-tailed macaques in eight forest fragments in IGWS and nearby private forests[a].

Groups	Area (ha)	Group size	No. of adult male	% adult female	% Juve	Juve /Af	Af/ Am	Birth rate
Varagaliar-I	2,500	14	1	42.86	50.00	1.17	6.00	0.50
Varagaliar-II	2,500	10	1	50.00	50.00	0.80	5.00	0.20
Varagaliar-III	2,500	13	1	38.40	53.00	1.40	5.00	0.60
Akkamalai	2,000	11	1	36.36	54.54	1.50	4.00	0.75
Monampoly	1,000	12	1	41.67	50.00	1.20	5.00	0.60
Andiparai	185	26	2	50.00	42.31	0.85	6.50	0.46
Puthuthottam-I	65	36	2	52.78	41.67	0.79	9.50	0.26
Puthuthottam-II	65	11	2	36.36	45.45	1.25	2.00	0.50
Pannimedu	50	8	1	50.00	37.50	0.75	4.00	0.25
Korangumudi	35	21	2	52.38	38.09	0.73	5.50	0.27
Tata	24	15	2	46.67	40.00	0.86	3.50	0.14

[a] Af = adult female; Juve = juvenile; Juve/Af = juvenile/adult female; Af/Am = adult female/adult male

Table 8. Spearman rank correlation coefficients between different demographic parameters of 11 groups of lion-tailed macaque and habitat parameters of the respective fragments (probability values in parenthesis).

Parameters	Area	Tree density	Canopy height	Canopy cover	Basal area
Group size	-0.26	-0.34	0.04	0.19	-0.28
	(0.22)	(0.15)	(0.46)	(0.08)	(0.19)
% adult females per group	-0.36	-0.33	-0.23	-0.24	-0.47
	(0.12)	(0.16)	(0.24)	(0.24)	(0.07)
% juveniles per group	0.86	0.77	0.59	0.46	0.60
	(0.01)	(0.01)	(0.03)	(0.08)	(0.03)
Juvenile/adult female	0.44	0.35	0.32	0.46	0.57
	(0.09)	(0.15)	(0.17)	(0.08)	(0.04)
Adult female/adult male	0.33	0.24	0.21	0.23	-0.07
	(0.16)	(0.24)	(0.26)	(0.25)	(0.42)
Birth rate	0.49	0.51	0.28	0.32	0.63
	(0.07)	(0.05)	(0.47)	(0.17)	(0.02)

The larger fragments (>100 ha) had an average group size of 14.3 animals (SE = 2.40, N = 6), while the smaller fragments (<100 ha) had an average of 18.2 animals (SE = 4.95, N = 5), the difference being close to significance (M-W U = 12.5, p = 0.06). The group size had weak negative correlation with tree density (r_s = -0.34 p = 0.15) and basal area (r_s = -0.28, p = 0.12) and had weaker correlation with other habitat parameters (Table 8). There was a tendency for groups to become larger as fragments become smaller and more disturbed. The group size also seemed to vary considerably as fragments became smaller, with some groups being very small and others being very large. In the large fragments (>100 ha), group size had a considerably lower coefficient of variation (CV = 41.1%, N = 6) compared to the smaller fragments (60.9%, N = 5). As expected, group size became more variable when disturbance increased as indicated by other habitat parameters (e.g., tree density).

Lion-tails in 11 groups from eight fragments were classified into adult males (>6 years), adult females (>5 years), and juveniles (<5 years) (Table 7). Five out of six single-male groups were from large fragments, while four out of five two-male groups were from small fragments, the difference being close to significance (Fisher exact test p = 0.06). The number of adult males in a group increased with a decrease in fragment size and tree density (r_s = -0.70, N = 11, p = 0.008 and r_s = -0.76, p = 0.012, respectively) and was also significantly correlated with group size (r_s = 0.67, N = 11, p = 0.003). A multiple regression showed that only area was negatively correlated with the number of adult males per group (R^2 = 0.79). There is thus a tendency for groups to be larger in the smaller fragments and for them to have more adult males.

The number of adult females in a group ranged from four (in three groups Akkamalai, Puthuthottam-II, and Pannimedu) to 19 (in Puthuthottam-I) (Table 7). The percentage of adult females in a group ranged from 36.4% to 52.8%. The large fragments (>100 ha) had slightly fewer adult females (43.2%, N = 6) than the smaller fragments (47.6%, N = 5), but area and percentage of females in the group were not significantly correlated (r_s = -0.36, p = 0.14). The other habitat parameters such as tree density, basal area, canopy height, and canopy cover also were not related to percent of adult females in the group (Table 8).

The percent of juveniles was positively correlated with the area of the fragment (r_s = 0.86, N = 11, p = 0.001). The larger fragments had an average of 50% juveniles, while the smaller had 41%, the difference being highly significant (M-W U = 1, p<0.01). The percent of juveniles was also significantly correlated with tree density, basal area, and canopy height (Table 8). Multiple regression showed that area alone was strongly correlated with percentage of juveniles in the group (R^2 = 0.91).

Birth rate was estimated in May 1996 for 11 groups from eight fragments as the number of infants (<1 year old) per adult female (Table 7). The highest birth rate (0.75) was recorded for a group in a very large fragment (Akkamalai Shola) and the lowest (0.14) in a small fragment (Tata Estate). The birth rate decreased with decreasing area of forest fragment (Table 8), but it was better correlated with tree density and basal area (Table 8). There was a large difference in the birth rate between large (mean = 0.52, N = 6) and small (mean = 0.28, N = 5) fragments, but not statistically significant (χ^2 = 2.02, p>0.05) due to small sample size. The birth rate in small fragments had a slightly larger coefficient of variation (46.4%) than in the large fragments (35.8%).

The exponential growth (r) rate of groups was calculated for one year (from May 1995 to April 1996) in the five continuously monitored groups of Akkamalai Shola and

Andiparai Shola and Puthuthottam, Korangumudi, and Tata estates. Andiparai Shola had the highest growth rate of 0.314 (from 19 to 26 individuals), and the lowest growth rate of 0 (15 to 15) was in the Tata Estate, the smallest fragment. Growth rate of groups decreased with decreasing area (r_s = 0.88, $0.10>p<0.05$) and was also significantly correlated with tree density (r_s = 0.91, p = 0.01, N = 5).

4.6.2. Nilgiri Langur

Since the Nilgiri langurs were very shy in nature, it was not possible to identify the age/sex of the individuals or even to obtain reliable counts of group size. Hence, the demographic analysis is restricted to eight groups in the four fragments (Akkamalai Shola, Andiparai Shola, Tata Estate, and Korangumudi Estate) where the intensive ecological studies were carried out. Another eight groups from the same fragments are also included for group size analysis, but not age/sex analysis.

The largest (18 animals) and smallest (7 animals) groups were in a small fragment (Korangumudi Estate, Table 9). The larger fragments had a slightly higher group size (mean = 13.2, SE = 0.70, N = 9) than smaller fragments (mean = 11, SE = 1.34, N = 7), the difference not being significant (M-W U = 16.5, $p>0.05$). The groups were classified into two categories, a small group <13 individuals and a big group >13 individuals. The large and small fragments had an almost equal proportion of groups (six small and eight big groups, five small and eight big groups, respectively). The difference was not being significant (χ^2 = 1.2, df = 1, $p>0.05$).

The number of adult males in a group did not vary much, six groups had one adult male, and only two had two adult males (Table 9). The number of the adult females in a group ranged from 3 to 10, both from a small fragment (Korangumudi Estate). The percentage of adult females in a group varied from 42.9% to 58.3%, the difference between large and small fragments not being significant (M-W U = 7.5, $p>0.05$). It was also not correlated with any of the habitat parameters (Table 10).

The percentage of juveniles in a group ranged from 36.6% to 46.7%. The highest was in a large fragment (Andiparai Shola) and the lowest in the smallest fragment (Tata Estate). The larger fragments had a higher average (43.5%) than the smaller (<100 ha) fragments (39.2%), but not significantly so (M-W U = 2, p = 0.08). The percentage of juveniles was most correlated with area, tree density, and canopy cover, with correlation approaching significance levels (Table 10).

Table 9. Demographic parameters of eight groups of Nilgiri langur in four forest fragments.

Group	Group size	No. of adult male	% adult female	% juvenile	Juve /Af[a]	Af/ Am[b]	Birth rate
Akkamalai-I	17	1	52.94	41.18	0.78	9	0.33
Akkamalai-II	10	1	50.00	40.00	0.80	5	0.40
Andiparai-I	15	1	46.67	46.67	1.00	7	0.29
Andiparai-II	13	1	46.15	46.15	1.00	6	0.33
Korangumudi-I	18	1	55.56	38.89	0.70	10	0.30
Korangumudi-II	7	2	42.86	42.80	1.00	3	0.00
Tata-I	12	1	58.33	36.57	0.57	7	0.14
Tata-II	13	2	46.15	38.46	0.83	3	0.33

[a] Juve/Af = juvenile/adult female; [b] Af/Am = adult female/adult male

Table 10. Spearman rank correlation between demography of Nilgiri langur and habitat parameters.

Parameters	Area	Tree density	Basal area	Canopy cover	Canopy height
Group size	0.15	0.15	0.07	0.15	-0.07
	(0.36)	(0.36)	(0.43)	(0.36)	(0.43)
% adult females in the group	0.10	0.10	0.07	0.19	0.22
	(0.40)	(0.40)	(0.43)	(0.32)	(0.28)
% juveniles in the group	0.59	0.59	0.29	0.59	-0.29
	(0.07)	(0.07)	(0.24)	(0.07)	(0.25)
Juvenile/adult female	0.20	0.20	0.10	0.20	-0.10
	(0.30)	(0.30)	(0.34)	(0.30)	(0.40)
Adult female/adult male	0.22	0.22	0.22	0.07	-0.20
	(0.29)	(0.29)	(0.29)	(0.43)	(0.30)
Birth rate	0.55	0.55	0.35	0.45	0.10
	(0.08)	(0.08)	(0.19)	(0.13)	(0.40)

There was no major difference in the percentages of the three age/sex classes between small and large fragments ($\chi^2 = 0.36$, df = 2, $p>0.05$). However, the ratio of infants and juveniles as a whole were greater in the larger fragments, though not significantly so.

Birth rate was estimated as the ratio of infants (<1 yr) to adult females at the time of the survey in May 1996. Birth rate was highest (0.4) in the very large fragment (Akkamalai Shola) and lowest (0.0) in a small fragment (Korangumudi Estate). Even though statistically not significant, the birth rate appeared to decrease with the decreasing fragment area ($r_s = 0.58$, N = 8, $p = 0.07$) and tree density ($r_s = 0.55$, N = 8, $p = 0.07$). There was a difference between birth rate in large fragments (mean = 0.34) and small fragments (0.19), even though this was not statistically significant ($\chi^2 = 0.4$, df = 1, $p>0.05$). The birth rate was also more variable in small fragments (CV = 79%) than the large fragments (CV = 14%).

The exponential growth rate of groups was calculated for one year (May 1995 to April 1996) for four continuously monitored groups in Korangumudi, Tata Estates, and Akkamalai and Andiparai Sholas. The groups in Akkamalai Shola had the highest growth rate of 0.19 (14 to 17 animals). The lowest rate of 0.15 was in the large fragment, Andiparai Shola. The other two groups (Korangumudi and Tata Estates) each had a growth rate of 0.17 (16 to 19 and 11 to 13 animals, respectively).

5. DISCUSSION

The occurrence and abundance of the primates varied between forest fragments. There was a high degree of multiple correlation among the different landscape and habitat parameters of the forest fragments, showing that the changes due to fragmentation, including those due to human impact, are often interrelated. Thus, the occurrence and abundance of primates was related to several parameters. Even though area had the most apparent impact, canopy height and tree density were the best predictors of occurrence, the former for the lion-tailed macaque and the latter for the

Nilgiri langur. The abundance of both these species became highly variable as fragment area declined.

Of the two primates that we examined, the lion-tailed macaque showed the largest difference in occurrence and abundance among the fragments. If the fragments at the time of its formation were a random sample of the contiguous forest, the small fragments would have had a very low probability of harboring the lion-tailed macaque when they were formed. The Nilgiri langur occurs in much greater densities than the lion-tailed macaque, probably with home ranges of less than 10 ha, even though no estimates are available. Therefore, the Nilgiri langur is more likely to have been initially present in small fragments. If the lion-tailed macaques were present when the fragments were formed, their subsequent survival seems to be more related to habitat variation than area. Most of the isolation-mediated habitat variations (such as vegetation structure) are highly correlated to each other and with area as indicated by this study and reported elsewhere (Abensperg-Traun et al., 1996).

Many recent studies show that the persistence of a species in a fragmented habitat is related to isolation-mediated habitat variations, within and around the fragments, that are consistent with the ecology and behavior of the species. The current occurrence of Nilgiri langur was more related to habitat parameters, tree density being the best predictor. This was also true for lion-tailed macaques. Unlike the lion-tailed macaque, the Nilgiri langur feeds mainly on young foliage and unripe fruits. These secondary compounds, rather than energy or protein content, dictate its food selection (Oates et al., 1980). The Nilgiri langur fed on a number of food plant species every day, even though the total number of food plant species that they used over the study period was considerably lower than those of the lion-tailed macaque (Umapathy and Kumar, 2000). Thus, it is the availability of a variety of plant species, rather than the abundance of foliage or fruits *per se*, that is important for the Nilgiri langur. Tree density, the best predictor of the occurrence of the Nilgiri langur, is perhaps the best indicator of the variety of plant species in forest fragments.

A· major factor that substantially affected the occurrence and abundance of both primates was the nature of vegetation that immediately surrounded the fragments. The very large fragments (mostly owned by Forest Department) had teak, eucalyptus, and tea estates immediately around them. The large fragments (owned by Forest Department and tea estates) were also surrounded by similar vegetation. Many privately owned small fragments had coffee plantation and orchards on the edges. Coffee beans, fruits of shade trees in coffee plantations, and many of the fruit trees in the orchards were important food items of primates, especially the lion-tailed macaque. The availability of these food resources to a large extent enhanced macaque occurrence and abundance in the small fragments.

The major demographic changes due to fragmentation in the lion-tailed macaque were a reduction in birth rate and proportion of juveniles in the group, an increase in group size, number of adult males, and variability in adult sex ratios. These correlated with decreasing fragment area and increasing disturbance levels. In the Nilgiri langur, there was a similar decrease in birth rate and proportion of juveniles; however, there was no change in group size and adult sex ratios. Growth rate of groups also decreased with area and disturbance levels. Kumar (1995b) found that birth in the lion-tailed macaque decreased with increasing group size due to increasing competition. In *Macaca sinica* in Sri Lanka, Dittus (1979) found an extensive reduction in birth rate and juvenile survival and very little change in adult survival during a period of acute food shortage. Drastic reduction in the proportion of juveniles has also been reported in primates following selective logging due to mortality following disturbance (Marsh et al., 1987; Johns, 1987). Dobson and Lyles (1989) suggested that primates respond to perturbations over several inter-birth intervals with a rapid decline in birth rates, while number of adults declined more slowly.

Reduction in survival, especially of juveniles, may also occur from increased predation. Increased densities of diurnal and nocturnal raptorial predators have been reported from logged forests in Malaysia (Johns, 1985). Similarly, in logged forests in Africa, a more open canopy made primates more vulnerable to predation by eagles. Such increased predation by birds has also been reported from forest fragments in Sweden (Andrén, 1992). Greater canopy discontinuity also forces the monkeys to spend more time on the ground, thereby increasing the risk of predation from terrestrial predators. Even though there is no data on ground predators, the number of feral or domestic dogs was often high in small fragments. A few cases of predation of lion-tailed macaque juveniles by dogs have been reported by local people. Thus, a reduction in birth rates and juvenile survival both due to resource shortage and the latter also due to increased predation is a major demographic consequence of habitat disturbance in primates. Although no predator was observed, this is very similar to the pattern found in the lion-tailed macaque.

In contrast, the Nilgiri langur did not show any variation in demographic parameters. The difference between the two species is related to the langurs' ability to use and disperse through the matrix between fragments and their particular social systems. Both males and females were often seen to move across treeless vegetation either natural or human made. They were also seen on isolated trees on the roadside far away from any forest. In Puthuthottam Estate, which had no Nilgiri langur, males were seen on a few occasions even though the fragment was totally surrounded by tea estates. In contrast, though the lion-tailed macaque spent more than 20% of their time on the ground in highly disturbed fragments, they were never seen to move into areas that were totally devoid of trees, such as tea estates that often surrounded the fragments. Even though adult and subadult males in contiguous forests emigrate from and immigrate into groups (Kumar, 1987), this was not observed in any of the groups that were intensively studied. Thus, male migration most probably does not occur between forest fragments for lion-tailed macaques.

Increased demographic stochasticity is a typical property of small populations (Soulé, 1986). This was reflected in the greater coefficient of variation in group size, adult sex ratios, and birth rate of lion-tailed macaque in the small fragments. As populations become small, the sampling variation inherent in the discrete nature of births and deaths can lead to distorted adult sex ratios and a high variability. In contiguous forests these are compensated for by adult male migration. The lack of male migration between fragments in lion-tailed macaques therefore causes a greater number of adult males per group in some fragments, reducing the adult sex ratio and increasing the variability in adult sex ratio. Since macaques live in female-bonded groups, females seldom migrate between groups (Wrangham, 1980; van Schaik, 1983). Thus, female dispersal is also unlikely to compensate for distorted sex ratios. Group fission and subsequent dispersal as by a matriline (Chepko-Sade and Sade, 1979) is also unlikely because of the inability of the lion-tailed macaques to use the treeless vegetation that often surrounds fragments. Thus, the larger group size, greater number of adult males per group, and greater variability in adult sex ratio in the smaller fragments is due to the absence of dispersal between fragments for lion-tailed macaques. In primates where dispersal between fragments is greater, such as with the Nilgiri langur, sex ratios are largely unaffected by fragmentation.

6. SUMMARY

The objectives were to examine the occurrence, abundance, and demographic consequences of population fragmentation in the lion-tailed macaque and Nilgiri langur in wet evergreen forest fragments in relation to several habitat and landscape parameters. In total, 25 forest fragments were identified and surveyed in the IGWS and surrounding areas in 1994–1996. The occurrence and abundance of both primates was estimated from 400 km of transects along existing forest trails. The habitat parameters were estimated from 350 circular plots of 5-m radius. Data on landscape parameters were obtained from historical documents, district and estate administration, and field observations. The parameters examined were group size, age/sex composition, birth rate, and population growth rate in relation to the various habitat parameters. Demographic parameters of the lion-tailed macaque were estimated for 11 groups in eight fragments, and that of the Nilgiri langur from 8 groups in four fragments during January to May 1996. The exponential growth rate of groups was calculated for the four main study groups of lion-tailed macaque and Nilgiri langur for one year (from March 1995 to February 1996).

The fragments differed considerably in landscape and habitat parameters, but variation showed a high level of multi-collinearity. Small fragments were more likely to be privately owned, had lower tree density, basal area, canopy height, and canopy cover and were more disturbed. The large fragments were more likely to be owned by Forest Department, had greater tree density, basal area, canopy height, and canopy cover. Vegetation around the fragment and connectivity were also related to area and ownership. The lion-tailed macaque was the most affected, being present in only 10 of 25 fragments, while Nilgiri langur was absent from six. The abundance of the lion-tailed macaque and Nilgiri langur showed a high variability in the smaller fragments. Because of multi-collinearity among landscape and habitat parameters, it was difficult to interpret bivariate or multivariate correlations between occurrence and abundance of animals and landscape and habitat parameters. Canopy height was the best predictor of the occurrence of the lion-tailed macaque and tree density for the Nilgiri langur. These are likely related to feeding ecology of the species. It is concluded that area may be an important predictor of the occurrence of the primates only when the fragments are very small, the initial occurrence depending on the relative densities at which the species occur. Once initially present, the continued occurrence and abundance are better predicted by habitat variation, consistent with the ecology of the species.

Demographic changes in the lion-tailed macaque included a reduction in birth rate and proportion of juveniles in the group and an increase in group size, number of adult males, and variability in adult sex ratios, with decreasing area and increasing disturbance levels. In the Nilgiri langur also, there was a similar decrease in birth rate and proportion of juveniles; however, there was no change in group size and adult sex ratio. Growth rate of groups also decreased with area and disturbance level in the lion-tailed macaque. The decrease in birth rate and survival in the lion-tailed macaque might be due to reduction in diet quality. The latter might also be due to falls from trees and greater predation in the small fragments. Lack of dispersal between fragments may be the reason for the fluctuation in adult sex ratios and larger group size. In the Nilgiri langur, there was a reduction in birth rate and juvenile survival, probably due to the same reasons as in the case of the lion-tailed macaque. There was no difference in sex ratios and group size of the Nilgiri langur because of their ability to disperse between fragments.

7. ACKNOWLEDGMENTS

This study was funded by the Ministry of Environment and Forests, Government of India. We thank Tamil Nadu Forest Department, Tata-Tea (Pvt) Ltd., Puthuthottam and Korangumudi Estates for permitting us to conduct this research and Dr. V. S. Vijayan, Director, Salim Ali Centre for Ornithology and Natural History for providing facilities.

8. REFERENCES

Abensperg-Traun, M., Smith, G. T., Arnold, G. W., and Steven, D. E., 1996, The effect of habitat fragmentation and livestock grazing on animal communities in remnants of gimlet *Eucalyptus salubris* woodland in the Western Australian wheatbelt, I. Arthropods, *J. of Applied Ecol.* **33**:1,281–1,301.
Adler, G. H., and Wilson, M. L., 1985, Small mammals on Massachusetts Islands: The use of probability functions in clarifying biogeographic relationships, *Oecologia* **66**:178–186.
Ali, S., and Ripley, D. S., 1987, *Compact Handbook to the Birds of India and Pakistan*, Oxford University Press, Bombay, India.
Andrén, H., 1996, Population responses to habitat fragmentation: Statistical power and random sample hypothesis, *Oikos* **76**:235–242.
Arnold, G. W., Steven, D. E., Weeldenburg, J. R., and Smith, E. A., 1993, Influences of remnant size, spacing pattern, and connectivity on population boundaries and demography in *Euros macropus robustus* living in a fragmented landscape, *Biol. Cons.* **64**:219–230.
Bierregaard, R. O., Lovejoy, T. E., Kapos, V., Aidos-Santos, A., and Hutchings, R. W., 1992, The biological dynamics of tropical rainforest fragments, *BioScience* **42**:859–866.
Bright, P. W., and Morris, P. A., 1996, Why are dormice rare? A case study in conservation biology, *Mam. Rev.* **26**:157–187.
Chepko-Sade, B. D., and Sade, D. S., 1979, Patterns of group splitting within matrilineal kinship groups, *Behav. Ecol. Sociobio.* **5**:67–80.
Congreve, C. R. T., 1938, *The Anamalais*, Madras.
Daniels, R. J. R., 1997, *A Field Guide to the Birds of Southwestern India*, Oxford University Press, New Delhi, India.
Decker, B. S., 1994, Effects of habitat disturbance on the behavioral ecology and demographics of the Tana River Bed Colobus (*Colobus badius rufomitratus*), *Int. J. Primatol.* **15**:703–737.
Dittus, W. P. J., 1979, The evolution of behaviours regulating density and age-specific sex ratios in a primate population, *Behaviour* **69**:265–302.
Dobson, A. P., and Lyles, A. M., 1989, The population dynamics and conservation of primate populations, *Cons. Bio.* **3**:362–380.
Downes, S. J., Handasyde, K. A., and Elgar, M., 1997, The use of corridors by mammals in fragmented Australian eucalypt forests, *Cons. Bio.* **11**:718–726.
Fitzgibbon, C. D., 1997, Small mammals in farm woodlands: The effects of habitat, isolation, and surrounding land-use patterns, *J. App. Ecol.* **34**:530–539.
Gilpin, M. E., and Soulé, M. E., 1986, Minimum viable populations: Process of species extinction, in: *Conservation Biology: The Science of Scarcity and Diversity*, M. E. Soulé, ed., Sinauer, Sunderland, Massachusetts, pp. 19–34.
Green, S. M., and Minkowski, K., 1977, The lion-tailed macaque and its south Indian rain forest habitat, in: *Primate Conservation*, G. H. Bourne and H. S. H. Rainer, eds., Academic Press, New York, pp. 289–337.
Hanski, I., and Gilpin, M. E., 1991, Metapopulation dynamics: Brief history and conceptual domain, *Biol. J. Linn. Soc.* **40**:3–16.
Hill, C. J., 1995, Linear strips of rain forest vegetation as potential dispersal corridors for rain forest insects, *Cons. Bio.* **9**:1,559–1,566.

Hladik, C. M., 1978, Adaptive strategies of primates in relation to leaf eating, in: *The Ecology of Arboreal Folivores*, G. G. Montgomery, ed., Smithsonian Institution Press, Washington, D.C., pp. 373–394.

Hosmer, D. W., and Lemeshow, S., 1989, *Applied Logistic Regression*, John Wiley and Sons, New York.

IUCN, 1996, *1996 IUCN Red List of Threatened Animals*, IUCN, Gland, Switzerland.

Johns, A. D., 1987, The use of primary and selectively logged rainforest by Malaysian hornbills (*Bucerotidae*) and implications for their conservation, *Biol. Cons.* **40**:179–190.

Johns, A. D., and Skorupa, J. P., 1987, Responses of rain-forest primates to habitat disturbance: A review, *Int. J. Primatol.* **8**:157–191.

Johns, A. D., 1988, Effects of "selective" timber extraction on rain forest structure and composition and some consequences for frugivores and folivores, *Biotropica* **20**:31–37.

Khoshoo, T. N., 1995, Census of India's biodiversity: Tasks ahead, *Current Science* **69**:14–17.

Klein, B. C., 1989, Effects of forest fragmentation on dung and carrion beetle communities in central Amazonia, *Ecology* **70**:1,715–1,725.

Kozakiewicz, M., 1985, The role of isolation in formation of structure and dynamics of the bank vole population, *Acta Theriologica* **30**:93–209.

Kozakiewicz, M., 1993, Habitat isolation and ecological barriers: The effect on small mammal populations and communities, *Acta Theriologica* **38**:1–30.

Kumar, A., 1987, The ecology and population dynamics of the lion-tailed macaque (*Macaca silenus*) in South India, Ph.D. thesis, Cambridge University, UK.

Kumar, A., 1995a, The life history, ecology, distribution, and conservation problems in the wild, in: *The Lion-tailed Macaque: Population and Habitat Viability Assessment Workshop*, A. Kumar, S. Molur, and S. Walker, eds., Zoo Outreach Organization, Coimbatore, India.

Kumar, A., 1995b, Birth rate and survival in relation to group size in the lion-tailed macaque, *Macaca silenus, Primates* **36**:1–9.

Kumar, A., and Kurup, G. U., 1993, The demography of the lion-tailed macaque in the wild, in: *Proceedings of the IVth International Symposium on the Lion-tailed Macaque*, 11-14 October 1994, Madras.

Kurup, G. U., 1973, Present status of the Nilgiri langur, *Presbytis johnii*, in the Anamalais, Western Ghats, *Indian Forester* **99**:518–521.

Lanly, J. P., 1982, Tropical forest resources, FAO Forestry Paper 30, FAO, Rome, Italy.

Laurance, W. F., 1990, Comparative responses of five arboreal marsupials to tropical forest fragmentation, *J. Mammal.* **71**:641–653.

Laurance, W. F., 1993, The pre-European and present distributions of *Antechinus godmani* (Marsupialia: Dasyuridae), a restricted rainforest endemic, *Australian Mammal* **16**:23–27.

Laurance, W. F., 1994, Rainforest fragmentation and the structure of small mammal communities in tropical Queensland, *Biol. Cons.* **69**:23–32.

Laurance, W. F., and Gascon, C., 1997, How to creatively fragment a landscape, *Cons. Biol.* **11**:577–579.

Levin, R., 1969, Some demographic and genetic consequences of environmental heterogeneity for biological control, *Bull. Entomol. Soc. Amer.* **15**:237–240.

Lindenmayer, D. B., and Possingham, H. P., 1995, The conservation of arboreal marsupials in the montane ash forests of the central highlands of Victoria, southeastern Australia-VII. Modelling the persistence of Leadbeater's possum in response to modified timber harvesting practices, *Biol. Cons.* **73**:239–257.

Marsh, C. W., Johns, A. D., and Ayres, J. M., 1987, Effects of habitat disturbance on rain forest primates, in: *Primate Conservation in the Tropical Rain Forest*, C. W. Marsh and R. A. Mittermeier, eds., Alan R. Liss, New York, pp. 83–107.

Martin, J., Gaston, A. J., and Hitier, S., 1995, The effects of island size and isolation on old growth forest habitat and bird diversity in Gwaii Haanas (Queen Charlotte Islands, Canada), *Oikos* **72**:115–131.

May, R. M., 1975, Island biogeography and the design of wildlife preserves, *Nature* **245**:177–178.

Menon, S., 1993, Ecology and conservation of the endangered lion-tailed macaque (*Macaca silenus*) in the landscape mosaic of the Western Ghats, Ph.D. dissertation, Ohio State University, Columbus.

Menon, S., and Bawa, K. S., 1997, Applications of geographic information systems, remote-sensing, and a landscape ecology approach to biodiversity conservation in the Western Ghats, *Current Science* **73**:134–144.

Menon, S., and Poirier, F. E., 1996, Lion-tailed macaques (*Macaca silenus*) in a disturbed forest fragment: Activity patterns and time budget, *Int. J. Primatol.* **17**:967–985.

Myers, N., 1991, Tropical forests: Present status and future outlook, *Climatic Change* **19**:3–32.

Myers, N., 1992, *Tropical Forests and Climate*, Kluwer, Academic Publishers, Dordrecht.

Nair, S. C., 1991, *The Southern Western Ghats: A Biodiversity Conservation Plan*, Indian National Trust for Art and Cultural Heritage, New Delhi.

Newmark, W. D., 1991, Tropical forest fragmentation and the local extinction of understory birds in the Eastern Usambara Mountains, Tanzania, *Cons. Biol.* **5**:67–77.

Norusis, M. J., 1990, SPSS *Inc.* SPSS Release 4.0 for Unisys 6000, Chicago, Illinois, USA.

Noss, R. F., 1987, Protecting natural areas in fragmented landscapes, *Nat. Areas J.* **7**:2–13.

Oates, J. F., 1979, Comments on the geographical distribution and status of the South Indian black leaf-monkey (*Presbytis johnii), Mammalia* **43**:485–493.

Oates, J. F., Waterman, P. G., and Choo, G. M., 1980, Food selection by a South Indian leaf-monkey, *Presbytis johnii,* in relation to leaf chemistry, *Oecologia* **45**:45–56.

Opdam, P. F. M., 1991, Metapopulation theory and habitat fragmentation: A review of holarctic breeding bird studies, *Lands. Ecol.* **5**:93–106.

Poirier, F. E., 1970, Dominance structure of the Nilgiri langur, *Presbytis johnii,* of South India, *Folia Primatol.* **12**:161–18.

Powell, A. H., and Powell, G. V. N., 1987, Population dynamics of male euglossine bees in Amazonian forest fragments, *Biotropica* **19**:176–179.

Prabhaker, A., 1999, Impact of forest fragmentation on terrestrial small mammal communities in western Ghats, India, Ph.D dissertation submitted to Bharathiar University, Coimbatore, India.

Prater, S. H., 1980, *The Book of Indian Animals, 3rd Ed.,* Bombay Natural History Society, Oxford University Press, Bombay.

Punttila, P., 1996, Succession, forest fragmentation, and the distribution of wood ants, *Oikos* **75**:291–298.

Quinn, J. F., and Hastings, A., 1987, Extinction in sub-divided habitats, *Cons. Bio.* **1**:198–208.

Robinson, G. R., Holt, R. D., Gaines, M. S., Hamburg, S. P., Johnson, M. L., Fitch, H. S., and Martinko, E. A., 1992, Diverse and contrasting effects of habitat fragmentation, *Science* **257**:524–526.

Russell-Smith, J., and Bowman, D. M. J. S., 1992, Conservation of monsoon rainforest isolates in the Northern Territory, Australia, *Biol. Cons.* **59**:51–63.

Saltz, D., 1996, Minimizing extinction probability due to demographic stochasticity in a reintroduced herd of Persian fallow deer, *Biol. Cons.* **75**:27–33.

Saunders, D. A., and Hobbs, R. J., 1991, *Nature Conservation 2: The Role of Corridors,* Surrey Beatty and Sons, Chipping Norton, NSW.

Saunders, D. A., Hobbs, R., and Ehrlich, P. R., 1993, *Nature Conservation 3: Reconstruction of Fragmented Ecosystems,* Surrey Beatty and Sons, Australia.

Saunders, G. W., Arnold, G. W., Burbidge, A. A., and Hopkins, J. M., 1987, *Nature Conservation: The Role of Remnants of Native Vegetation,* Surrey Beatty and Sons, Chipping Norton, Australia.

Shaffer, M. L., 1981, Minimum population sizes for species conservation, *BioScience* **31**:131–134.

Shaffer, M. L., 1987, Minimum viable populations: Coping with uncertainty, in: *Viable Populations for Conservation,* M. E. Soulé, ed., Cambridge University Press, Cambridge, England, pp. 69–86.

Siegel, S., and Castellan, N. J., Jr., 1988, *Nonparametric Statistics for the Behavioral Sciences, 2nd Ed.* McGraw-Hill Book Co., USA.

Soulé, M. E., 1986, Conservation biology and real world, in: *Conservation Biology: The Science of Scarcity and Diversity,* M. E. Soulé, ed., Sinuar Associates, Inc., Sunderland, MA, pp. 1–12.

Soulé, M. E., 1987, Introduction, in: *Viable Populations for Conservation,* M. E. Soulé, ed., Cambridge University Press, Cambridge, England, pp. 1–10.

Soulé, M. E., and Simberloff, D., 1986, What do genetics and ecology tell us about the design of nature reserves? *Biol. Cons.* **35**:19–49.

Srivastava, K. K., Zacharias, V. J., Bhardwaj, A. K., Joseph, P., and Joseph, S., 1996, Some observations on troop structure, activity budget, and food habits of the Nilgiri langur (*Presbytis johnii*) in Periyar during monsoon (June - August), *Indian Forester* **122**:946–950.

Stouffer, P. C., and Bierregaard, R. O., 1994, Effects of forest fragmentation on understory hummingbirds in Amazonian Brazil, *Cons. Bio.* **9**:1,085–1,094.

Sugiyama, Y., and Ohsawa, H., 1982, Population dynamics of Japanese macaques with special reference to the effect of artificial feeding, *Folia Primatol.* **39**:238–263.

Sundararaju, R., 1987, *Management Plan for Indira Gandhi Wildlife Sanctuary, Pollachi (for the period of 1987-88 to 1992-93),* Office of the Chief Wildlife Warden, Chennai.

Terborgh, J., 1976, Island biogeography and conservation: Strategy and limitations, *Science* **193**:1,029–1,030.

Terborgh, J., 1986, Keystone plant resources in the tropical forest, in: *Conservation Biology: The Science of Scarcity and Diversity,* M. E. Soulé, ed., Sinauer Associates, Sunderland, MA, pp. 330–344.

Terborgh, J., 1992, *Diversity and the Tropical Rainforest,* Scientific American Library, New York.

Terborgh, J., and Winter, B., 1980, Some causes of extinction, in: *Conservation Biology: An Evolutionary-Ecological Perspective,* M. E. Soulé and B. A. Wilcox, eds., Sinauer Associates, Sunderland, MA.

Turner, I. M., and Corlett, R., 1996, The conservation value of small, isolated fragments of lowland tropical rain forest, *TREE* **11**:330–333.

Umapathy, G., and Kumar, A., 2000, Impacts of habitat fragmentation on time, budget, and feeding ecology of lion-tailed macaque (*Macaca silenus*) in rain forest fragments of Anamalai Hills, South India, *Primate Report* **58**:67–82.

van Dorp, D., and Opdam, P. F. M., 1987, Effects of patch size, isolation, and regional abundance on forest bird communities, *Lands. Ecol.* **1**:59–73.

van Schaik, C. P., 1983, Why are diurnal primates living in groups? *Behaviour* **87**:120–144.

Waterman, P. G., 1984, Food acquisition and processing by primates as a function of plant chemistry, in: *Food Acquisition and Processing by Primates*, D. J. Chivers, B. A. Wood, and A. Bilsborough, eds., Plenum Press, New York, pp. 177–211.

Whitmore, T. C., 1997, Tropical forest disturbance, disappearance, and species loss, in: *Tropical Forest Remnants: Ecology, Management, and Conservation of Fragmented Communities*, W. F. Laurance and R. O. Bierregaard, Jr., eds., University of Chicago Press, Chicago, pp. 3–12.

Wilcove, D. S., McLellan, C. H., and Dobson, A. P., 1986, Habitat fragmentation in the temperate zone, in: *Conservation Biology: The Science of Scarcity and Diversity*, M. E. Soulé, ed., Sinauer Associates, Sunderland, MA, pp. 237–256.

Wilcox, B. A., and Murphy, D. D., 1985, Conservation strategy: The effects of fragmentation on extinction, *Am. Nat.* **125**:879–887.

Williams, M., 1990, Forests: The earth as transformed by human action, in: *Global and Regional Changes in the Biosphere Over the Past 330 Years*, B. L. Turner, W. C. Clark, R. W. Kates, J. F. Richards, J. T. Mathews, and W. B. Meye, eds., Cambridge University Press, Cambridge, England, pp. 179–201.

Willson, M. F., Sabag, C., and Armesto, J. J., 1994, Avian communities of fragmented south-temperate rainforests in Chile, *Cons. Bio.* **8**:508–520.

Wittenberger, J. F., 1980, Group size and polygyny in social mammals, *Am. Nat.* **115**:197–221.

Wrangham, R. W., 1980, An ecological model of female-bonded primate groups, *Behaviour* **75**:262–300.

WRI, 1990, *World Resources* 1990-1991, Oxford University Press, Oxford.

POPULATION SIZE AND HABITAT USE OF SPIDER MONKEYS AT PUNTA LAGUNA, MEXICO

Gabriel Ramos-Fernández and Bárbara Ayala-Orozco[*]

1. INTRODUCTION

The Yucatán Peninsula is currently a mosaic of forest in various stages of secondary succession, mostly attributed to human-induced disturbances. No more than 50 years ago, approximately 86,000 km² of the peninsula were covered by medium, semi-evergreen forest (Rzedowski, 1978). Today only a few unprotected fragments of this vegetation type remain larger than 1,000 km², and deforestation continues at a rate of 8,000 km² per year (Challenger, 1998). How does this disturbance affect spider monkey populations? Do spider monkeys modify their behavior in any way to survive in a fragmented habitat? If so, what is the minimum size of a forest fragment that can support a healthy population of spider monkeys? How can the remaining populations be protected?

Habitat destruction is certainly the strongest threat to the survival of the extant primates of the Yucatán Peninsula—the black-handed spider monkey (*Ateles geoffroyi*) and the Central American black howler monkey (*Alouatta pigra*). Most of their habitat has been destroyed by slash-and-burn agriculture and accidental fires associated with it (Challenger, 1998). This form of agriculture is practiced mainly at a subsistence level by around 200,000 farmers who live on less than $3.00 (US) per day (National Institute of Statistics, Geography, and Information Science, 2000). Each farmer cuts down trees in 0.02 to 0.04 km² of either primary or secondary forest no less than 10 years old. Fields are used for up to three consecutive years. When the soil is no longer rich in nutrients, fast-growing species invade. During the dry season, farmers use fire to clean up the land just before planting corn and beans as the rains begin. Using fire involves the risk of burning large portions of the surrounding forest by accident. This risk is especially high after a hurricane when a large amount of dead vegetation lying on the ground favors the spread of fire. For example, after hurricane Gilbert in 1989, 1350 km² of medium semi-evergreen forest were burned by accident in the eastern Yucatán Peninsula (López-Portillo et al., 1990).

[*] Gabriel Ramos-Fernández, Pronatura Península de Yucatán, A.C., Calle 17 #188A × 10, Col. Garcia Gineres, Mérida Yucatán, México 97070. Bárbara Ayala-Orozco, Department of Environmental Studies, University of California, Santa Cruz, CA, 95064, USA. Correspondence to G. Ramos-Fernández (email: ramosfer@sas.upenn.edu).

Primates in Fragments: Ecology and Conservation
Edited by L. K. Marsh, Kluwer Academic/Plenum Publishers, 2003

Meffe and Carroll (1994) conclude that species that occur naturally in low numbers, occupy wide ranges, have low fecundity rates, and depend on patchy or unpredictable resources are particularly vulnerable to extinction from habitat fragmentation. Spider monkeys fulfill these characteristics: first, they have been found at low densities in forests in Latin America, including Peru (Symington, 1987), Suriname (van Roosmalen and Klein, 1987), Costa Rica (Chapman, 1990), Brazil (Nunes, 1995), and Colombia (Klein and Klein, 1977). Second, home ranges for a group of 18 to 55 individuals have been reported from 1.5 to 3.9 km^2 (Klein and Klein, 1977; van Roosmalen, 1985; Symington, 1987; Chapman, 1990; Nunes, 1995; Wallace, 1998). Third, a female spider monkey will give birth to one infant every 3 to 4 years, a low fecundity rate compared to other primate species (Robinson and Janson, 1987). Finally, spider monkeys are highly frugivororus, with 80% to 90% of the time feeding on the fruit of 50 to 150 species of trees, while the rest of the time feeding on leaves, flowers, and bark (van Roosmalen and Klein, 1988). Because of these characteristics, spider monkey populations are more vulnerable to the effects of fragmentation than populations of other sympatric species. For the same reason, spider monkeys can be monitored to detect the early effects of fragmentation on species diversity (e.g., Dirzo and Miranda, 1990).

This chapter reports the results of a three-year study on the behavioral ecology of spider monkeys living in a fragmented habitat in the northeastern Yucatán Peninsula. The main goal of the study was to evaluate the conservation status of the population. This was achieved by estimating the size of the population and by describing the habitat characteristics, diet, use of space, and social organization in two groups that were studied continuously for three years. The comparison with other populations of *Ateles* in non-fragmented habitats may prove useful for understanding how habitat fragmentation affects the permanence of viable populations of the species. These results can also be used as guidelines for the design of conservation strategies that help ensure the permanence of other remaining populations throughout the Yucatán.

2. STUDY AREA

The study was carried out in the area of forest around Punta Laguna Lake, in the state of Quintana Roo on the Yucatán Peninsula, México (Figures 1 and 2). This 2-km-wide lake and the neighboring smaller lakes are surrounded by a forest fragment of approximately 2 km^2 of relatively undisturbed, semi-evergreen medium forest with trees up to 25 m in height and an abundance of ramon (*Brosimum alicastrum*). Local communities have protected this fragment from slash-and-burn agriculture because of their interest in the permanence of the spider monkey population as a source of income from tourist visits. Several other bodies of water, including small sinkholes and lagoons, are also surrounded by medium forest. The study fragment was 0.6 km^2 of medium forest. Most of the vegetation in the area consists of 30- to 50-year-old successional forest that local communities use for slash-and-burn agriculture. Cornfields and grasslands constitute a smaller proportion of the area (Figure 3).

The results discussed below constitute the basic information used for the design of a protected area, the *Otoch Ma'ax Yetel Kooh* sanctuary for spider monkeys ("House of the Monkey and the Puma," in Yucatec Maya; Figure 3). Currently, this sanctuary is managed by the local communities, Yucatec Mayas with a minimum level of formal education, who initiated a conservation effort that led to the declaration of the area as

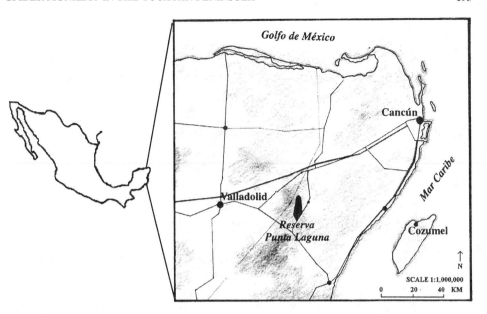

Figure 1. Location of the study area in the Yucatán Peninsula.

Figure 2. Aerial view of the Punta Laguna Lake, showing fragments of undisturbed forest around it and patches of recently disturbed forest (upper right corner).

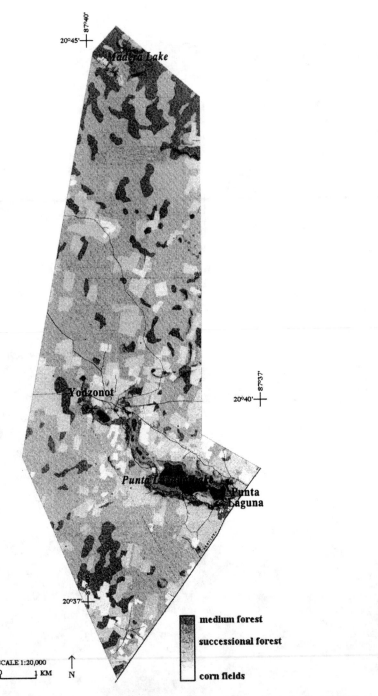

Figure 3. Vegetation map of the *Otoch Ma'ax Yetel Kooh* sanctuary for spider monkeys, showing the different vegetation types in the area, including the medium forest surrounding Punta Laguna and Madero lakes. The total area shown is 53.3 km^2 (see text for area of each vegetation type). Map by John Kelly, Pronatura Peninsula de Yucatán.

federally protected in June 2002. Conservation work by Pronatura Península de Yucatán and other non-government organizations continues.

3. POPULATION DENSITY

As an initial approach to evaluating the conservation status of the population, a line transect census was carried out during one year from September 1997 to September 1998. Two observers walked a total distance of 307 km in five transects 2.5 to 7.4 km in length. These transects were located in the areas around Punta Laguna Lake covering 67% of successional forest, 21% of medium forest, and 12% of other vegetation types, such as grassland and cornfields. All individual spider monkeys found during these walks were counted and their age and sex was recorded. Density was calculated by the transect-to-animal distance method outlined by the National Research Council (1981).

Spider monkeys were found at much higher densities in the medium forest than in any other vegetation type. A mean density of 89.5 individuals per km^2 ($\pm$ 2.4 confidence interval at 95%) was found for the medium forest, compared to a density of 6.3 individuals per km^2 ($\pm$ 5.6 confidence interval at 95%) for the 30- to 50-year-old successional forest. No monkeys were found in the cornfields and grasslands. Comparing between vegetation types is justified since the lengths of transects in each vegetation type and each vegetation type in the total area are proportional (National Research Council, 1981).

Assuming a total area of medium forest in the whole sanctuary of 7.7 km^2, the size of the population would be 648 individuals. This figure includes the fragment of medium forest around Punta Laguna Lake as well as the fragment around Madera Lake, 10 km north of Punta Laguna (see Figure 3). Assuming a total area of successional forest in the whole sanctuary of 29 km^2, an additional 183 individuals may be present. Clearly, because spider monkeys utilize these vegetation types in different ways, this is only an approximate figure of the monkeys found in the successional forest.

4. VEGETATION

Once we determined that spider monkeys could be found in both medium and successional forest, we carried out a vegetation survey of these two forest types. A total of 110 quadrats of 400 m^2 (for a total of 0.043 km^2) were set up at random on the trail system shown in Figure 4. Sixty-four of these quadrats were in medium forest, and 46 were in successional forest. Following the methodology of Higgins et al. (1994), four people counted all trees in the quadrats with a diameter at breast height larger than 10 cm. Tree species were identified by our local field assistants who, in the very rare cases when they did not know the species' Mayan name, would climb the tree and take a sample to show to the community's elders. Only one of the species' Mayan names was unknown. Mayan names were then looked up in the existing literature (Flores and Espejel, 1994; Sosa et al., 1985) or in the herbarium of the Autonomous University of Yucatán.

New quadrats were sampled until the relationship between the accumulated area and the estimated density of the most abundant species reached an asymptote. The distinction between the two forest types was made by the local assistants who defined the medium forest as "an old forest where nobody has ever made a field," where trees normally

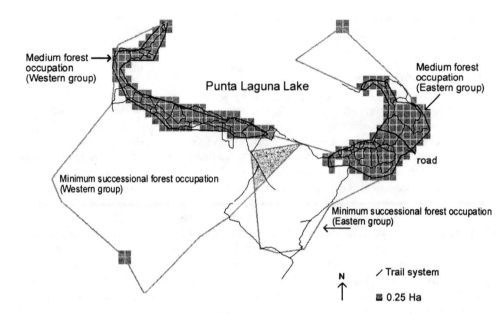

Figure 4. Home ranges of the two study groups. Squares represent the areas of medium forest around the lake.

exceeded 15 m in height, and successional forest as "a place where someone made a field less than 40 years ago" where trees normally did not exceed 15 m. This forest also had a high abundance of chakah (*Bursera simaruba*). In contrast, it contained very few trees of *Brosimum alicastrum*, a species that clearly dominated the medium forest.

Figure 5 shows the relative abundance of the most common tree species in both forest types. While *Brosimum alicastrum* is clearly abundant in the medium forest (comprising 43% of the total sample of trees at a density of 288 individuals per hectare), it occurs at much lower densities in the successional forest (less than 1 individual per hectare). Other important differences between both forest types include the relatively high abundance of *Bursera simaruba*, *Lysiloma bahamensis,* and *Piscidia piscipula* in the successional forest (at densities of 73 to 144 individuals per hectare) and a higher abundance of *Metopium brownei* in the successional forest compared to the medium forest (44 vs 2.7 individuals per hectare). *Ficus cotinifolia* and *F. ovalis* occur at similar low densities in both vegetation types (2 to 3 individuals per hectare).

5. DEMOGRAPHY AND SOCIAL ORGANIZATION

Two groups of spider monkeys living around the Punta Laguna Lake were studied continuously from January 1997 to December 2000. During this time, information on individually identified monkeys was obtained by means of an instantaneous scan sample procedure (Martin and Bateson, 1993). In each sample, taken every 20 minutes during all daylight hours, the location and composition of the subgroup was registered, as well as the activity of all independently moving individuals. If the majority was feeding, the tree species and part of the tree were also noted.

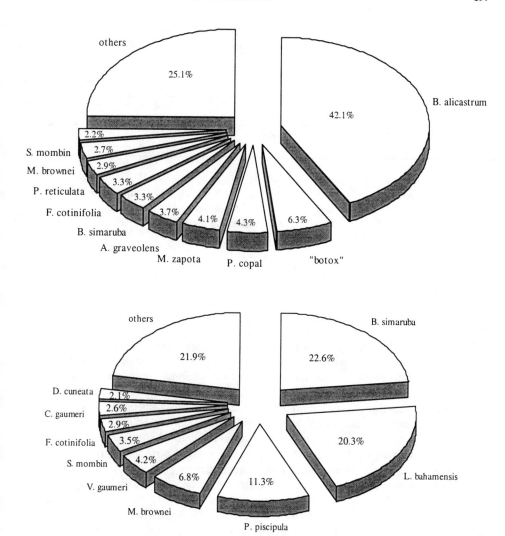

Figure 5. Relative abundance of tree species in the two vegetation types where spider monkeys were found in the study. (a) medium semi-evergreen forest; (b) successional forest. See Table 3 for full species names.

By December 1999, the two groups contained a total of 16 and 41 individuals, respectively. Table 1 shows the composition of the groups in terms of age and sex classes.

In the three-year period between January 1997 and December 1999, a total of 16 births and six disappearances were registered. Two of these disappearances were confirmed deaths and four were subadult females that possibly emigrated from their natal group. The mean per capita birth rate for four adult females that gave birth twice during that period is 0.48 births per year (± 0.12 SD).

Table 1. Age and sex class composition of the two study groups by December 1999.

Eastern group	Infants	Juveniles[a]	Adults	Total
Females	3	2	5	10
Males	1	2	3	6
Western group				
Females	6	7	15	28
Males	4	3	6	13

[a] Juveniles were independently moving monkeys that had not yet reached adult size.

Spider monkeys are typically organized in a fission-fusion society (Symington, 1990). Figure 6 shows the relative frequency with which subgroups of different sizes were observed in the four years of study. Median subgroup size for the smaller, Eastern group was 4.4 (range 1 to 13), while the median subgroup size for the larger, Western group was slightly smaller, 3.3 (range 1 to 15).

6. HOME RANGES

Ranging patterns were analyzed based on data collected from January 1997 to December 2000. The home range area of a group was defined as the area within a minimum polygon enclosing all locations where monkeys were observed (sighting sites). Medium forest occupation was defined as the area within a polygon enclosing all sighting sites in medium forest, while the minimum successional forest occupation referred to the area within a polygon enclosing all sighting sites in successional forest. Areas were measured by superimposing a grid of 0.25 ha over the respective polygons. Each individual did not use the entire group's home range, but rather restricted most of its activity to portions of it. The individual core area was defined as the area in which each individual spent 80% of the total observations. Core areas were calculated for all adult individuals observed for at least 50 hours (16 individuals).

Daily ranging lengths, defined as the distance that an individual traveled in a single day, were analyzed based on instantaneous scan samples taken every five minutes during all daylight hours. These samples were collected from September to November 1999. Daily paths were calculated from the length of a line joining all locations at which instantaneous scan samples were recorded.

By the end of the third year of study, the Eastern group used a total area of 0.95 km^2, from which 0.29 km^2 corresponds to medium forest occupation and 0.66 km^2 to minimum successional forest occupation. The Western group ranged in a total area of 1.66 km^2. Their medium forest occupation and minimum successional forest occupation was estimated at 0.29 and 1.37 km^2, respectively (Figure 4).

Individual core areas differ between males and females. In both groups males used larger areas than females (Eastern group: males mean = 0.15 km^2, females mean = 0.08 km^2; Western group: males mean = 0.21 km^2, females mean = 0.12 km^2; Table 2), but only in the Western group was a significant difference found (Mann-Whitney U = 24, P = 0.024).

The daily path length varied between 1,182 m to 3,872 m, with a mean of 2,302 m. In both groups, males traveled longer distances than females (Eastern group: males mean

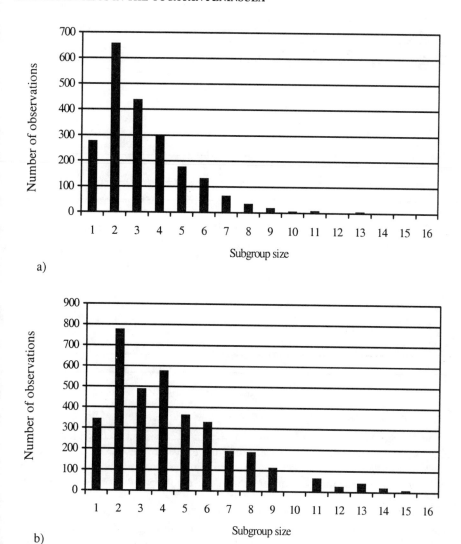

a)

b)

Figure 6. Frequency of observation of subgroups of different size. (a) Western group; (b) Eastern group. Total number of observations is 3,555 for (a) and 2,136 for (b).

= 3,327 m, females mean = 2,573; Western group: males mean = 2,226, females mean = 1,900).

7. DIET

To identify the most important food species in the spider monkey diet, we analyzed the instantaneous scan samples in which a subgroup was feeding and calculated a

Table 2. Core areas for all adult individuals observed for at least 50 hours.

Individuals	Sex[a]	Group[b]	Core area (ha)	N	Mean	SD
CE	F	E	8.75			
CH	F	E	8.25			
VE	F	E	8.25			
CL	F	E	8.25			
FL	F	E	8.50	5	8.40	0.22
PA	M	E	15.00			
DA	M	E	15.25	2	15.13	0.18
Total				7	10.32	3.29
BT	F	W	8.25			
MO	F	W	14.00			
OC	F	W	8.50			
IS	F	W	13.50			
LU	F	W	15.75			
R I	F	W	10.00	6	11.69	3.16
RO	M	W	23.75			
EN	M	W	21.00			
AJ	M	W	19.25	3	21.33	2.27
Total				9	14.89	5.56

[a] F = female; M = male.
[b] E = Eastern group; W = Western group.

proportion of feeding time on each item out of the total time spent feeding. Even though the spider monkeys in both groups under study traveled into the successional forest on many of the observation days, it was not possible to obtain quantitative data in this type of vegetation. Therefore, this section only includes data from the observations in the medium forest.

Table 3 shows the most common tree species found in medium and successional forest with their Mayan name used for identification in the field and the scientific name as found in the existing literature (Flores and Espejel, 1994; Sosa et al., 1985; Ogata et al., 2000) or in the herbarium at the Autonomous University of Yucatán. These species comprise 75% of the trees found in each forest type.

Figure 7 shows the proportion of time spent feeding on different items, relative to the total amount of time spent feeding. The figure only shows the top 10 items that in each study group accounted for 85% of the feeding sample. Fruit and leaves of *Brosimum alicastrum* comprise a large proportion of the diet in both groups, as well as the fruit of *Ficus cotinifolia, F. ovalis,* and *Manilkara zapota.* Other items that comprise a large proportion of the diet at least in one of the two groups include the fruit of *Sideroxylon capiri, Guazuma ulmifolia,* and *Metopium brownei* (Figure 8). The last two species are more abundant in successional than in medium forest.

The data summarized above do not consider the wide temporal variation shown in species consumption. To illustrate this, Figure 9 shows the proportion of the total monthly feeding time accounted for by the four main species, *Brosimum alicastrum,*

Table 3. List of tree species found in the forest types that spider monkeys use.

Common in:	Maya name	Scientific name
Medium forest	Ramon	*Brosimum alicastrum*
	Botox	Not identified
	Pom	*Protium copal*
	Ya'	*Manilkara zapota*
	K'ulim che'	*Astronium graveolens*
	Chakah	*Bursera simaruba*
	Kopo'	*Ficus cotinifolia*
	Zapotillo	*Pouteria reticulata*
	Chechem	*Metopium brownei*
	Ju'jub	*Spondias mombin*
	Caracolillo	*Sideroxylon capiri*
	Pixoy	*Guazuma ulmifolia*
	Pich	*Enterolobium cyclocarpum*
	Alamo	*Ficus ovalis*
Successional forest	Chakah	*Bursera simaruba*
	Tsalam	*Lysiloma bahamensis*
	Ha'bin	*Piscidia piscipula*
	Chechem	*Metopium brownei*
	Ya'axnik	*Vitex gaumeri*
	Ju'jub	*Spondias mombin*
	Kopo'	*Ficus cotinifolia*
	Kitam che'	*Caesalpinia gaumeri*
	Tsilil	*Diospyros cuneata*
	K'aan chunuub	*Thouinia paucidentata*
	Pixoy	*Guazuma ulmifolia*

Ficus cotinifolia, F. ovalis, and *Manilkara zapota*. There are large changes in the consumption of these species from month to month and in their proportion of the total diet. Most notably, the fruit of *Brosimum alicastrum* can account for up to half of the monthly diet (Figure 9). Similarly, the fruit of *Manilkara zapota* and *Ficus cotinifolia* sometimes accounted for more than a third of the monthly diet (Figure 10). Conversely, in some months, these species together do not constitute more than 10% of the monthly diet.

To evaluate the importance of these components of the spider monkey diet in terms of consumption and abundance, we calculated the selectivity index (following Wrangham et al., 1996) for each of the frequently consumed species shown in Figure 7. This index is an approximation to the degree with which spider monkeys prefer or select these species out of all the possible food species in their environment. We calculated this index by dividing the relative consumption of each species (shown in Figure 7) by the relative abundance of that species (shown in Figure 5). Therefore, if the selectivity index for a given species is lower than or equal to 1, monkeys were not necessarily selecting that species but were consuming it in the same proportion in which they encountered it. A high the selectivity index implies that they selected foods in a higher proportion than that in which they found them. Figure 11 shows the selectivity indices for the top species in each group's diet.

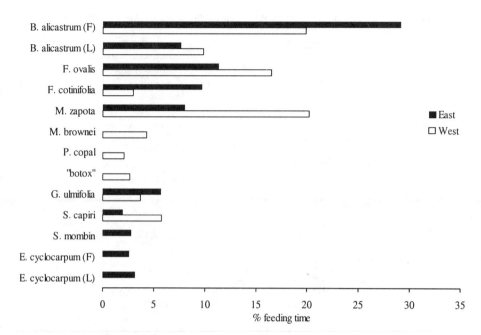

Figure 7. Proportion of scan samples in which subgroups were observed feeding on different items out of the total number of samples. See Table 3 for full species names. F = fruit and L = leaves.

Figure 8. Mother with subadult daughter and infant son feeding on *Guazuma ulmifolia* fruit within an area of successional forest. Photo by David M. Taub.

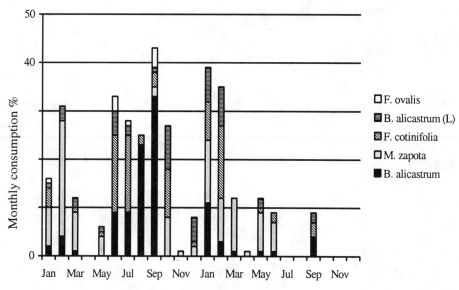

a)

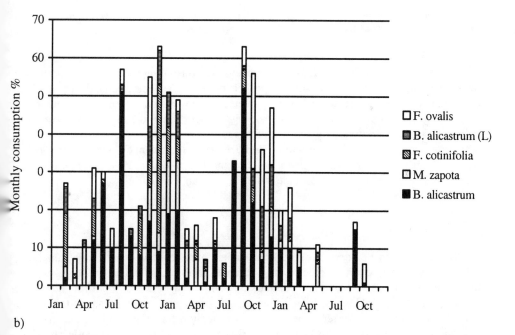

b)

Figure 9. Proportion of scan samples in each month in which subgroups were observed feeding on five main diet items. (a) Western group, data from January 1998 to October 1999. (b) Eastern group, data from February 1997 to October 1999. Months with no bars contain no observations. See Table 3 for full species names (L = leaves).

Figure 10. Female spider monkey with young infant from the Eastern group, feeding on *Manilkara zapota* fruit within a fragment of medium forest. Photo by Emilio Ramos-Fernández.

As can be seen in Figure 11, spider monkeys are consuming both the fruit and leaves of *Brosimum alicastrum* in a similar proportion to the species' abundance in their environment. Given its relatively high abundance, *Brosimum alicastrum* accounts for a significant proportion of their diet. Both species of *Ficus* seem to be preferred, given that their abundance is relatively low compared to the rate at which spider monkeys consume their fruit. Other species that seem to be preferred by spider monkeys include *Manilkara zapota*, *Sideroxilon capiri*, *Guazuma ulmifolia,* and *Enterolobium cyclocarpum*. From *E. cyclocarpum*, spider monkeys consume more leaves than fruit.

8. DISCUSSION

In the Yucatán Peninsula, four large protected areas contain most of the remaining original vegetation cover: medium, semi-evergreen forest (Challenger, 1998). These are the protected areas of Laguna de Términos (containing around 7,000 km² of this forest type), Calakmul (5,800 km²), Sian Ka'an (1,320 km²), and Yum Balam (1,000 km²). Outside of these areas, the remaining patches of medium forest are no larger than 1,000 km² and are in grave danger of being lost to the pressures of agriculture, cattle ranching, and logging. Even within the protected areas, management plans currently allow several productive activities in the areas of medium forest, including logging and low-impact agriculture. Managers of these protected areas and state and local authorities that are

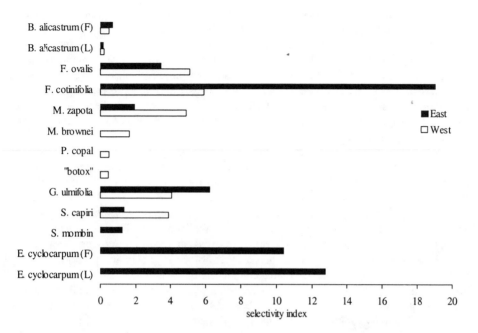

Figure 11. Selectivity indices for the 10 most consumed species in both study groups. An index lower than or equal to one shows little selectivity; the higher the index, the more preference or selectivity is shown for that item as compared to that species abundance. See Table 3 for full species names. F = fruit and L = leaves.

interested in conserving primate populations outside protected areas could benefit from an estimation of minimum habitat requirements.

This study reports a high density of spider monkeys in a relatively small fragment of medium forest. The results suggest that this fragment is large enough to support a viable population of spider monkeys. Given that the area has recently been declared as protected, its successful management will need to incorporate the results of this study.

Because of the difficulty in following spider monkey subgroups in the successional forest, it was not possible to obtain quantitative data on diet and movements when monkeys were in this forest type. It is intriguing, however, that two of the selected food species are more abundant in successional than in medium forest (*Metopium brownei* and *Guazuma ulmifolia*). In fact, during the few systematic observations achieved in successional forest, spider monkeys were feeding on these two species, as well as on isolated trees of *Manilkara zapota* (Ramos-Fernández, personal observations). Whether the reason they venture out into the successional forest is the higher abundance of these species cannot be said at this point. Further studies will allow the identification of the most important resources that spider monkeys are using in this forest type and whether they use this forest increasingly more as it regenerates.

This is the first detailed study of spider monkeys in the Yucatán Peninsula, a region with unusual soil and vegetation characteristics (Duch, 1988; Flores and Espejel, 1994). The northeastern region of the peninsula constitutes the northernmost range of the distribution of all Neotropical primates. Because of this, it is remarkable that we find several similarities in social organization, use of space, and diet between this population and others of *Ateles* studied in different parts of the Neotropics.

Spider monkeys in the two study groups were organized in the typical fission-fusion pattern that has been found in other studies (Symington, 1990; Chapman, 1990). Compared to these two studies, frequency and distribution of observed subgroups and size and sex ratios were similar to our study. Males traveled more per day and ranged over larger areas than females (Ayala-Orozco, 2001). As in other studies of *Ateles*, males showed higher association indices and affiliative behaviors between them than females did (Ramos-Fernández, 2001).

Home ranges for both groups in this study, including successional forest (0.95 and 1.66 km^2), are close to the lower limit of those reported in other studies of *Ateles* (Klein and Klein, 1977, for *A. belzebuth*: 2.6 and 3.9 km^2 for two groups; van Roosmalen, 1985, for *A. paniscus*: 2.2 km^2; Symington, 1987, for *A. chamek*: 2.3 and 1.5 km^2 for two groups; Chapman, 1990, for *A. geoffroyi*: 1.7 km^2; McDaniel, 1994, for *A. geoffroyi*: 1.4 km^2; Nunes, 1995, for *A. belzebuth*: 3.2 km^2; and Wallace, 1998, for *A. chamek*: 2.9 km^2). The proportion of fruit in the diet is similar across studies, ranging from 83% to 90% (van Roosmalen and Klein, 1987). The fact that home range size and proportion of fruit in the diet do not vary across study sites that clearly differ in tree species composition and diversity suggests that the flexibility in grouping patterns in this species allows them to successfully exploit a wide variety of habitats. *A. geoffroyi*, in particular, appears to be less restrictive in its habitat requirements (van Roosmalen and Klein, 1987).

The density of spider monkeys in medium forest found in this study is higher than reported in other studies (Freese, 1976; White, 1986; McDaniel, 1994). This could be attributed to the fact that, even though their habitat spans both medium and successional forest, monkeys are spending more time in the former than in the latter, thereby increasing the probability of being found there during a transect census.

The source of the primate disturbance in the study area may have been a fire that occurred more than 30 years ago, after Hurricane Beulah. According to the local peoples' reports, this fire damaged most of the medium forest existing in the area, except for the forest surrounding bodies of water. It is possible that after being drastically reduced in size by this disturbance, as the successional forest regenerates, it provides increasingly more fruit for spider monkeys. Longitudinal observations on identified individuals allowed us to estimate a per capita birth rate of approximately one offspring every two years, as well as an increase in the combined size of the two groups under study from 47 to 57 individuals in three years. These might indicate that the population overall is in a stage of expansion. Strier (1991) found a similar situation in a group of muriquis (*Brachyteles arachnoides*) in Minas Gerais, Brazil, which after an unknown disturbance grew in size from 22 to 43 individuals in eight years of study.

Water could be a limiting factor in this latitude for the distribution of spider monkeys. During the dry season (December–April), subgroups of spider monkeys were seen on several occasions drinking water from the lake (Ramos-Fernández, personal observations). One by one, individuals in a subgroup would come down and drink, while the rest remained quiet and wary, presumably from being so close to the ground. It is likely that their normal sources of water in tree holes and bromeliads up in the canopy dry up during this season and monkeys are forced to come down. If this is the case, then populations of this species would only survive around permanent bodies of water such as lakes and sinkholes.

It is helpful to compare this situation to what occurs in other areas of the tropics where the forest is converted to cattle ranches (e.g., Lovejoy et al., 1986 in Manaus, Brazil; Chiarello and de Melo, 2001 in the Atlantic forest in Brazil; and Estrada and Coates-Estrada, 1996 in Los Tuxtlas, México). In these areas, small fragments of old-growth forest remain isolated within large extensions of pasture that cannot be utilized by spider monkeys. There, spider monkey populations go extinct in fragments of less than 1 km^2. In the Yucatán Peninsula, cattle ranching is not profitable because grass cannot grow easily on limestone (Challenger, 1998). Therefore, human disturbances to forested areas consist mainly of mosaics of different aged forest produced by slash-and-burn agriculture. Local people that practice this form of agriculture set aside areas of forest in different stages of secondary succession so the soil can regain its nutrient content. Spider monkeys can travel through these areas of medium forest, thereby increasing the effective population size. In our study area, spider monkeys in Punta Laguna and in Madero Lake, 10 km north, might actually be part of the same population. This would provide an excellent opportunity to implement conservation measures for the permanence of the spider monkey population in the context of continuous human use. An analysis of the genetic composition of the population could provide information on the history of its reduction and overall health.

As of now, the proposed management plan for the *Otoch Ma'ax Yetel Kooh* Sanctuary for spider monkeys (53.3 km^2 total area) contains the total protection of two nucleus zones: one around Punta Laguna Lake and adjacent lakes (8.5 km^2) and another around Madero Lake (2.8 km^2). In the rest of the area, a percentage of relatively old successional forest should be kept at all times, which should suffice for maintaining a single population in these two areas. Assuming that no deleterious genetic effects of past reductions in population size have already accumulated, this implies that practicing slash-and-burn agriculture at low rates with short regenerating cycles would be compatible with the permanence of the local population of spider monkeys. This result adds to those of others (e.g., Smith et al., 1999 in the Peruvian Amazon), which suggest that the successional forest used in slash-and-burn agriculture has a high value for the conservation of forest biodiversity.

The main incentive for local communities to protect the spider monkey population is ecotourism. Only 100 km from Punta Laguna is Cancún, the largest tourist destination in México, and the associated tourist corridor, stretching 200 km to the south near the Caribbean coast. Tourism agencies are eager to find new activities for visitors, and Punta Laguna is currently receiving an average of 10,000 people per year (Ramos-Fernández, unpublished data). This has already had an impact on the living conditions of several families, whose members work as tour guides and sell crafts to the visitors. We hope that the results of this study will provide useful information for managing activities that help conserve spider monkeys while improving the living conditions of the people.

9. SUMMARY

A study of the population size, habitat characteristics, ranging, and foraging behavior of spider monkeys in the northeastern Yucatán Peninsula was carried out to evaluate its conservation status. A high density of spider monkeys (89 individuals per km^2) was found in the largest fragment of undisturbed forest (0.6 km^2). The home range of two groups that were studied continuously for three years includes all 0.6 km^2 of medium forest around Punta Laguna Lake, but also includes large areas of 30- to 50-year-old successional forest, where monkeys fed and traveled in about one-third of the observation days. In both of these forest types, spider monkeys were found feeding on the fruit and leaves of around 50 species of trees, although 10 species comprised three-quarters of their total diet. *Brosimum alicastrum*, a species that clearly dominates the medium forest in this region, fulfilled a major part of the monkeys' diet, especially during the months of July–September. Species that were selected, given their consumption by the monkeys relative to their abundance, included two species of *Ficus* and *Sideroxylon capiri, Manilkara zapota, Metopium brownei,* and *Guazuma ulmifolia.* High per capita birth rates and a net increase in the size of the groups under study suggest that the population is currently expanding. An area of 53.3 km^2, containing two fragments of undisturbed forest within a large area of successional forest in different stages used for slash-and-burn agriculture, has been declared as a reserve by federal authorities. These results, together with the incentive from ecotourism that the local communities have for protecting the spider monkey population, provide an excellent opportunity to ensure the permanence of the population of spider monkeys in the context of continuous human use of their habitat.

10. ACKNOWLEDGMENTS

Many thanks to Eulogio and Macedonio Canul, from the communitiy of Punta Laguna, for their assistance in the field, without which this study would not have been feasible. David M. Taub and Laura G. Vick initiated the study of the spider monkeys in Punta Laguna. John Kelly provided the maps of the study area. The personnel at Pronatura Peninsula de Yucatán, especially their president, Joann Andrews, helped at all times during field work. Daniel Janzen, Dorothy Cheney, W. John Smith, Salvador Mandujano, and Alfredo Cuaron provided comments on earlier versions of this manuscript. The study was financed by a grant from the National Commission on Biodiversity Knowledge and Use (CONABIO, project M120). G. R. F. was supported by a Ph.D. grant from the National Council for Science and Technology (CONACYT).

11. REFERENCES

Ayala-Orozco, B., 2001, Uso de hábitat por dos grupos de monos araña (*Ateles geoffroyi yucatanensis*) en Punta Laguna, Yucatán, México, Honors Thesis, UNAM, México.

Challenger, A., 1998, *Utilización y Conservación de los Ecosistemas Terrestres de México: Pasado, Presente y Futuro*, CONABIO, UNAM, Sierra Madre, México.

Chapman, C. A., 1990, Association patterns of spider monkeys: The influence of ecology and sex on social organization, *Behav. Ecol. Sociobiol.* **26**:409–414.

Chiarello, A. G., and de Melo, F. R., 2001, Primate population densities and sizes in Atlantic forest remnants of northern Espirito Santo, Brazil, *Int. J. Primatol.* **22**:379–396.

Dirzo, R., and Miranda, A., 1990, Contemporary Neotropical defaunation and forest structure, function, and diversity, a sequel to John Terborgh, *Cons. Bio.* **4**:444–447.

Duch, J., 1988, *La Conformación Territorial del Estado de Yucatán: Los Componentes del Medio Físico*, Universidad Autonoma de Chapingo, Centro Regional de la Península de Yucatán.

Estrada, A., and Coates-Estrada, R., 1996, Tropical rainforest fragmentation and wild populations of primates at Los Tuxtlas, Mexico, *Int. J. Primatol.* **17**:759–783.

Flores, J., and Espejel, I., 1994, Tipos de vegetación en la península de Yucatán, *Etnoflora Yucatanense* **3**.

Freese, C., 1976, Censusing *Alouatta palliatta, Ateles geoffroyi*, and *Cebus capucinus* in the Costa Rican dry forest, in: *Neotropical Primates: Field Studies and Conservation*, R. W. Thorington, and P. G. Heltne, eds. National Academy of Sciences, Washington D.C.

Higgins, K. F., Oldemeyer, J. L., Jenkins, K. J., Clambey, G. K., and Harlow, R. F., 1994, Vegetation sampling and measurement, in: *Research and Management Techniques for Wildlife and Habitats*, T. A. Bookhout, ed., The Wildlife Society, Bethesda, MA.

Klein, L. L., and Klein, D. J., 1977, Feeding behaviour of the Colombian spider monkey, in: *Primate Ecology* T. H. Clutton-Brock, ed., Academic Press, London.

López-Portillo, J., Keyes, M. R., González, A., Cabrera, E. C., and Sánchez, O., 1990, Los incendios de Quintana Roo: Catástrofe ecologica o evento periódico? *Ciencia y Desarrollo* 16:43–57.

Lovejoy, T. E., Bierregaard, R. O., Jr., Rylands, A. B., Malcolm, J. R., Quintela, C. E., Harper, L. H., Brown, K. S., Jr., Powell, A. H., Powell, G. V. N., Schubart, H. O. R., and Hays, M. B., 1986, Edge and other effects of isolation on Amazon forest fragments, in: *Conservation Biology: The Science of Scarcity and Diversity*, M. E. Soulé, ed., Sinauer Associates, Sunderland, MA.

McDaniel, P., 1994, The social behavior and ecology of the black-handed spider monkey (*Ateles geoffroyi*), Ph.D. dissertation, University of Saint Louis.

Martin, P., and Bateson, P. B., 1993, *Measuring Behaviour: An Introductory Guide, 2nd edition*, Cambridge University Press.

Meffe, G. K., and Carroll, C. R., 1994, *Principles of Conservation Biology*, Sinauer Associates, Sunderland, MA.

National Institute of Statistics, Geography, and Information Science, 2000, Results of the population census of 2000, http://www.inegi.gob.mx.

National Research Council, 1981, *Techniques for the Study of Primate Population Ecology*, U.S. Committee on Nonhuman Primates, Subcommittee on conservation of natural populations.

Nunes, A., 1995, Foraging and ranging patterns in white-bellied spider monkeys, *Folia Primatol.* 65:85–99.

Ogata, N., Gómez-Pompa, A., Aguilar-Meléndez, A., Castro-Cortés, R., and Plummer, O. E., 2000, Arboles tropicales del área maya, Sistema de Información Taxonómica, University of California, Riverside.

Ramos-Fernández, G., 2001, Patterns of association, feeding competition, and vocal communication in spider monkeys, *Ateles geoffroyi*, Ph.D. dissertation, University of Pennsylvania.

Robinson, J., and Janson, C. H., 1987, Capuchins, squirrel monkeys, and Atelines: Socioecological convergence with Old World primates, in *Primate Societies*, B. B. Smuts, D. L. Cheney, R. M. Seyfarth, R. W. Wrangham, and T. T. Struhsaker, eds., University of Chicago Press, Chicago.

Rzedowski, J., 1978, *Vegetación de México*, Limusa, México.

Sosa, V., Flores, J. S., Rico-Gray, V., Lira, R., and Ortiz, J. J., 1985, Lista floristica y sinonomia Maya, Fasciculo 1 de Etnoflora Yucatanense, INIREB.

Strier, K. B., 1991, Demography and conservation of an endangered primate, *Brachyteles arachnoids*, *Cons. Bio.* 5:214–218.

Symington, M. M., 1987, Ecological and social correlates of party size in the black spider monkey, *Ateles paniscus chamek*, Ph.D. thesis, Princeton University.

Symington, M. M., 1990, Fission-fusion social organization in *Ateles* and *Pan*, *Int. J. Primatol.* 11:47–61.

van Roosmalen, M. G. M., 1985, Habitat preferences, diet, feeding strategy, and social organization of the black spider monkey (*Ateles p. paniscus*) in Surinam, *Acta Amaz.* 15:12–38.

van Roosmalen, M. G. M., and Klein, L. L., 1987, The spider monkeys, Genus *Ateles*, in: *Ecology and Behavior of Neotropical Primates*, R. A. Mittermeier and A. B. Rylands, eds., World Wildlife Federation, Washington, D.C.

Wallace, R. B., 1998, The behavioural ecology of black spider monkeys in northeastern Bolivia, Ph.D. thesis, University of Liverpool.

White, F., 1986, Census and preliminary observations on the ecology of the black-faced black spider monkey (*Ateles paniscus chamek*) in Manu National Park, Peru, *Am. J. Primatol.* 11:125–132.

Wrangham, R. W., Chapman, C. A., Clark-Arcadi, A. P., and Isabirye-Basuta, G., 1996, Social ecology of Kanyawara chimpanzees: Understanding the costs of great ape groups, in: *Great Ape Societies*, W. C. McGrew, L. F. Marchant, and T. Nishida, eds., Cambridge University Press.

CHANGES IN FOREST COMPOSITION AND POTENTIAL FEEDING TREE AVAILABILITY ON A SMALL LAND-BRIDGE ISLAND IN LAGO GURI, VENEZUELA

Marilyn A. Norconk and Brian W. Grafton[*]

1. INTRODUCTION

Fragmentation of tropical forests affects the viability of primate populations worldwide. A recent assessment of habitat loss in Latin America has estimated that 9.7% of extant forest was lost between 1980 and 1995 (Chapman and Peres, 2001). Forest fragmentation has many causes (e.g., human encroachment for settlements, agricultural practices, logging, and flooding, Alvarez et al., 1986; Cosson et al., 1999; Chapman and Peres, 2001), but these causes share a common phenomenon. Disruption of contiguous forest creates disjunct patches of forest separated by different types of land use, vegetation, or water, in the case of flooding (Alvarez et al., 1986; Saunders et al., 1991; Terborgh et al., 1997; Cosson et al., 1999). Forest remnants are both smaller, when compared to contiguous forest, and isolated from other forest patches (Saunders et al., 1991). The nature of the surrounding modified habitats—or matrix—imposes a variety of novel (and often detrimental) effects on the plant and animal species still residing within a given fragment (Cosson et al., 1999). In the case of land-bridge islands, water as a barrier has a powerful effect, both in terms of limiting dispersal of resident species and providing an unusable habitat for those species (Turner, 1996; Terborgh et al., 1997; Cosson et al., 1999).

Whatever the barrier, fragmentation causes an 'ecological disruption' because of a combination of edge effects, and fragment size, shape, and location (Saunders et al., 1991). Interactions between the forest interior and the adjacent habitat along the edge include abiotic effects (changes in such environmental conditions as air temperature, humidity, light intensity), direct biological effects (changes in species distribution and abundance due to the modifications of environmental conditions), and indirect biological

[*] Marilyn A. Norconk and Brian W. Grafton at the Department of Anthropology and Biological Anthropology Program, School of Biomedical Sciences, Kent State University, Kent, Ohio, 44242, USA. Correspondence to M. A. Norconk (email: mnorconk@kent.edu).

Primates in Fragments: Ecology and Conservation
Edited by L. K. Marsh, Kluwer Academic/Plenum Publishers, 2003

effects (higher-order changes resulting from alterations in species interactions) (Murcia, 1995; Turton and Freiburger, 1997; Sizer and Tanner, 1999).

Studies of forest fragments have resulted in the discovery of a variety of problems associated with microhabitat changes. For example, increases in windthrow and ambient temperature, as well as reduced humidity near fragment boundaries, result in a sharp increase in tree mortality, tree damage, and the formation of canopy gaps. Such changes can bring about an increase in plant species adapted to gap and disturbed habitats, resulting in an associated decrease in old-growth canopy trees (Lovejoy et al., 1986; Kapos, 1989; Leigh et al., 1993; Kapos et al., 1997; Laurance et al., 1998a; Mesquita et al., 1999). Turner (1996) found that alterations in microclimate may limit the usefulness of the forest to residents, further reducing the size of the useable area and causing both an increase in forest plant mortality rates and a reduction in their recruitment near the edge. The fact that a species is present in a forest fragment immediately after its isolation does not ensure that it will continue to persist; successful reproduction and recruitment are required (Saunders et al., 1991). A lack of required pollinators and seed dispersers for some plant species can seriously affect the future reproduction of those species and can have far-reaching effects for the future integrity of a given forest fragment (Howe, 1984; Powell and Powell, 1987; Pannel, 1989; Turner, 1996).

2. PRIMATES IN FOREST FRAGMENTS

How do primates respond to habitat fragmentation? Two major characteristics of primates—home range size and the degree of frugivory in the diet of a species— influence the ability of different species to live in forest fragments (Tutin and White, 1999; Onderdonk and Chapman, 2000; Estrada and Coates-Estrada, 1996; Lovejoy et al., 1986). The interaction between fragment size, home range size, and diet type is complex; the limited area resulting from fragmentation reduces the diversity of plant species and the number of food plants available to consumers (Tutin and White, 1999). Fruit as a resource is highly heterogeneous in terms of its spatial and temporal distribution, and larger frugivorous primates usually require large tracts of forest to provide enough resources to support viable populations (Johns and Skorupa, 1987; Turner, 1996; Onderdonk and Chapman, 2000).

Fragmentation of contiguous habitat can 1) exclude a primate species from residence in a given forest fragment, effectively causing localized extinctions (Lovejoy et al., 1986), 2) alter the group sizes and population densities of species still able to inhabit the fragmented landscape (Milton, 1982; Estrada and Coates-Estrada, 1988, 1996; Terborgh et al., 1997; Tutin and White, 1999; Tutin, 1999), 3) alter the dietary strategies of species able to reside in fragments (Johns and Skorupa, 1987; Tutin, 1999), and 4) affect gene flow among resident populations (Estrada and Coates-Estrada, 1996; Pope, 1996; Cosson et al., 1999; Gravitol et al., 2001).

The ability of primates to deal with the challenges of living in a fragmented habitat is obviously variable and clear patterns that can characterize this response have yet to be found (Onderdonk and Chapman, 2000). The purpose of this paper is to examine changes in the vegetation of a 15 ha island in Lago Guri, Venezuela, using data collected in 1988-89 (Parolin, 1992, 1993; Peetz and Parolin, unpublished) and 2001-02. Since changes in the diversity of plant species and in the numbers of individuals of plant species used as food sources can negatively affect the viability of primates in fragments, we will also examine the frequency of white-faced saki (*Pithecia pithecia*) feeding trees occurring in the sample plots.

White-faced sakis (*Pithecia* spp.) are the smallest (c. 1.4 to 3.1 kg: Hershkovitz, 1987a) members of the Pitheciini that includes bearded sakis (*Chiropotes* spp.: 2.5 to 3.2 kg: Hershkovitz, 1985) and uacaris (*Cacajao* spp.: 2.7 to 3.4 kg: Hershkovitz, 1987b). Hershkovitz (1987a) recognized two groups of *Pithecia* sakis, the smaller-bodied, strongly sexually dichromatic Guianan group (*P. pithecia*, 2 ssp.:Figure 1a and b) and the larger-bodied and more subtly dichromatic Amazonian group (*P. monachus, P. irrorata, P. aequatorialis*, and *P. albicans*). The larger bearded sakis/uacaris are allopatric, but *Pithecia* spp. overlap much of the range of *Cacajao* (primarily west of the Rio Branco in Brazil and tributaries of the left bank of the Rio Amazonas) and *Chiropotes* (primarily east of the Rio Branco and Rio Madeira, both north and south of the Rio Amazonas) (Hershkovitz, 1985). All of the saki/uacaris share dental adaptations for opening hard fruit–robust, laterally flaring canines, procumbant incisors, crenulated molars of low cusp relief, and robust jaws (Kinzey, 1992). While they ingest a variety of food types including leaves, pith, flowers, and insects, they are primarily seed predators with a preference for large, multiseeded fruit of the Chrysobalanaceae, Lecythidaceae, Sapotaceae, and Bignoniaceae plant families (van Roosmalen et al., 1988; Norconk, 1996; Stevenson, 2001).

3. METHODS

3.1. Study Site

This report is part of a primate behavioral ecology study conducted in Lago Guri, Bolívar State, Venezuela, from 1987 to 2002. The Embalse de Guri or Lago Guri (Figure 2) is the catchment basin (3,919 km²: CVG-EDELCA, 1997) for the Raúl Leoni hydroelectric plant constructed in the company town of Guri (7° 45' N, 56° 10' W). The site is 90 km upriver of the confluence of the Caroní and Orinoco Rivers at Puerto Ordaz.

Hydroelectric plant construction and maintenance is under the auspices of the company EDELCA (Electrificación del Caroní), a subsidiary of CVG (Corporación Venezolana de Guayana). Planning and construction of the Raúl Leoni/Guri dam began in 1968 (Roo, 1987). Dam building and flooding of the sparsely populated, mildly hilly terrain between Guri and Puerto Ordaz continues with ground-breaking for the fourth dam (Tacoma) in 1999. This region of Venezuela was considered to be potentially valuable for the production of hydroelectric energy as early as 1912 (Roo, 1987). Vegetation on this northernmost rim of the Guianan Shield consists of tropical savannah and gallery forest, but perhaps most important, the Rio Caroní is a black-water river. The low level of suspended organic material in black-water rivers minimizes turbine damage and equipment maintenance required. Of the four hydroelectric plants on the Caroní River, the Guri plant is the most productive, providing energy to major cities in Venezuela, as well as to portions of Columbia and Brazil bordering southern Venezuela.

Two rivers, the Caroní and the Caura, are the largest in eastern Venezuela. Both are black-water rivers draining the precambrian deposits of the Guianan Shield. The Caroní is by far the better known from the perspective of tourism and accessibility. This region is well known for its mineral deposits and stunning geological formations, the characteristic table-top mountains or tepuis.

Figure 1. Adult female white-faced saki (left) and adult male white-faced saki (right). Photos by Ken Glander.

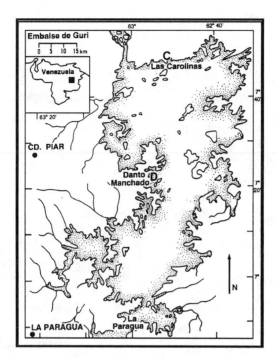

Figure 2. The "Embalse de Guri," or Caroní River basin was formed by inundation behind the Guri dam at the site of the Raúl Leoni hydroelectric plant. The location of the dam and hydroelectric plant are indicated by the narrow portion of the lake in the extreme northwest. The study island (Isla Redonda) was located in the portion of the lake referred to as "Las Carolinas." Only a few of the more then 200 islands are represented in this figure. Map by Gerardo Aymard.

There are four primate species in this region of Venezuela. We surveyed islands for the presence of howler monkeys (*Alouatta seniculus*) in 1988 and found individuals on almost every island that supported a patch of forest (Kinzey et al., 1989). We characterized the capuchins (*Cebus nigrivittatus*) as "widely distributed" in the lake from north to south, but less abundant than howlers.

The two sakis species, *Pithecia pithecia* and *Chiropotes satanas*, are not sympatric in Venezuela. Before flooding, saki distributions were separated by the Caroní River with white-faced sakis on the right bank and bearded sakis on the left bank. We have limited and anecdotal evidence of primate "migrations" between islands. In the early years of the study when rising and falling water levels were still fresh in the minds of our Venezuelan colleagues, we were repeatedly told stories of howlers drowning during high water periods, still hanging by their tails as water levels receded. Recently, dispersal of howlers between islands or between island and mainland, has occurred and was attributed to low water levels exposing land masses that were previously separated by water (Terborgh, personal communication). Nevertheless, it is not a common occurrence for howlers and we have no evidence that sakis have moved among water-bound islands.

Research for this study was conducted on a 12.8-hectare island (medium-sized, following Terborgh et al., 1997) called Isla Redonda (Figure 3). The island is a remnant hilltop that has been isolated from the nearby mainland since at least 1981 (CVG-

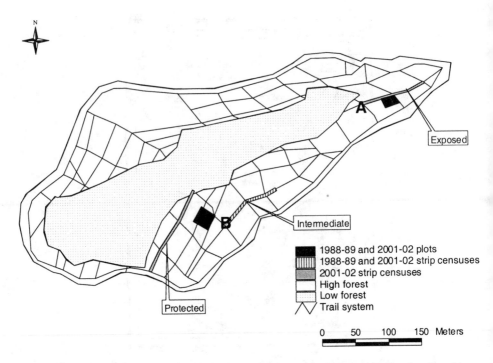

Figure 3. Trail and vegetation map of Isla Redonda. Three plots were measured in 1988-89 and replicated in 2001-02, labeled "protected," "intermediate," and "exposed." The protected and exposed plots measured 25 m^2. The intermediate plot measured 100 m long and 4 m wide.

EDELCA map, 1981) after the first stage of flooding (1963-1978) (CVG-EDELCA, 1997). The second stage of flooding began in 1978 and was completed in 1986. Parolin collected her first vegetation samples in 1988, approximately 10 years post isolation. Ghost forest (sensu Terborgh et al., 1997) surrounds the present day island, but does not connect the island to other islands or to the mainland even when the lake level is extremely low (personal observation, 2001).

Vegetation in the northern portion of the Caroní basin was characterized as low to medium deciduous non-flooded forests (5 to 25 m in height) and shrub savannas by Huber (1986). This description fits well with more detailed work on the island by Aymard et al. (1997). They described the vegetation of Isla Redonda as a dry tropical forest with trees of medium height (maximum height 23 m), consisting of predominantly small stems, growing on rocky quartzite and clayey ferruginous soils. Parolin (1992) characterized distinct regions of high, low, or mixed forest on the island, and Brush (2000) used the terms high or low forest with variable degrees of rockiness, canopy cover, and undergrowth density. Parolin's 1988-89 samples found the Leguminosae to be the best represented family with three genera of Caesalpinioideae and one genus of Mimosoideae in the top 10 most abundant genera (Parolin, 1992). Annual rainfall averages 1,100 mm (CVG-EDELCA, unpublished).

3.2. Sampling Methods

Parolin measured and identified every stem ≥5.0 cm DBH (diameter at breast height) in 16, 25-m squared plots placed randomly around the island (Parolin, 1992). She and A. Peetz (unpublished) also conducted a strip sample 100 m long and 4 m wide on the south side of the island. In 2001-02, we replicated the measurements she took on two plots on the south side of the island and the strip sample (see Figure 3). In addition, we assigned each stem a leaf phenology score of 0 to 4, where 0 = no leaves, stem presumed dead, to 4 = fully leafed. In addition, we added two new strip samples in 2001 (100 m long and 4 m wide) in the vicinity of the two established plots (see Figure 3). All stems ≥0.5 cm maximum diameter (trees, lianas, saplings, and understory shrubs) were measured in contiguous 2 m^2 blocks along three trails for a total of 250 blocks. Stems smaller than a maximum diameter of 0.5 cm were counted and identified to species when possible. We refer to these data below as evidence for recruitment.

We characterized the plots and strip samples as "protected," "exposed," or "intermediate" using both windthrow data and distance from the windward (southeastern) edge of the island. The exposed plot was within 50 m of the south edge and approximately 140 m from the southeastern edge of the island. The protected plot was within 100 m of the edge and approximately one-third up the slope to the crest of the island and 250 m from the southeastern end of the island. The intermediate strip sample was approximately 25 m and only about 4 m in elevation from the southern edge and also approximately 250 m from the southeastern end of the island (see Figure 3).

Windthrow and temperature were estimated in 2001 using a Kestrel® 2000 Pocket Thermo Wind Meter. Average and maximum wind speed (in knots) were collected at two points on the island, observation point A—on the northwestern corner of the exposed plot—and observation point B—the western end of the strip sample, approximately the mid-point on the southeast side of the island (see Figure 3). The wind meter was hand held 2 m above the ground and pointed toward the direction of maximum wind speed as indicated on the read-out. Maximum wind speed (measured as 3-second gusts) as well as average wind speeds and temperature readings were recorded 60 seconds after the meter was activated. Paired samples from points A and B were taken approximately 15 minutes apart.

Data on primate feeding trees were collected in 1991-92 during a 12-month study of the group of nine white-faced sakis (*Pithecia pithecia*) (see methods in Norconk, 1996). Feeding trees totaling 3,570 stems were identified, labeled, and measured (DBH, height, canopy breadth) and ranked by total feeding minutes. We use feeding tree rank as a reflection of dietary preference of the sakis, with the caveat that feeding tree preferences do fluctuate with annual variation in fruit availability (Norconk, unpublished).

3.3. Analysis

We estimated within year (1988-89 and 2001-02) structural variation among vegetation plots using the non-parametric analysis of variance, Kruskal-Wallis test (SPSS, v. 7.0). The DBH distributions were truncated at 5 cm as the minimum DBH thus the distributions were not normal. DBH differences in each plot were compared between time samples using Mann-Whitney tests for independent samples. Feeding tree species were ranked by total feeding minutes over a 12-month study in 1991-1992 and we used those ranks to correlate feeding trees with stem abundance in the sample plots using Pearson's R correlation. We set α values at 0.05 and all tests were two-tailed.

4. RESULTS

4.1. Wind Data

Northwesterly winds affect predominantly the southeastern portion of the island. Windthrow speed averaged 2.3 knots (N = 15) with gusts to 6.6 knots at the southeast end of the island (observation point A). Wind speed was lower at observation point B—1.8 knots (N = 15) with gusts to 3.9 knots. Paired samples demonstrated that the windier the conditions, the larger the difference between the two observation points. When maximum wind speed was ≥4.0 knots, the difference between the two points averaged 2.4 knots compared with a difference of 0.5 knots when maximum wind readings were <4.0 knots.

4.2. Structural Changes in the Forest

The 1988-89 baseline sample consisted of two plots sampled by Parolin (1992) and one strip sample measured by Parolin and Peetz in 1989 (unpublished). The Kruskal-Wallis test detected significant differences in the average DBH of protected, intermediate, and exposed samples of 1988-89 (χ^2 = 9.8, df = 2, ρ = 0.007). No significant differences were found in the 2001-02 samples (χ^2 = 2.53, df = 2, ρ = 0.28). The data, as plotted in Figure 4, do not suggest a significant difference either within or between samples, despite the statistically significant result in the 1988-89 data. Nevertheless, we did find significant differences in the between year samples for the exposed and the intermediate plots, but not the protected plot (exposed: Z = -2.53, p = 0.011; intermediate: Z = - 3.14, p = .002; protected:, Z = -1.5, ns). Furthermore, all three measures of central tendency increased in the exposed area in 2001-02 suggesting that losses were more frequent in smaller stems (means compared in Figure 5).

To obtain a better sense of the selective nature of tree loss, we examined the size and frequency of dead trees in the 2001-02 samples (Figure 6). Dead trees were often still standing so that we could measure DBH, but we also measured the maximum diameter of dead trees that had fallen to the ground. Almost three-quarters (73.5%) of the trees in the small category (5.0 to 9.9 cm DBH) were leafless and apparently dead in the exposed plot, compared with a third of the total trees in the intermediate plot and a fifth in the protected plot. Trees ≥10 cm DBH represented about 40% of the sample in the exposed plot, of which a third (14) were dead. This compared with 37% of trees in the protected plot measuring ≥ 10 cm DBH with 7% tree loss and 6.4% in the intermediate strip sample. Dead trees now represent a considerable portion of the exposed plot, with the largest trees surviving better than the smallest ones.

The finding that small trees are suffering high losses in the exposed plot was supported by the additional data we collected from the strip samples of 2001. For this sample, we added the category of stems ≤0.5 cm (greatest diameter). We tallied these very small stems for each 2 m² block of the sample strips. Small stems averaged 1.31 and 3.46 stems/2 m² on the exposed and intermediate samples, respectively, and 11.71/2 m² in the protected area.

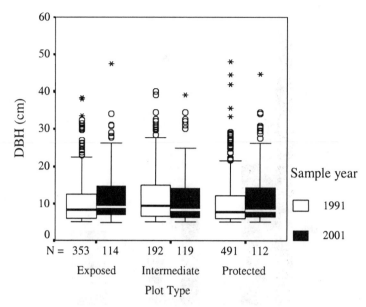

Figure 4. Box plots comparing DBH values measured in the three plots between the two years. The "1991" sample refers to the 1988-89 collection and the "2001" sample refers to the 2001-02 collection. The box plots indicate the median as a heavy line within the boxes, themselves representing the 25th to 75th percentiles of the distributions. Outliers are identified as circles above the largest value (horizontal line) that is not an outlier. Extremes are represented by asterisks.

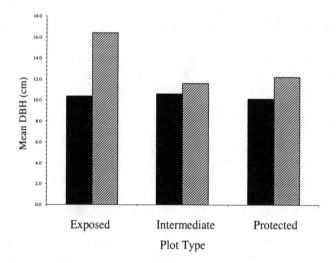

Figure 5. Mean DBH values compared between years. Solid bars are the 1988-89 sample and hatched bars represent data from 2001-02. The average DBH increased from 10.4 ± 6.4 to 16.8 ± 9.8 in the exposed plot; from 10.1 ± 6.2 to 12.2 ± 8.4 in the intermediate plot; and from 9.5 ± 7.2 to 11.2 ± 7.4 in the protected plot.

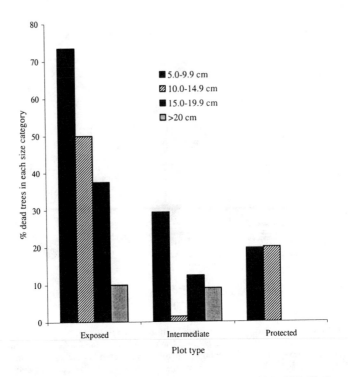

Figure 6. Percentage of dead trees by size category for each of the 2001-02 plots.

4.3. Plant Species Diversity and Impact on Primate Feeding

We found a total of 31 species in the three 2001-02 plots. Of these, 51.6% (16 species) were found only in one plot, 25.8% (8 species) were found in two plots, and 22.6% (7 species) were found in all three plots. Since these three plots are within a radius of 150 m, more than half the species found only in one plot could be considered to be rare. Of the more common species (i.e., 16 species that were found in at least two plots), the species representation in the intermediate plot was positively correlated with both exposed (r = 0.868, p < 0.001) and protected plots (r = 0.573, p < 0.025), but not between exposed and protected plots (r = 0.368, ns). The intermediate plot was not only spatially intermediate between the exposed and protected plots, but was also more diverse in species number than either of the other two plots (Figure 7).

We used leaf phenology as an indicator of stem health. None of the species examined here were deciduous, so scores of three or four suggested to us that the tree was doing well and scores of 1 or 0 indicated that the tree was declining or dead, respectively. The assessment of tree death was somewhat subjective, however. For example, *Erythroxylum steyermarkii* trees had zero scores for all plots. We have not seen *E. steyermarkii* trees with leaves, flowers, or fruit on northern islands in the lake since 1998, and we suspected a population-wide crash for this species (Figure 8). As of 2001, however, new shoots were beginning to regenerate from the bases of many "dead" trees. *E. steyermarkii* aside, there appear to be more stems of low phenology values in the exposed plot than the other

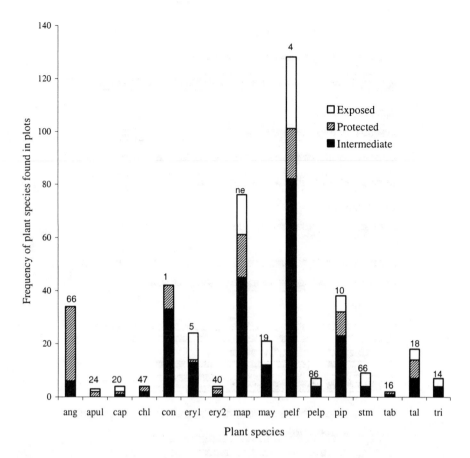

Figure 7. Frequency of feeding tree species in 2001-02 plots and their relative rank (number above the column) in terms of feeding preference by the white-faced sakis. Species are ang = *Angostura trifoliata*, apul = *Apuleia molaris*, cap = *Capparis muco*, chl = *Chrysophylum lucentifolium*, con = *Connarus venezuelanus*, ery1 = *Erythroxylum steyermarkii*, ery2 = *Erythroxylum* sp. 2, map = *Maprounea guianensis*, may = *Maytenus guianensis*, pelf = *Peltogyne floribunda*; pelp = *Peltogyne paniculata*, pip = *Piptadenia leucoxylon*, stm = *Strychnos mitscherlichii*, tab = *Tabebuia serratifolia*, tal = *Talisia retusa*, tri = *Trichilia lepidota*.

two plots. *Maytenus guianensis, Strychnos mitcherlichii, Maprounea guianensis, Piptadenia leucoxylon*, and *Peltogyne floribunda* appeared to be struggling in the exposed plots compared to either the intermediate or protected plots. A few species were doing as well or better in the exposed plot, however. *Apuleia molaris, Erythroxylum* sp. 2, *Peltogyne paniculata, Talisia retusa*, and *Trichilia lepidota* may be well adapted to dryer conditions and eventually result in a changed flora in that area of the forest.

Comparing the intermediate and protected plots, we noted a number of phenology "reversals" with some species having higher phenology scores in some plots and lower scores in others. Individual variation in this small sample is difficult to account for, but there were not marked differences in scores for same tree species in these two plots (Figure 8).

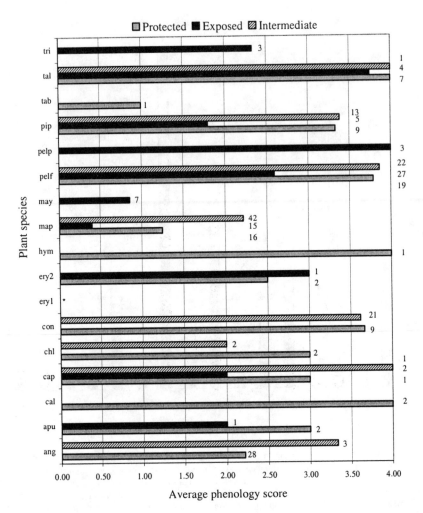

Figure 8. Average phenology scores for stems measured in each of the three sample plots: protected, exposed and intermediate. Scores ranged from 0 (no leaves) to 4 (fully leafed). Plant codes are the same as those used in Figure 7. Sample sizes are given at the end of each bar. * In the case of "eryl" *(Erythroxylum steyermarkii)*, there were 10 stems in the exposed plot, one stem in the protected plot, and no stems in the intermediate plot. All eryl stems had a phenology score of zero.

5. DISCUSSION

5.1. Edge Effects on Forest Composition on Isla Redonda

We compared the change in forest structure (using DBH) and species composition in three plots on the windward side of Isla Redonda. The original sample was collected by Parolin (1992) and was compared with samples collected in 2001-02. The average DBH increased in all three forest plots, significantly in the exposed and intermediate plots between the 1988-89 and 2001-02 samples. We interpreted the increase in the mean value to represent a reduction in small stems. That was confirmed when we calculated the stem losses for each plot, with about three-quarters of the small stems (≤5.0 cm DBH) dead (and often still standing) in the exposed plot. Finally, we assessed recruitment in the number of very small (0.5 cm maximum diameter) stems and found the exposed plot to have several magnitudes fewer stems than the protected or intermediate plots. Only at this level of size comparison did we see that the intermediate forest resembled the exposed more than the protected forest.

We anticipated that we would find heavy losses in larger stems as other studies have done and this may have been the case if our plots were closer to the edge. Laurance et al. (1998a, b) and Ferriera and Laurance (1997) found increases in tree mortality within 100 m of fragment edges caused by increased wind turbulence and microclimatic changes. On Isla Redonda, the effect of wave action is apparent at the immediate edge on the south and southeast side of the island, but in recent years engineers at the hydroelectric plant have manipulated the level of the lake by regulating the flow of water through the dam so that we have not seen high lake levels for about four years. While direct erosion has affected the edge of the island, wind and windthrow appear to have had a more pervasive effect on vegetation survival and establishment.

Our interpretation that abiotic factors (wind and dessication) were primary in the habitat changes we observed on this medium-sized island is at odds with a recent study by Terborgh et al. (2001). They compared tree plots on windward and leeward sides of both large and small islands in the southern portion of Lago Guri and found land mass to be the significant variable contributing to stem loss (i.e., all small islands had lower tree densities than large islands). No significant differences in stem density were found between plots on the windward or leeward sides of islands. Our findings may be different since we compared relatively closely spaced plots, all on the windward side of Isla Redonda. The observed differences may be due to differences in the scale of the two studies. All small islands are at higher risk than large islands, but the process of stem loss may be accelerated in areas closest to the edge.

Our findings are consistent with other studies (e.g., Benitez-Malvido, 1998) that indicate that seedling density of mature-phase, shade-tolerant species declines in forest fragments due to reductions in seedling establishment rate and/or increases in seedling mortality. The finding that the 27 individuals of *Peltogyne floribunda* in the exposed plot may be declining in health (based on phenology scores) is of concern since this species represents the predominant cover in that plot. The relative success of other species in this plot may be dependent on shade provided by *P. floribunda*. Alternatively, the loss of some species may provide opportunities for others. We were struck by the number of very small stems of *Connarus venezuelanus* in the exposed plot, relative to the adult stems in that area. If these small stems are successful, it would be good news for the saki monkeys. *C. venezuelanus* has an asynchronous fruit cycle so that fruit was found in fruit traps in 13 out of 15 months of a previous feeding study, in the phenology sample 12 out of 15 months, and in the feeding records every month (Norconk, 1996).

Important plant species for the saki/uakaris (Pitheciins) include Sapotaceae, Chyrsobalanaceae, Lecythidaceae, Euphorbiaceae and Leguminosae (Soini, 1986; Ayres, 1989; Kinzey and Norconk, 1990; Peres, 1993; Boubli, 1999, Stevenson, 2001). Despite the small size of the island, the sakis use a very diverse set of resources. For example, the top five feeding species represented 56.65% of the diet in Lago Guri compared well with 49.85% of the white-faced saki diet in French Guiana (Vié et al., 2001). Nevertheless, some species have emerged as particularly important to Lago Guri's island-bound *Pithecia* sakis. *Connarus venezuelanus* (mentioned above) may be increasing its distribution on the island to include the "exposed" area. In contrast, *Piptadenia leucoxylon* and *Erythroxylum steyermarkii* have both suffered losses in the exposed area. The 2001 phenology scores of large *Piptadenia* in the exposed area averaged 2.67 (N = 16) on a scale of 0 to 4 compared with an average score of 3.35 (N = 20) in the intermediate sample. Selective leaf loss could indicate that individual trees are declining, perhaps from wind and desiccation. *Erythroxylum steyermarkii* suffered heavy losses in what appeared to have been a population-wide collapse and apparent death of all mature *E. steyermarkii* trees on Isla Redonda and other islands in the northern portion of the lake by 1996 (personal observation). However, we are encouraged by viable new growth on a number of apparently dead stems in the year 2000 and fruit appearing on a few stems in 2001. Both *E. steyermarkii* and *P. leucoxylon* were important feeding tree species in the 1991-92 sample, ranking 5th and 10th, respectively.

6. SUMMARY

We compared the forest structure and plant species composition in three areas on the windward side of a medium-sized island in Lago Guri, Venezuela. The early sample (two 25 m^2 plots and a 100-m strip sample) was collected in 1988-89 and demonstrated significant structural variation among the three samples. In 2001-02, we remeasured the trees in the three original samples and added three 100-m-long strip samples. The principle findings were two-fold: we noted a drastic decline in the number of surviving small stems in the range of 5 cm DBH, as well as heavy losses for trees of all sizes in the exposed plot, and a low level of recruitment of very small stems (from 5.0 cm to 0.5 cm maximum diameter). The stem losses appear to be related to proximity to the edge and relatively high levels of windthrow compared with the more protected samples. The picture of plant species diversity, particularly as it relates to feeding trees used by white-faced sakis, is more complex and appears to be changing more slowly than the structural changes. However, if recruitment continues to proceed as indicated by the survivorship of small stems, the exposed area may eventually include a high proportion of *Connarus venezuelanus*, the highest ranking feeding tree for white-faced sakis even though this may be accompanied by an overall loss in dietary diversity as other important resources are lost.

While edge effects will continue to degrade the forest on the south side of the island, demographic changes occurring in the saki group may outpace changes occurring in the vegetation. Approximately 20 years after isolation, it is clear that the foremost pressure on the monkeys is not dietary. The changes in forest composition are proceeding relatively slowly compared to the failure of infants to survive since 1995 (Norconk, unpublished). Infant deaths in combination with a few adult deaths and lack of immigration have seriously impaired replacement. Thus, we feel that the primates are at much higher risk of local extinction due to stochastic events and lack of gene flow than they are of starvation due to changing forest composition.

7. ACKNOWLEDGMENTS

We are very grateful to Pia Parolin and Angela Peetz for sharing unpublished research during their 1988-89 tree study. TSU Luis Balbás and his staff in Estudios Básicos-Guri provided essential logistical support during the entire project. We also thank EDELCA-Guri for permission to work on islands in the lake. Venezuelan botanist, Gerardo Aymard, from the Herbarium of the Universidad Nationál Experimentál de los Llanos Occidentales Ezequiel Zamora (UNELLEZ), visited the site several times, collected samples, and trained us to identify plant species on the island. We are grateful to Jean Engle and Steve Ruhl who assisted with plot measurement in the 2002 sample. The late Warren G. Kinzey was the prime mover in early stages of the project. We dedicate this paper to Warren and to Jesus (Quique) Pacheco, both who died far too young. Research was supported by NSF (BNS 87-19800 and SBR 98-07516) and CUNY Research Foundation. We thank Laura Marsh for asking us to think about island dynamics and saki feeding in Lago Guri.

8. REFERENCES

Alvarez, E., Balbás, L., Massa, I., and Pacheco, J., 1986, Aspectos ecológicos del Embalse Guri, *Interciencia* **11**:325–333.

Aymard, G., Norconk, M., and Kinzey, W., 1997, Composición florística de comunidades vegetales en islas en el embalse de Guri, Rio Caroní, Estado Bolívar, Venezuela, *BioLlania Edición Esp.* **6**:195–233.

Ayres, M. M., 1989, Comparative feeding ecology of the uakari and bearded saki, *Cacajao* and *Chiropotes, J. Hum. Evol.* **18**:697–716.

Benitez-Malvido, J., 1998, Impact of forest fragmentation on seedling abundance in a tropical rain forest, *Cons. Bio.* **12**:380–389.

Boubli, J. P., 1999, Feeding ecology of black-headed uacaris (*Cacajao melanocephalus melanocephalus*) in Pico da Neblina National Park, Brazil, *Int. J. Primatol.* **20**:719–749.

Brush, J. A., 2000, Sleeping ecology of white-faced saki monkeys in Lago Guri, Venezuela, Master's thesis, Kent State Univerisity, Kent, OH.

Chapman, C. A., and Peres, C. A., 2001, Primate conservation in the new millennium: The role of scientists, *Evol. Anthrop.* **10**:16–33.

Cosson, J. F., Ringuet, S., Claessens, O., de Massary, J. C., Dalecky, A., Villiers, J. F., Granjon, L., and Pons, J. M., 1999, Ecological changes in recent land-bridge islands in French Guiana, with emphasis on vertebrate communities, *Biol. Conser.* **91**:213–222.

CVG -EDELCA (Corporación Venezolana de Guayana, Electrificación del Caroní), 1997, Guri, "Raul Leoni" Hydroelectric Central, *CVG-EDELCA*, Caracas.

Estrada, A., and Coates-Estrada, R., 1988, Tropical rain forest conversion and perspectives in the conservation of wild primates (*Alouatta* and *Ateles*) in Mexico, *Am. J. Primatol.* **14**:315–327.

Estrada, A., and Coates-Estrada, R., 1996, Tropical rain forest fragmentation and wild populations of primates at Los Tuxtlas, Mexico, *Int. J. Primatol.* **17**:759–783.

Ferriera, L. V., and Laurance, W. F., 1997, Effects of forest fragmentation on mortality and damage of selected trees in Central America, *Cons. Bio.* **11**:797–801.

Gravitol, A. D., Ballou, J. D., and Fleischer, R. C., 2001, Microsatellite variation within and among recently fragmented populations of the golden lion tamarin (*Leontopithecus rosalia*), *Conser. Gen.* **2**:1–9.

Hershkovitz, P., 1985, A preliminary taxonomic review of the South American bearded saki monkeys genus *Chiropotes* (Cebidae, Platyrrhini), with the description of a new subspecies, *Fieldiana, Zoology n.s.* **27**:1–46.

Hershkovitz, P., 1987a, The taxonomy of South American sakis, genus *Pithecia* (Cebidae, Platyrrhini): A preliminary report and critical review with the description of a new species and a new subspecies, *Am. J. Primatol.* **12**:387–468.

Hershkovitz, P., 1987b, Uacaris, New World monkeys of the genus *Cacajao* (Cebidae, Platyrrhini): A preliminary taxonomic review with the description of a new subspecies, *Am. J. Primatol.* **12**:1–53.

Howe, H. F., 1984, Implications of seed dispersal by animals for tropical reserve management, *Biol. Cons.* **30**:261–281.

Huber, O., 1986, La vegetación de la cuenca del Rio Caroní, *Interciencia* **11**:301–310.

Johns, A. D., and Skorupa, J. P., 1987, Responses of rain-forest primates to habitat disturbance: A review, *Int. J. Primatol.* **8**:157–191.

Kapos, V., 1989, Effects of isolation on the water status of forest patches in the Brazilian Amazon, *J. Trop. Ecol.* **5**:173–185.

Kapos, V., Wandelli, E., Camargo, J. L., and Ganade, G., 1997, Edge-related changes in environment and plant responses due to forest fragmentation in Central Amazonia, in: *Tropical Forest Remnants: Ecology, Management, and Conservation of Fragmented Communities*, W. F. Laurance and R. O. Bierregarrd, eds., University of Chicago Press, Chicago, pp. 33–44.

Kinzey, W., 1992, Dietary and dental adaptations in the Pitheciinae, *Am. J. Phys. Anth.* **88**:499–514.

Kinzey, W. G., Norconk, M. A., and Alvarez-Cordero, E., 1989, Primate survey of eastern Bolívar, Venezuela, *Primate Cons.* **9**:66–70.

Kinzey, W., and Norconk, M., 1990, Hardness as a basis of fruit choice in two sympatric primates, *Am. J. Phys. Anth.* **81**:5–15.

Kinzey, W. G., and Norconk, M. A., 1992, Physical and chemical properties of fruit and seeds eaten by *Pithecia* and *Chiropotes* in Surinam and Venezuela, *Int. J. Phys. Anth.* **14**:207–227.

Laurance, W. F., Ferreira, L. V., Rankin-DeMerona, J. M., and Laurance, S. G., 1998a, Rain forest fragmentation and the dynamics of Amazonian tree communities, *Ecol.* **79**:2,032–2,040.

Laurance, W. F., Ferreira, L. V., Rankin-DeMerona, J. M., Laurance, S. G., Hutchings, R. W., and Lovejoy, T. E., 1998b, Effects of forest fragmentation on recruitment patterns in Amazonian tree communities, *Cons. Bio.* **12**:460–464.

Leigh, E. G., Jr., Wright, S. J., Herre, E. A., and Putz, F. E., 1993, The decline of tree diversity on newly isolated tropical islands: A test of a null hypothesis and some implications, *Evol. Ecol.* **7**:76–102.

Lovejoy, T. E., Bierregaard, R. O., Jr., Rylands, A. B., Malcolm, J. T., Quintela, C. E., Harper, L. H., Brown, K. S., Jr., Powell, A. H., Powell, G. V. N., Schubart, H. O. R., and Hays, M. B., 1986, Edge and other effects of isolation on Amazon forest fragments, in: *Conservation Biology: The Science of Scarcity and Diversity*, M. E. Soulé, ed., Sinauer, Assoc. Inc, Sunderland, MA, pp. 257–285.

Mesquita, R. C. G., Delamonica, P., and Laurance, W. F., 1999, Effect of surrounding vegetation on edge-related tree mortality in Amazonian forest fragments, *Biol. Cons.* **91**:129–134.

Milton, K., 1982, Dietary quality and demographic regulation in a howler monkey population, in: *Ecology of a Tropical Forest: Seasonal Rhythms and Long-term Changes*, E. G. Leigh, Jr., A. S. Rand, and D. M. Windsor, eds., Smithsonian Institution Press, Washington, D.C., pp. 273–289.

Murcia, C., 1995, Edge effects in fragmented forests: Implications for conservation, *TREE* **10**:58–62.

Norconk, M. A., 1996, Seasonal variation in the diets of white-faced and bearded sakis (*Pithecia pithecia* and *Chiropotes satanas*) in Guri Lake, Venezuela, in: *Adaptive Radiations of Neotropical Primates*, M. A. Norconk, A. L. Rosenberger, and P. A. Garber, eds., Plenum Press, New York.

Onderdonk, D. A., and Chapman, C. A., 2000, Coping with forest fragmentation: The primates of Kibale National Park, Uganda, *Int. J. of Primatol.* **21**:587–611.

Pannell, C. M., 1989, The role of animals in natural regeneration and the management of equatorial rain forests for conservation and timber production, *Com. For. Rev.* **68**:309–313.

Parolin, P., 1992, Characterization and classification of the vegetation in an island of Lake Guri, Venezuela, Masters Thesis, University of Bielefeld, Germany.

Parolin, P., 1993, Forest inventory in an island of Lake Guri, Venezuela, in: *Animal-Plant Interactions in Tropical Environments*, W. Barthlott, C. M. Naumann, K. Schmidt-Loske, and K. L. Schuchmann, eds., Zoologisches Forschungsinstitut und Museum Alexander Koenig, Bonn, pp.139–147.

Peres, C. A., 1993, Notes on the ecology of buffy saki monkeys (*Pithecia albicans*, Gray 1860): A canopy seed-predator, *Am. J. Primatol.* **31**:129–140.

Pope, T. R., 1996, Socioecology, population fragmentation, and patterns of genetic loss in endangered primates, in: *Conservation Genetics: Case Histories from Nature*, J. Avise and J. Hamrick, eds., Kluwer Academic Publishers, Norwell, MA, pp. 119–159.

Powell, A. H., and Powell, G. V. N., 1987, Population dynamics of male euglossine bees in Amazonian forest fragments, *Biotropica* **19**:176–179.

Roo, H., 1987, El desarrollo hidroeléctrico en Venezuela, *EDELCA*, CVG, Año XII, Segunda Epoca, Edición Especial, pp. 5–10.

Saunders, D. A., Hobbs, R. J., and Margules, C. R., 1991, Biological consequences of ecosystem fragmentation: A review, *Cons. Bio.* **5**:18–32.

Sizer, N., and Tanner, E. V. J., 1999, Responses of woody plant seedlings to edge formation in a lowland tropical rainforest, Amazonia, *Biol. Cons.* **91**:135–142.

Soini, P., 1986, A synecological study of a primate community in the Pacaya-Samiria National Reserve, Peru, *Primate Cons.* **7**:63–71.

Stevenson, P. R., 2001, The relationship between fruit production and primate abundance in Neotropical communities, *Biol. J. Linn. Soc.* **72**:161–178.

Terborgh, J., Lopez, L., Tello, J., Yu, D., and Bruni, A. R., 1997, Transitory states in relaxing ecosystems of land bridge islands, in: *Tropical Forest Remnants: Ecology, Management, and Conservation of Fragmented Communities*, W. F. Laurance and R. O. Bierregarrd, eds., University of Chicago Press, Chicago, pp. 256–274.

Terborgh, J., Lopez, L., Nuñez V. P., Rao, M., Shahabuddin, G., Orihuela, G., Riveros, M., Ascanio, R., Adler, G. H., Lambert, T. D., and Balbas, L., 2001, Ecological meltdown in predator-free forest fragments, *Science* **294**:1,923–1,926.

Turner, I. M., 1996, Species loss in fragments of tropical rain forest: A review of the evidence, *J. App. Ecol.* **33**:200–209.

Turton, S. M., and Freiburger, H. J., 1997, Edge and aspect effects on the microclimate of a small tropical forest remnant on the Atherton Tableland, Northern Australia, in: *Tropical Forest Remnants: Ecology, Management, and Conservation of Fragmented Communities*, W. F. Laurance and R. O. Bierregarrd, eds., University of Chicago Press, Chicago, pp. 45–54.

Tutin, C. E. G., 1999, Fragmented living: Behavioural ecology of primates in a forest fragment in the Lopé Reserve, Gabon, *Primates* **40**:249–265.

Tutin, C. E. G., and White, L., 1999, The recent evolutionary past of primate communities: Likely environmental impacts during the past three millenia, in: *Primate Communities*, J. G. Fleagle, C. Janson, and K. E. Reed, eds., Cambridge University Press, Cambridge, MA, pp. 220–236.

van Roosmalen, M. G. M., Mittermeier, R. A., and Fleagle, J. G., 1988, Diet of the northern bearded saki (*Chiropotes satanas chiropotes*): A neotropical seed predator, *Am. J. Primat.* **14**:11–35.

Vié, J. C., Richard-Hansen, C., and Fournier-Chambrillon, C., 2001, Abundance, use of space, and activity patterns of white-faced sakis (*Pithecia pithecia*) in French Guiana, *Am. J. Primat.* **55**:203–222.

FORAGING STRATEGY CHANGES IN AN *ALOUATTA PALLIATA MEXICANA* TROOP RELEASED ON AN ISLAND

Ernesto Rodríguez-Luna, Laura E. Domínguez-Domínguez, Jorge E. Morales-Mávil, and Manuel Martínez-Morales [*]

1. INTRODUCTION

The distribution of primates has been diminishing during the last decades because of fragmentation and disappearance of habitat (Estrada and Coates-Estrada, 1994). As a consequence, the geographic distribution is no longer continuous and is comprised of areas relatively inaccessible to people. Fragmented habitat is where some primates manifest a great behavioral elasticity as an adaptive response to changes in their environment (Chivers, 1991; Garcia-Orduña, 1996; Rodríguez-Luna, 2000). The renewed interest for the study of primates under different environmental conditions is attributed to a change in perspectives of theory and methodology for research. Chivers (1986) suggested that, as a complement to the study of the survival of primates in disturbed habitat, work must be developed in different-sized fragments of tropical forest.

Howler (genera *Alouatta*) populations are found from tropical deciduous forests to tropical rain forest. Howlers occupy the largest variety of habitats, even very disturbed ones. Eisenberg (1979) considers howlers a pioneer species, being able to adapt to diverse habitats. This adaptability allows them to survive, even when the habitats are degraded. This adaptability allows them to prosper while habitat regenerates.

Howler habitat is not homogeneous. We must consider that, for any species, the distribution of food resources offered by the ecosystem reflects the cycle of seasonal change, in time as well as in space. These changes can affect the behavior of the primates, for example, in their patterns of daily activities or foraging strategies, whether in a continuous or fragmented habitat. Some references have mentioned that howlers

[*] Ernesto Rodríguez-Luna, Parque de Flora y Fauna Silvestre Tropical, Instituto de Neuroetología, Universidad Veracruzana, Jalapa, Veracruz, México, 91000. Laura E. Domínguez-Domínguez, Parque de Flora y Fauna Silvestre Tropical, Instituto de Neuroetología, Universidad Veracruzana, Jalapa, Veracruz, México, 91000. Jorge E. Morales-Mávil, Parque de Flora y Fauna Silvestre Tropical, Instituto de Neuroetología, Universidad Veracruzana, Jalapa, Veracruz, México, 91000. Manuel Martínez-Morales, Maestría en Inteligencia Artificial, Universidad Veracruzana, Jalapa, Veracruz, México, 91000. Correspondence to E. Rodríguez-Luna (email: errodriguez@uv.mx).

Primates in Fragments: Ecology and Conservation
Edited by L. K. Marsh, Kluwer Academic/Plenum Publishers, 2003

need at least 30 to 60 ha of tropical forest for their survival (Estrada, 1984; Estrada and Coates-Estrada, 1993; 1994). However, we observed Mexican howlers (*Alouatta palliata mexicana*) in fragments within Los Tuxtlas for 15 years with just four or five individuals living in 8 to 20 ha maximum (Silva-López et al., 1988; García-Orduña, 1996; Rodríguez-Luna, 2000). This could indicate that the most important reasons for the survival of the howlers in the fragmented landscape of Los Tuxtlas are the availability of resources and the degree of human pressure more than the size of the fragments.

   This study proposes an explanation for the process of environmental adjustment of a population of Mexican howler monkeys on an island within a lake. We test the hypothesis that if a group of howler monkeys increases in population within a fragment, then their foraging strategy (search and ingestion of food) will change at both individual and group levels because of greater demand for food sources. As a result, their pattern of daily activities and their feeding preferences will also change.

1.1. Daily Activity Pattern

   Howler monkeys are strictly diurnal. Their activities tend to take place in a standard sequence, alternating between resting episodes, locomotion, and feeding (Serio-Silva, 1992; Carrera-Sánchez, 1993). In a behavior study by Mendel (1976), these three activities take up 95% of the total focussed observations of the study. In almost all studies about the daily pattern these activities are in about the same time proportions: between 64% and 80% resting, about 10.5% to 24% feeding, and 9.5% to 12% for locomotion (Chivers, 1969; Richard, 1970; Milton, 1980; Serio-Silva, 1992). Nocturnal activity is not present, although, in some places, howling at night has been reported (Horwich and Lyon, 1988). At the beginning of the day, howlers can be found in the same tree they were last observed in at dusk (Altmann, 1959).

1.2. Feeding Preferences

   Howlers are completely herbivorous. Once classified as folivorous, today they are classified as folivorous-frugivorous or primarily folivorous (Chivers and Hladik, 1980). Some studies have shown that howlers eat considerable amounts of fruits and leaves; typically, they eat more younger, greener leaves than older ones, and the matured fruits are preferred, even though the green ones are still eaten. However, howlers have been observed to be up to 95% frugivorous (Altmann, 1959). Table 1 shows the percentage of time dedicated in diverse habitats to the consumption of different parts of plants.

Table 1. Percentage of time *A. palliata* spend consuming various plant parts. Other consists of bark, stems, petioles, fungus, or insects.

Author	Area	Leaves	Fruits	Flowers	Other
Milton (1980)					
Old Forest	31.7 ha	53.4%	36.9%	9.3%	-----
Lutz Ravine	31.1 ha	43.6%	46.7%	9.6%	-----
Estrada (1984)	60 ha	49.3%	49.9%	0.2%	0.6%
Serio-Silva (1992)	8.3 ha	21.2%	58.6%	0.2%	20.0%
Juan et al. (1999)	3.6 ha	57.0%	38.0%	1.0%	5.0%
Juan et al. (2000)	35 ha	44.0%	44.1%	10.7%	1.2%

Some fruits in the diet are high in fiber, such as figs (*Ficus*), which represent an important portion of the howler's food (Estrada and Coates-Estrada, 1986; Serio-Silva, 1996; Serio-Silva et al., 1997). Other fruits are of a consistency similar to that of leaves, and they require a similar digestive process (Crockett and Eisenberg, 1986). Carnivorous consumption is insignificant and includes insects (ants, termites, and wasps) eaten mostly by accident (Crockett and Eisenberg, 1986).

Estrada (1984) reported that in continuous habitat of Los Tuxtlas, howlers used 120 trees of 27 different species during one year; eight of these species added up to 78% of the trees used during 87% of the feeding time (Table 1). He pointed out that howlers consumed similar proportions of leaves and fruits. Seasonal changes influenced the choice of these foods and the different parts of the plants. Estrada and Coates-Estrada (1984) mention that 80% of the fruits eaten were mature and from a few species, especially species from Moraceae and Lauraceae.

1.3. Home Range

Napier and Napier (1985) recognized that the use of the space available to the primates depends on the resources and whether the space is a defined territory or an undefined home environment. Of these resources, food is considered the most important one, as the primates depend on whatever seasonal vegetation is available.

Field data do not support the conclusion that mature howlers are territorial (Crockett and Eisenberg, 1986). However, Milton (1980) and Sakulik (1982) argue that howlers cannot be considered territorial because their ranges considerably overlap. However, home range can vary for individuals and groups of howlers from year to year. Freese (1976) reported that howlers moved between locations depending on wet or dry climate. Napier and Napier (1985) report that howler populations establish ranges based on experience about fruiting cycles and travel corridors.

1.4. Feeding Habits and Foraging Strategies

The composition of the primate's environment is such that the monkeys will usually find their food distributed in "spots" (Oates, 1986), meaning in areas of high concentration of food, separated by areas of low concentration. There is also a great variation in the size of these "spots" and the separation between them in time and space, as well as the density and availability of the preferred parts of the plants.

The spatial distribution of food is a strong influence on pauses in the locomotion of the primates. However, it is not easy to discern the relationship between the distribution of the food and the locomotion of the monkeys. In general, it is assumed that the monkeys exploit food resources that are highly dispersed and unpredictable. Compared to species feeding in areas with abundant resources, wide dispersal, and predictable availability, they travel more every day and they cover a larger area per year. The distance between the feeding spots and the costs of the trip might be influencing the size of the groups of primates. If the size of feeding spots remains constant, the number of spots needed in a determined amount of time will increase with the size of the troop, but the energetic restrictions will limit the amount of spots that they can actually visit (Oates, 1986).

Because these feeding spots have a very complex distribution in time and space, the primates have to make decisions about when and where to eat. This also involves making decisions about the direction, distance, and speed of the locomotion. Oates (1986) makes reference to some studies that give evidence that the primates can make movements highly directed towards particular feeding spots and other resources, which strongly suggests a path to the fulfillment of a goal and the existence of mental maps.

2. METHODOLOGY

2.1. Group Study

This study was conducted during two observation periods of a troop of howler monkeys introduced to Agaltepec Island (Catemaco Lake, Veracruz, Mexico) (Figure 1). The first group was captured in the wild in 1987 at Mirador Pilapa. A second group was captured on a ranch next to the San Juan Evangelista River (both places to the south of Veracruz, Mexico) (Rodríguez-Luna et al., 1993). During the captures, samples were taken from the monkeys for clinical analysis (hematic biometry, bacteriological, and coproparasitoscopic analysis), as well as weight and morphology. In general, the health of the monkeys was considered poor (Canales-Espinosa, 1992).

They remained in captivity for 17 months, and in October of 1988 five howler monkeys were released on Agaltepec Island (four adult females and one adult male), however, the male died a few days later. After one week, one infant was born on the island. In April 1989, another five individuals (four adult females and one adult male) were released. Before they were released, there was a new evaluation of their health; most of them had shown an improvement in various clinical parameters (Canales-Espinosa, 1992). The two groups joined to form a single group with one adult male, eight females, and one infant.

In the first period of study (1989-1990), the troop was composed of the 10 founder individuals. In the second period (1997), the group had 57 monkeys (12 adult males, 23 adult females, 16 young, and 6 infants). This implied an important and notable increase in population density from 1.2 individuals per hectare to 6.9 individuals per hectare.

2.2. Study Area

Agaltepec Island (Figure 1) has a surface area of approximately 83,719 m^2 (8.3 ha). It is located at coordinates 18° 24' – 18° 25' N and 95° 05' W. Elevation is between 380 to 432 m. The island is located in the climate region Am(e)gw, which corresponds to a warm, humid climate with an annual precipitation of 1,980 mm (Gonzalez-Capistrán, 1991). The maximum, medium, and low temperatures are 36.7, 20.8, and 12.9 °C, respectively. The flora has been characterized into four kinds: lowland forest, riparian vegetation, secondary vegetation (acahual), and grassland (López-Galindo and Acosta-Pérez, 1998).

Parallel to the capture and handling of the monkeys, there was a botanical study of the island to ecologically evaluate the area for release. There were 1,605 trees recorded during the first stage of the sampling and 2,079 in the second one. All trees sampled had a diameter above 30 cm at breast height, and 65 tree species were identified corresponding to 32 families and 58 genera. The number of trees per hectare and relative density were also estimated (see Appendix).

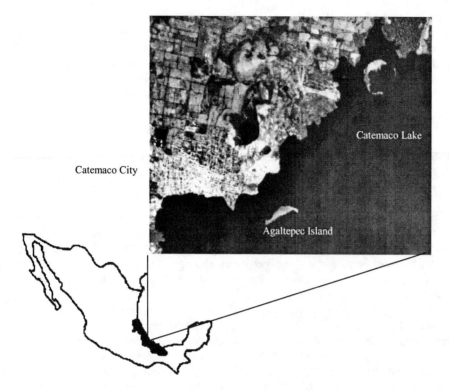

Figure 1. Location of Agaltepec Island.

2.3. Sampling and Recording Methods

Systematic observations were made during two stages of 10 months each: the first one (Stage 1) from November 1989 to August 1990; the second one (Stage 2) from January to July and from October to December 1997. In both, focal-primate sampling was combined with continuous recording (Martin and Bateson, 1991).

For Stage 1, 10 hours of daily observation took place (from 0600 to 1600 hours), for a total of 1,500 hours. For Stage 2, observations were for five hours a day in an alternating schedule—in the morning (from 0700 to 1200 hours) and in the afternoon (from 1200 to 1700 hours)—for a total of 630 hours. In both stages, the daily selection of the monkey to be focused on was made at random.

The frequency and duration of the three relevant behavior categories for the study—rest, feeding, and locomotion—were recorded (Glander, 1975; 1981). During feeding, additional information was taken about the plant in use, including feeding time, height of the trees (tall, medium, low), where the monkey was feeding, the part of the plant ingested [fruits, leaves, and others (buds, flowers, stems, petioles, bark), among other biotypes], number of parts consumed, and number of monkeys who ate at the same time on the same tree.

For the analysis, we calculated respective percentages for the frequencies and duration of the recorded activities, and for the parts of the plants that were eaten. The

Student t-test, the Mann-Whitney U-test, and the Medium Comparison Test were used (with a confidence level of 99%), with the parameters time dedicated for the execution of each behavior and the time of consumption of each part of the plant. A comparative analysis of proportions (Freedman et al., 1980) was performed to determine the differences in the number of tree species consumed during both sampling stages. Finally, the Concordance Index (Daniel, 1978) was calculated to establish similarities among the stages, in the order of preference of the 10 most used tree species.

3. RESULTS

3.1. Pattern of Daily Activities

The time dedicated by the howlers to the three basic behaviors during the two research stages was similar (from larger to smaller percentage: rest, feeding, and locomotion) (Figure 2). In general we can see significant differences in the behavior times when we compare the three behaviors during each stage (Chi-square; 2 g.l., $p<0.01$), and each behavior in both stages combined (Mann-Whitney, $p<0.01$). The average times are shown in Table 2.

We theorized that when the size and density of the population increased, the competition for the feeding sources would also increase, forcing the monkeys to make longer foraging trips. Therefore, the time dedicated to rest would decrease, but this reduction was not significant (Mann-Whitney U = 401, 136.5, $p>0.01$). We found that the time dedicated to locomotion increased a small percentage, but was significant statistically during Stage 2 (Mann-Whitney U = 274,746.0, $p<0.001$) (Table 2). This regularity in the proportions in the pattern of daily activities, even with the change in the monkey's population, coincides with proportions shown by groups that live in

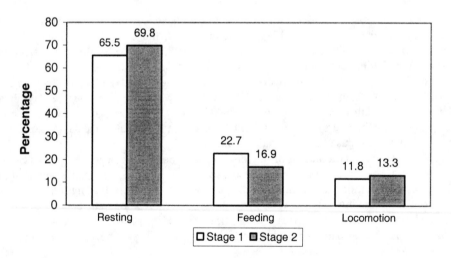

Figure 2. Percentage of time dedicated to each behavior during each stage of the research.

Table 2. Analysis of the pattern of daily activities (minutes).

a) Feeding

Statistics	Stage 1			Stage 2		
	Number of episodes (per day)	Total duration (per day)	Mean duration (per episode)	Number of episodes (per day)	Total duration (per day)	Mean duration (per episode)
Minimum	0	0	0	0	0	0
Maximum	12	196	67	8	132	113
Mean	3.9	67.6	19.4	3.5	51.1	16.9
Median	4	69	18.8	3.5	50.5	14.8
SD	2.2	35.9	7.7	1.9	31.6	11.6
Total	626	10,965	3,108	437	6,382	2,194.8

b) Resting

Statistics	Stage 1			Stage 2		
	Number of episodes (per day)	Total duration (per day)	Mean duration (per episode)	Number of episodes (per day)	Total duration (per day)	Mean duration (per episode)
Minimum	1	66	11.3	1	73	12.2
Maximum	11	294	261	12	300	300
Mean	4.3	196.3	61.2	4.8	209	60.8
Median	4	178	51	4	209	47.3
SD	2.2	44.1	41.1	2.5	45.6	51
Total	693	31,626	9,762.4	620	26,401	7,532.9

c) Locomotion

Statistics	Stage 1			Stage 2		
	Number of episodes (per day)	Total duration (per day)	Mean duration (per episode)	Number of episodes (per day)	Total duration (per day)	Mean duration (per episode)
Minimum	0	0	0	0	0	0
Maximum	9	133	80	11	119	41.2
Mean	3	35.3	11.9	4.4	40	9.4
Median	3	36	12	4	32	9.4
SD	1.9	26.3	10.2	2.5	29.9	6
Total	484	5,675	1,902.5	563	5,017	1,153.8

environments with different area, diverse flora, and degree of disturbance (e.g., Chivers, 1969; Richard, 1970; Mittermeiner, 1973; Milton, 1980; and Estrada, 1982; 1984). According to our results, the change in the foraging of a growing, isolated population of monkeys is not obtained by changing the pattern of daily activities. The monkeys showed a generally similar behavior during both stages; at least the behaviors seem to extend in similar proportions.

The frequency and duration of the feeding episodes were analyzed. They were significantly shorter during Stage 2 than Stage 1(median time for Stage 1 was 19.4 minutes and 16.9 for Stage 2; Comparative Measurements Test: T = 1774.5, $p<0.01$). The frequency of feeding (number of episodes per day) was similar in both stages (3.9 for Stage 1 and 3.5 for Stage 2). For Stage 1, the feeding time was 67.6 minutes and for

Stage 2 it was 51.1 (Comparative Measurements Test: T = 0.34, $p>0.01$; Table 2). The duration of the locomotion episodes in Stage 2 was significantly shorter than the duration in Stage 1. This was surprising, as we expected the competition for food would make the searching periods longer for finding fruits (average for each episode was 13.6 minutes for Stage 1 and 10 minutes for Stage 2, T = -6.67. $p<0.05$; Table 3).

We expected the feeding episodes to be shorter (on average) and more frequent during Stage 2 because of the lack of spots with resources and the increase in demand for food. However, our predictions were mistaken and the feeding episodes remained similar in both research stages. We did however see an increase in the total number of locomotion episodes during Stage 2 (from 484 in Stage 1 to 563), which increased the daily average of episodes from 3 to 4.4 and may correspond to the expected increase in the search for food. We also saw a change in the correlation between locomotion and feeding (Table 2). There was a greater amount of locomotion episodes compared to feeding in Stage 2, which makes us suppose the monkeys were spending more time looking for feeding sources. In general, they were traveling more for shorter overall durations to search for food. This may indicate familiarity of the site in Stage 2 as compared to Stage 1.

We believe that the effort (locomotion episodes and duration combined) in the search for food was increased when the population became larger, and that the lack of equilibrium between availability and demand (abundance and availability of resources on the island) became greater.

Table 3. Descriptive statistics of travel to feeding locations (minutes).

a) Stage 1

	Locomotion Episodes (Fruits)	Average Duration	Locomotion Episodes (Leaves)	Average Duration
Number of Cases	86	86	50	50
Minimum	1	2	1	2
Maximum	4	88	4	53
Mean	1.3	13.6	1.34	11.9
Median	1	9	1	8
SD	0.6	14.1	0.7	10.8
Total	115	1,171.9	67	597

b) Stage 2

	Locomotion Episodes (Fruits)	Average Duration	Locomotion Episodes (Leaves)	Average Duration
Number of Cases	35	35	59	59
Minimum	1	2	1	1
Maximum	3	50	3	45
Mean	1.3	10.0	1.42	10.0
Median	1	8	1	4
SD	0.5	9.3	0.7	9.3
Total	47	349.8	84	588.9

3.2. Feeding Preferences

We proposed that the number of species used as a food source would be larger during Stage 2, however, no significant differences were found in the number of species used in both stages (Proportion comparison test, $p<0.01$). The monkeys used 32 species as food sources during Stage 1 and 30 during Stage 2 (Tables 4 and 5). Twenty-four of the species used in Stage 1 were common during Stage 2, while only eight of the ones used in Stage 2 (26.7%) were not even recorded in Stage 1. An increase in the number of trees used as a food source was expected. In Stage 1, 117 trees were used, while during Stage 2, 153 were used (Tables 4 and 5).

The tree species preference by the howlers was similar in both stages. Seven of 10 species most consumed in the first stage remained among the most favorite in the second stage (*Ficus pertusa, F. cotinifolia, F. maxima, F. insipida, Bursera simaruba, Andira galeottiana,* and *Mastichodendron capiri*). In Stage 2 *Ficus pertusa* was the most consumed species. *Gliricidia sepium,* a species little used by the monkeys in Stage 1, was in the top 10 most consumed species in Stage 2 (Tables 4 and 5). The comparison between Stage 1 and Stage 2 for the 10 most used food species reveals that during Stage 2, for five of these species, the number of trees used increased; one species remained with the same number of individuals used, and four showed a decrease in the number of trees used (Table 6). Thirty six percent of the trees used in Stage 1 were also used during Stage 2.

The fruits were a more limited resource than leaves, therefore, when the population increased during Stage 2, we proposed a larger demand and greater competition among individuals as well an increase in the consumption of leaves. The results that confirm this prediction are shown in Figure 3. There was a noticeable increase in both the consumption of other plant species during Stage 2, as well as the ingestion of other parts of plants (Figure 3). The increase in the consumption of leaves can be interpreted as an opportunistic response when faced with the food the fragment offers. A notable difference between the plant species used as food during Stage 1 and Stage 2 was the increase in the percentage of the consumption of lianas and vines (Figure 3). This percentage increased from 8.8 during Stage 1 to 21.0 during Stage 2. According to Carmona-Diaz et al. (1999), the howlers on the island foraged on 38 species of lianas, vines, epiphytes, and parasitic. Of those, the consumption of lianas the largest (47.3%). In a previous study, Carmona-Diaz et al. (1997) reported that the howlers of Agaltepec used as a food source 38.5% of the total species of lianas recorded (N = 52) on the island, eating the leaves (young and old), stems, sprouts, tendrils, flowers, and fruits.

Carmona-Diaz (personal communication) also considers that within recent years there has been an increase in the epiphytes and hemiepiphytes on the island as a consequence of the environmental conditions resulting from greater disturbance. This could be a reason for the increased consumption of these resources by the monkeys. Putz (1983, 1984) documented that the proliferation of lianas in tropical forest is a response to the established environmental conditions associated with disturbance (this is common at the forest's edge, or in the interior, because of tree fall gaps and blocked sunlight by a discontinuous arboreal roof), although decreased in abundance with time since last disturbance. The increase in the consumption of epiphytes and hemipiphytes recorded with the larger monkey population (Stage 2) leads us to consider adaptation of *A. p. mexicana* towards disturbed habitat. This might also explain the prolonged stay of the

Table 4. Species and number of individuals used by howlers for feeding during Stage 1 where percent of usage is the percent of individual trees per species howlers consumed out of the total on the island.

Species	% of Time (total 89.7)	Individuals Used	Existing Individuals	% of Usage
Ficus pertusa	34.0	10	21	47.6
Protium copal	7.7	12	129	9.3
Mastichodendron capiri	6.8	7	20	35.0
Ficus maxima	6.7	7	12	58.3
Ficus sp.	6.4	2	4	50.0
Ficus cotinifolia	5.2	4	24	16.7
Ficus insipida	5.1	3	3	100.0
Andira galeottiana	4.1	10	52	19.2
Bursera simaruba	3.2	16	347	4.6
Ficus obtusifolia	2.6	5	15	33.3
Dendropanax arboreus	1.8	5	81	6.2
Chlorophora tinctoria	1.7	4	23	17.4
Erythrina folkersii (-)	0.5	1	5	20.0
Spondias mombin	0.5	5	42	11.9
Chrysophyllum mexicanum (-)	0.4	1	11	9.1
Cordia dodecandra (-)	0.4	2	13	15.4
Astronium graveolens	0.4	3	43	7.0
Brosimum alicastrum	0.3	2	17	11.8
Leucaena leucocephala	0.3	1	2	50.0
Nectandra coriacea	0.3	1	62	1.6
Scheelea liebmannii	0.2	1	1	100.0
Lonchocarpus cruentus	0.2	2	33	6.1
Spondias radlkoferi	0.1	1	7	14.3
Diospyrus verae-crucis	0.1	1	23	4.3
Tapirira mexicana (-)	0.1	2	11	18.2
Zanthoxylum caribaeum	0.1	1	7	14.3
Pisoania aculeata (-)	0.1	1	12	8.3
Heliocarpus donnell-smithii (-)	0.1	1	13	7.7
Gliricidia sepium	0.1	3	85	3.5
Compositae (-)	0.05	1	30	3.3
Pachira aquatica	0.02	1	45	2.2
Cupania dentata (-)	0.02	1	9	11.1
Trees not identified	0.8			
Other biological entities (*)	9.5			
TOTAL	**100**	**117**	**1,202**	**9.7**

(-) Arboreal species not used during Stage 2.
(*) Generic name for plants with distinct biological forms other than trees (creeping plants, lianas, epiphytes, parasitic), which were not taxonomically determined for this study.

monkeys in highly altered fragments with small dimensions that are found in the region of Los Tuxtlas (Silva-Lopez, 1987; Garcia-Orduña, 1996), mainly in the humid period, since a relationship between climber abundance and rainfall has been suggested (Putz and Chai, 1987; Balfour and Bond, 1993).

Table 5. Species and number of individuals used by howlers for feeding during Stage 2, % of usage as in Table 4.

Species	% of Time (total 75.5)	Individuals Used	Existing Individuals	% of Usage
Ficus pertusa	21.7	9	27	33.3
Ficus cotinifolia	6.4	6	24	25.0
Ficus maxima	6.0	3	14	21.4
Bursera simaruba	5.9	28	445	6.3
Andira galeottiana	5.9	16	56	28.6
Mastichodendron capiri	5.5	11	24	45.8
Gliricidia sepium	2.7	13	186	7.0
Chlorophora tinctoria	2.6	7	25	28.0
Ficus insipida	2.5	2	3	66.7
Leucaena leucocephala	2.3	1	2	50.0
Ficus obtusifolia	2.0	7	19	36.8
Spondias mombin	1.7	3	43	7.0
Protium copal	1.6	10	157	6.4
Ficus sp.	1.4	2	6	33.3
Dendropanax arboreus	1.3	7	104	6.7
Brosimum alicastrum	1.1	3	17	17.6
Lonchocarpus cruentus	1.0	4	45	8.9
Guazuma ulmifolia (+)	0.7	1	14	7.1
Astronium graveolens	0.6	4	70	5.7
Diospyrus verae-crucis	0.5	2	27	7.4
Plumeria rubra (+)	0.5	4	79	5.1
Ceiba aesculifolia (+)	0.3	2	45	4.4
Pachira aquatica	0.3	1	46	2.2
Tabernaemontana alba (+)	0.3	1	10	10.0
Inga vera (+)	0.2	1	8	12.5
Nectandra coriacea	0.1	1	68	1.5
Cedrela odorata (+)	0.1	1	15	6.7
Spondias radlkoferi	0.1	1	13	7.7
Zanthoxylum caribaeum	0.1	1	9	11.1
Scheelea liebmannii	0.1	1	1	100.0
Trees not specified	4.0			
Other biological entities (*)	20.5			
TOTAL	**100**	**153**	**1,602**	**9.6**

(+) Species not used during Stage 1.
(*) Generic name for plants with distinct biological forms other than trees (creeping plants, lianas, epiphytes, parasitic), which were not taxonomically determined for this study.

To explain the decrease in the consumption of fruits, we must consider the following: a) the amount of fruits produced by the trees is similar from year to year. Data obtained by López-Galindo et al. (in prep.) recorded regularity in the phenology of the different arboreal fruiting species on the island, except for *Ficus* b) The reduction may be a consequence of less access to individual resources. This restriction is attributed to a larger number of monkeys during Stage 2 taking advantage of the same feeding sources, and so, every individual would have had less fruits available c) If there were more fruits

Table 6. Number of trees of the 10 most eaten species.

Species	Trees used during Stage 1	Trees used during Stage 2	Trees used during both Stages
Ficus pertusa	10	9	6
Protium copal	12	10	1
Mastichodendron capiri	7	11	3
Ficus maxima	7	3	3
Ficus sp.	2	2	1
Ficus cotinifolia	4	6	2
Ficus insipida	3	2	2
Andira galeottiana	11	17	4
Bursera simaruba	21	28	3
Ficus obtusifolia	6	7	5

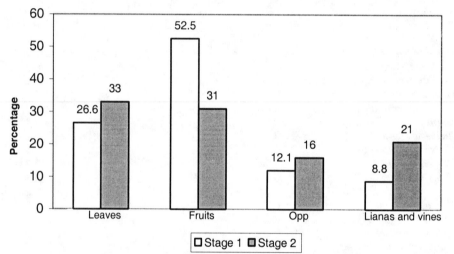

Figure 3. Percentage of time dedicated to the consumption of the different parts of the plant per stage. OPP = other parts of plants.

available, it is assumed that the monkeys would actually take advantage, so that the decrease in the consumption is not due to an auto-restriction. The increase in the consumption of other food sources might be a response to the fruit deficit, even when the natural sources of fruits could not be substituted in totality by the other parts of the plants.

When we took into consideration the small size of the environment (8 ha) and the low arboreal diversity, we were expecting that the most often consumed flora during the first stage would also remain so during the second stage. A very high percentage of the species were common to both stages (63%). For Stage 1, the 10 most used species represented 81.8% of the food for the monkeys, and for this they used only 4.7% of the total trees on the island. For Stage 2, the 10 most used species of trees represented 61.5% of the food, and, to reach that level of consumption, the monkeys used only 4.6% of the trees on the island. These data accentuate the importance of a small number of species and individual trees in the monkey's diet.

Even under extreme environmental conditions such as the high population density of monkeys on the island, the population only used a fraction of the arboreal component of the area. This makes us suppose that there are factors that limit the availability of the food resources; factors such as the phenological variations of the arboreal components, the palatability and digestibility of the vegetal parts, and the accessibility to certain areas of the island. We can also say that there is a progressive reduction in food availability from the most important species in the monkey's diet, which is a consequence of increased competition. To fix this deficit, other food items have to be exploited.

The species *Brusera simaruba* requires a special commentary. It was the arboreal species with the highest density on the island (Stage 1= 21.61 and Stage 2 = 21.4). During Stage 2, there was increased consumption (from the ninth position in the preference chart to the fourth) and the number of trees utilized (from 10 to 28). It needs to be pointed out that *B. simaruba* is a species that prospers in the disturbed forest; it is fast growing and easily spreads. The howlers' ability to use these trees as a food source may be strategically of great importance when the monkeys are in a fragmented and disturbed habitat. The growth in the consumption of a species such as *B. simaruba* may also allow the growing population of monkeys to adjust to the differences in primary food availability in a forest fragment.

Another meaningful event during Stage 2 was the decrease in the use of *Ficus* (Tables 7 and 8). This species has been reported as key in the diet of howler monkeys (Milton, 1980; Serio-Silva et al., 1997; Serio-Silva and Rico-Gray, this volume). This decrease may be interpreted as an indication that the maximum use of that resource had been reached. It is also necessary to mention that the food offered by the *Ficus* trees varies according to the seasons (Milton, 1980). The *Ficus* species' phenological patterns are known as asynchronic even among trees of the same species and same location. For the *Ficus* of Agaltepec Island in particular, their fruiting has been monitored and visible patterns vary between the species and even among the same species within the same year (López-Galindo et al., in prep). The principal adjustment in the diet of the monkeys is identified as the increase in the use of species that we call "secondary food choices" (species such as *B. simaruba* and other species such as vines and lianas) associated with disturbed forest or regenerating secondary growth.

Table 7. Consumption of *Ficus* species on Agaltepec Island during Stage 1. See Table 4 for definition of percent and usage.

Species	% Consumption (minutes)	Number of individual trees used	% of Usage
Ficus pertusa	21.7	9	33.3
Ficus cotinifolia	6.4	6	25.0
Ficus maxima	6.0	3	21.4
Ficus insipida	2.5	2	66.7
Ficus obtusifolia	2.0	7	36.8
Ficus sp.	1.4	2	33.3
Total	40% of the total feeding time	19% of all the trees consumed	31.2% of all *Ficus* trees on Agaltepec

Table 8. Consumption of *Ficus* species on Agaltepec Island during Stage 2. See Table 4 for definition of percent of usage.

Species	% Consumption (minutes)	Number of individual trees used	% of Usage
Ficus pertusa	34.0	10	47.6
Ficus maxima	6.7	7	58.3
Ficus sp.	6.4	2	50.0
Ficus cotinifolia	5.2	4	16.7
Ficus insipida	5.1	3	100.0
Ficus obtusifolia	2.6	5	33.3
Total	60% of total feeding time	26.5% of all trees consumed	39.2% of all *Ficus* trees on Agaltepec

4. DISCUSSION

The fact that availability of trees from the most consumed species did not match use is very interesting. Why are some of the trees of the same species used as a food source and not others? Are not all the trees of the same species equal in their availability to the monkeys? Do some trees impose restrictions on the monkeys? It is also interesting that only some of the 10 most used arboreal species during Stage 1 were also used during Stage 2. Why are the other trees not being used during Stage 2? Have they suffered changes? Could those changes be independent of the foraging the monkeys exercised? Do they depend on the seasonal changes? All of these questions must be answered through another series of studies. Either way, considering all the trees of importance in the howlers' diet that are not used, one may believe that the island still has potential food sources for the howlers.

According to Ankel-Simmons (2000), the dynamics of the monkey population can be reviewed based on the analysis of the distribution and abundance of the food resources. In Agaltepec, we would expect that the problems the growing monkey populations might face could be solved by a change in their strategy to find, choose, and consume their resources. One might also see a change in the patterns in the use of the tree species and the key food sources. This change in the foraging strategy could imply an increase in the extension of the zones of intense use and also the magnitude of the use.

During the growth of the population, there were changes in the migration of the monkeys and the occupation of the island. In the first sampling stage, the monkeys moved in only one group in a cohesive way and in a linear progression. In the second stage, the monkeys moved in sub-groups, apparently unstable, to the length and width of the island and eventually the pattern of occupation changed. Within the island, there are some intense use zones that can be identified, which are simultaneously explored by the monkeys. Baldwin and Baldwin (1972) discuss overlapping home ranges when the population density increases. It is possible that the isolation of the population obliges it to divide into these sub-groups for foraging. If the physical barrier of water surrounding the fragment did not exist to impede migration, the population density would not remain high for such a small area.

Milton (1982) suggested that the availability of the food is the limiting factor for the population's growth. The central problem seems to be the (quantitative and qualitative) determination of the food availability in a fragment or on an island, assuming that the vegetative structure is not static, and recognizing that the monkeys do not keep this ecological relationship unchanged in time and space. This study suggests changes in the feeding preference as a response to increased demand for food by the population. Should this change in the foraging strategy be considered to determine a change in the capacity of the ecosystem? How can we recognize this new limit in the availability of food in the fragment? Either way, there will be a limit on the capacity to find new food sources in the fragment and, thus, other changes will be expressed (for example, an increase in infant perishing rate) that will limit the population's growth. Which are the population indicators that will allow us to recognize the overexploiting of the food sources in the environment? Is it possible to find these indicators in the environment itself (for example, changes in the arboreal structure)? The answers to these questions rest on modeling the ecological adaptation of the primate population in the fragmented habitat.

The fragmentation of the primate's habitat is determined by diverse human activities that have not been prevented, even in some protected natural areas. Therefore, it is necessary to increase our knowledge of the primate populations in fragmented habitats or on islands to establish actions to help their conservation. This will be possible as long as the mechanisms of ecological adaptation that operate in the populations in fragmented habitats are known. No doubt, we shall be more obliged to take on conservation activities in fragmented habitats (c.f., Marsh et al., this volume). This is the reason research such as this one has special current interest, since it will allow us to design better models and effective decisions for primate conservation. For example, with the base we have in this study, we obtain new criteria to evaluate the amount of food support a fragmented island habitat can offer a troop of howlers, and so we can orient the execution of the appropriate tasks for the reintroduction of the species.

We must point out the relevance of this analysis as a tool to understand the persistence of the howler populations in ecologically impoverished environments. In the scientific literature there are numerous reports on the populations of the genera *Alouatta*, but there are few with a long-term following of the populations in disturbed habitats. The importance of this work in a disturbed habitat reveals the mechanisms of ecological adaptation in a scale of a couple of years. It is accepted that the monkey population studied was analyzed under special ecological conditions (on an island), yet they were similar in many ways to the ones presented by the fragmented habitats where populations of this species have become isolated (García-Orduña, 1996). The similarity referred to is given mostly by the extension, arboreal composition, and degree of disturbance of the area the primates occupy, as well as their persistence under ecological conditions arguably adverse to them.

The analysis of this process of environmental adaptation gives us a relevant perspective for the comprehension of the extraordinary persistence of populations of howlers in severely disturbed and fragmented habitats. This knowledge can be helpful in designing tactics and strategies for the conservation of this species, for example, to evaluate the viability of the habitat and populations (Rodríguez-Luna et al., 1995). With a base pattern of daily activities and food preference, it is possible to model the foraging strategy of the howler monkeys and to solve the research problems needed to achieve a more complete understanding of environmental adaptation displayed by populations of monkeys in fragmented and disturbed habitat. This study contributes to the comprehension of the plasticity of the howler monkey, and the notable persistence of their populations in fragmented habitats.

5. SUMMARY

In order to analyze the changes in the foraging strategy of howler populations living in a fragmented habitat, as well as to generate relevant information for conservation efforts, we decided to place a group of monkeys on an island. In this study, we considered the island to be equivalent to a fragment in the "Tuxtlas" region. The howlers that were moved to the island were taken from two habitat fragments. The first group came from Mirador Pilapa Ejido, in "Sierra de Los Tuxtlas," and the second group was captured in a farm near the border of the San Juan Evangelista River, also in the "Tuxtlas" region. Both groups were confined separately under strict captivity. Two cages were built for this management on Totogochillo Island in Lake Catemaco. In captivity, we provided the animals with tree branches collected on Agaltepec Island, as well as cultivated fruit (banana, mango, pineapple). Four females and one male howler from Mirado Pilapa were released on Agaltepec Island in 1988. Six months later a group from San Juan Evangelista, of similar composition, was released at the same site. Both groups formed the founder troop on the island.

Agaltepec Island is one of the four islands on Lake Catemaco. This island has 8.3 ha covered with the following vegetation types; lowland forest, secondary forest, riparian vegetation and induced grassland, similar in composition to any tropical forest fragment in Los Tuxtlas. Before the release of the howlers, a taxonomic identification of the trees (diameter to the height of the chest bigger or equal to 20 cm) of the island was carried out. We collected behavioral data and found that in the first stage of our study, the daily activity pattern was similar to those reported in continuous habitat in the "Tuxtlas" region. The food preferences showed a decrease in the consumption of leaves and a significant increase in the consumption of fruits. We found a very high consumption of other biological forms, like vines and climbing plants, whose distribution and abundance is commonly associated with disturbed forests. This study shows that these biological forms can be an important alternative food source for the monkeys. The howlers were highly selective of foods in proportion to their availability. The population of monkeys in Agaltepec Island increased from ten individuals in 1990 (first stage of our study) to 57 individuals in 1997 (second stage). We saw an increase of population density from 1.2 to 6.9 individuals per ha.

Recognizing the population growth of howlers as well as the environmental limitations of the island, we expected the daily activity pattern and food preferences of the howlers to change, as reflected in a new foraging strategy. We thought that, as the group of howlers increased in population size, the foraging strategy (search and food ingestion) would change due to high demand on the food resources. However, even as the population density of howler monkeys increased, the daily activity patters did not change. This experience studying the howlers on the island illustrates the type of foraging strategy of populations that live in fragments of habitat. There appears to be an adjustment in behavior in order to have the scantiness of primary food resources available in the habitat. The comparative analysis between the two stages in the population growth of the howlers in the island leads us to consider the measuring of the carrying capacity in fragments of habitat. Without any doubt, this kind of study is relevant for conservation planning to support howler populations living in fragments of habitat. The study of howlers relocated to an island acts as a controlled experiment to learn about foraging strategies and population growth that may occur in fragments. This kind of study is important to understanding howlers in fragments throughout Mexico.

6. ACKNOWLEDGMENTS

The entire project was supported by Patronato Pro-Universidad Veracruzana, A.C., CONACYT (register number: 4316P-H9608), World Wildlife Fund, US Fish and Wildlife Service, and Wildlife Preservation Trust International. We would like to thank Juan Carlos Serio Silva, Arturo González Zamora, and César Arturo Pérez Moscoso for their assistance in the field work. We thank Adolfo López Galindo and Gustavo Carmona Díaz for identifying the species of trees and comments about lianas. We would like to thank Emiliano Salatino for the translation of the manuscript.

7. REFERENCES

Altmann, S. A., 1959, Field observations on a howling monkey society, *J. Mammal.* **40**:317–330.
Ankel-Simmons, F., 2000. *An Introduction to Primate Anatomy,* Academic Press, San Diego, CA, USA, pp. 349–380.
Baldwin, J. D., and Baldwin, J. I., 1972, Population density and use of space by howling monkeys (*Alouatta villosa*) in southwestern Panama, *Primates* **13**:371–379.
Balfour, D. A., and Bond, W. J., 1993, Factors limiting climber distribution and abundance in a southern Africa forest, *J. Ecol.* **81**:93–99.
Canales-Espinosa, D., 1992, Programa piloto de translocación del mono aullador (*Alouatta palliata*), Tesis de Licenciatura, Fac. de Medicina Veterinaria y Zootecnia, Universidad Veracruzana. Veracruz, Veracruz, México, p 97
Carmona-Díaz, G., 1997, Listado preliminar de lianas consumidas por el mono aullador (*Alouatta palliata*) en la isla Agaltepec, Catemaco, Veracruz, México, in: *Resumen VI Simposio Nacional de Primatología,* México, D.F. p 3.
Carmona-Díaz, G., Gómez-Marín, F. J., Asensio-Herrero, N., and Rodríguez-Luna, E., 1999, Forrajeo de lianas, enredaderas, epífitas y parásitas por *Alouatta palliata* en tres sitios de la región de Los Tuxtlas, Veracruz, in: *Resumen VII Simposio Nacional de Primatología,* Catemaco, Veracruz, México.
Carrera-Sánchez, E., 1993, Etograma del mono aullador (*Alouatta palliata mexicana* Merriam) en la isla de Agaltepec, Lago de Catemaco, Veracruz, Tesis de Licenciatura, Fac. de Biología, Universidad Veracruzana, Xalapa, Veracruz, México, 105 pp.
Chapman, C. A., and Balcom, S. R., 1998, Population characteristics of howlers; Ecological conditions of group history, *Int. J. Primatol.* **19(3)**:385–403.
Chivers, D. J., 1986, Current issues and new approaches in primate ecology and conservation, in: *Primate Ecology and Conservation,* J. G. Else and P. C. Lee, eds., Cambridge University Press, Cambridge, Great Britain.
Chivers, D. J., 1969, On the daily behavior and spacing of howling monkeys groups, *Folia Primatol.* **54**:1–15.
Chivers, D. J., and Hladik, L. M., 1980, Morphology of the gastrointestinal tract in primates: Comparisons with mammals in relation to diet, *J. Morph.* **166**:337–386.
Crockett, C. M., and Eisenberg, J. F., 1986, Howlers: Variations in group size and demography, in: *Primates Societies,* B. B. Smuts, D. L. Cheney, R. M. Seyfarth, R. W. Wrangham, and T. T. Struhsaker, eds., The University of Chicago Press, Chicago, Illinois, pp. 54–68.
Daniel, W. W., 1978, *Applied Nonparametric Statistics,* Houghton Mifflin Co., U.S.A.
Eisenberg, J. F., 1979, Habitat, economy, and society: Some correlations and hypotheses for Neotropical primates, in: *Primate Ecology and Human Origins,* I. S. Bernstein and E. O. Smith, eds., Garland Press, New York, pp. 215–262.
Estrada, A., 1982, Survey and census of howler monkeys in the rain forest of Los Tuxtlas, Veracruz, Mexico, *Am. J. Primatol.* **2(4)**:363–372.
Estrada, A., 1984, Resource use by howler monkeys (*Alouatta palliata*) in the rain forest of Los Tuxtlas, Veracruz, Mexico, *Int. J. Primatol.* **5**:105–131.
Estrada, A., and Coates-Estrada, R., 1984, Fruit eating and seed dispersal by howling monkeys (*Alouatta palliata*) in the tropical rain forest of Los Tuxtlas, *Am. J. Primatol.* **6**:77–91.
Estrada, A., and Coates-Estrada, R., 1986, Fruiting and frugivores at a strangler fig in the tropical rain forest of Los Tuxtlas, Mexico, *J. Trop. Ecol.* **2**:349–357.

Fox, B. A., and Cameron, A. G, 1992, *Ciencia de los alimentos, Nutrición y Salud*, Edit. Limusa, México, D.F., 457 pp.

Freedman, D., Pisani, R., and Purves, R., 1980, *Statistics*, W. W. Norton and Co., U.S.A.

Freese, C., 1976, Censusing *Alouatta palliata, Ateles geoffroyi,* and *Cebus capucinus* in the Costa Rican dry forest, in: *Neotropical Primates: Field Studies and Conservation*, R. W. Thorington and P. G. Heltne, eds., National Academy of Sciences, Washington, D.C., pp. 4–9.

García-Orduña, F., 1996, Distribución y abundancia del mono aullador *Alouatta palliata* y el mono araña *Ateles geoffroyi* en fragmentos de selva del municipio de San Pedro Soteapan, Veracruz, Tesis de Licenciatura, Fac. de Biología, Universidad Veracruzana, Xalapa, Veracruz, México, 48 pp.

Glander, K. E., 1975, Habitat and resource utilization: An ecological view of social organization in mantled howling monkeys, PhD Dissertation, University of Chicago, USA.

Glander, K. E., 1981, Feeding patterns in mantled howling monkeys, in: *Foraging Behavior*, A. C. Kamil and T. D. Sargen, eds., pp. 231–257.

González-Capistrán, M. E., 1991, Regionalización climática de la Sierra de Santa Martha y el Volcán de San Martín Pajapan, Veracruz, Tesis de Maestría, Fac. de Ciencias, UNAM, México, D.F., 91 pp.

Horwich, R. H., and Lyon, J., 1990, *A Belizean Rain Forest*, Orangutan Press, USA, 420 pp.

Juan, S. S., Ortiz, M., Estrada, T. J., and Coates-Estrada, R., 1999, Uso de plantas como alimento por *Alouatta palliata* en un fragmento de selva en Los Tuxtlas, México, *Neotrop. Primates* **7(1)**:8–11.

Juan, S., Estrada, A., and Coates-Estrada, R., 2000, Contrastes y similitudes en el uso de recursos y patrón general de actividades en tropas de monos aulladores (*Alouatta palliata*) en fragmentos de selva en Los Tuxtlas, México, *Neotrop. Primates* **8(4)**:131–135.

López-Galindo, A., and Acosta-Pérez, R., 1998, Listado florístico de la isla Agaltepec, Lago de Catemaco, Veracruz, *Foresta Veracruzana* **1(1)**:1–4.

López-Galindo, A., Morales-Mávil, J. E., Rodríguez-Luna, E., and García-Orduña, F., (In preparation) Fenología de la vegetación arbórea de la Isla Agaltepec, Lago de Catemaco, Veracruz, México.

Lovejoy, T. E., Bierregard, R. O., Rylands, A. B., Malcom, J. R., Quintela, C. E., Harper, L. H., Brown, K. S., Powell, A. H., Powell, G. V. N., Schubart, H. O. R., and Hays, M. B., 1986, Edge and other effects of isolation on Amazon forest fragments, in: *Conservation Biology: The Science of Scarcity and Diversity*, M. E. Soulé, ed., Sinnauer Associates, Massachusetts, USA, pp. 257–285.

Martin, P., and Bateson, P., 1991, *La Medición del Comportamiento*, Alianza Universitaria, Madrid, España, 237 pp.

Mendel, F., 1976, Postural and locomotor behavior of *Alouatta palliata* on various substrates, *Folia Primatol.* **26**:36–53.

Milton, K., 1980, *The Foraging Strategy of Howler Monkeys (A Study in Primate Economics)*, Columbia University Press, New York, 165 pp.

Milton, K., 1982, Dietary quality and population regulation in a howler monkey population, in: *The Ecology of a Tropical Forest: Seasonal Rhythms and Long-term Changes*, E. G. Leigh and D. M. Windosr, eds., Smithsonian Institution Press, Washington, D.C., pp. 273–290.

Mittermeier, R. A., 1973, Group activity and population dynamics of the howler monkeys on Barro Colorado Island, *Primates* **14**:1–19.

Napier, J. R., and Napier, P. H., 1985, *The Natural History of the Primates*, The MIT Press, Great Britain.

Ninomiya, Y., and Funakoshi, M., 1989, Peripheral neural basis for behavioral discrimination between glutamate and the four fasic taste substance in mice, *Comp. Biochem. Phybiol. A Comp. Physiol*, **92**:371–376.

Oates, J. F., 1986, Food distribution and foraging behavior, in: *Primates Societies*, B. B. Smuts, D. L. Cheney, R. M. Seyfarth, R. W. Wrangham, and T. T. Struhsaker, eds., The University of Chicago Press, Chicago, IL, pp. 197–209.

Putz, F. E., 1984, The natural history of lianas on Barro Colorado Island, Panama, *Ecology* **65(6)**:1,713–1,724.

Putz, F. E., and Chai, P., 1987, Ecological studies of lianas in Lambir National Park, Sarawak, Malaysia, *J. Ecol.* **75**:523–531.

Richard, A., 1970, A comparative study of the activity patterns and behavior of *Alouatta villosa* and *Ateles geoffroyi*, *Folia Primatol.* **12**:241–263.

Riney, T., 1982, *Study and Management of Large Mammals*, John Wiley and Sons, England.

Rodríguez-Luna, E., 2000, Cambios en la estrategia de forrajeo del mono aullador (*Alouatta palliata mexicana*); estudio de una población en un fragmento de selva, Tesis de Maestría, Instituto de Neuroetología, Universidad Veracruzana, 106 pp.

Rodríguez-Luna, E., García-Orduña, F., and Canales-Espinosa, D., 1993, Translocación del mono aullador *Alouatta palliata*: una alternativa conservacionista, in: *Estudios Primatológicos en México, Vol. I*, A. Estrada, E. Rodríguez-Luna, R. López-Wilchis, and R. Coates-Estrada, eds., Biblioteca de la Universidad Veracruzana, Xalapa, Veracruz, México, pp. 129–177.

Rodríguez-Luna, E., Martínez-Morales, M., Serio-Silva, J. C., and Domínguez-Domínguez, L. E., 1995, Forrajeo del mono aullador (*Alouatta   palliata*) en semilibertad, in: *Estudios Primatológicos en México, Vol. II*, E. Rodríguez-Luna, L. Cortés-Ortíz, and J. Martínez-Contreras, eds., Biblioteca de la Universidad Veracruzana, Xalapa, Veracruz, México, pp. 133–148.

Scott, R. T., 1990, Gustatory control of food selection, in: *Handbook of Behavioral Neurobiology, Vol. 10, Neurobiology of Food and Fluid Intake*, E. M. Stricker, ed., Plenum Press, New York, pp. 243–263.

Sekulik, R., 1982, Daily and seasonal patterns of roaring and spacing in four red howler (*Alouatta seniculus*) troops, *Folia Primatol.* **39**:22–48.

Serio-Silva, J. C., 1992, Patrón diario de actividades y hábitos alimenticios de *Alouatta palliata* en semilibertad, Tesis de Licenciatura, Fac. de Biología, Universidad Veracruzana, Córdoba, Veracruz, México, 66 pp.

Serio-Silva, J. C., 1996, Calidad del alimento consumido por *Alouatta palliata* en condiciones de semilibertad, Tesis de Maestría, Maestría en Neuroetología, Instituto de Neuroetología, Universidad Veracruzana, Xalapa, Veracruz, México, 59 pp.

Serio-Silva, J. C., Rodríguez-Luna, E., Hernández-Salazar, L. T., Espinosa-Gómez, R., and Rico-Gray, V., 1997, Nutritional and chemical considerations for the food selection by howler monkeys (*Alouatta palliata mexicana*) in Agaltepec Island, Catemaco, Veracruz, Mexico, Abstract, *Am. J. Primatol.* **42(2)**:147.

Silva-López, G., 1987, La situación actual de las poblaciones de monos araña (*Ateles geoffroyi*) y aullador (*Alouatta palliata*) en la Sierra de Santa Martha (Veracruz, México), Tesis de Licenciatura, Fac. de Biología, Universidad Veracruzana, Xalapa, Veracruz, México, 162 pp.

Appendix. Number of individuals and species of trees on Agaltepec in the two stages study

	Number of Individuals		Relative Density (%)	
	Stage 1	Stage 2	Stage 1	Stage 2
ANNONACEAE	2	2	0.12	0.10
Annona reticulata L.				
ANACARDIACEAE				
Astronium graveolens Jacq.	43	70	2.26	3.37
Spondias mombin L.	42	43	2.61	2.07
Spondias radlkoferii J:D Smith	7	13	0.43	0.62
Tapirira mexicana Marchand	11	18	0.68	0.87
APOCYNACEAE				
Plumeria rubra L.	71	79	4.42	3.80
Stemmadenia donnell-smithii (Rose) Woodson	3	4	0.18	0.19
Tabernaemontana alba Miller	8	10	0.49	0.48
ARALIACEAE				
Dendropanax arboreus (L.) Decne. & Planchon.	81	104	5.04	5.00
BURSERACEAE				
Bursera simaruba (L.) Sarg.	347	445	21.61	21.4
Protium copal (Schlechtendal y Cham.) Engel	129	157	8.03	7.55
BOMBACACEAE				
Ceiba aesculifolia (Kunth) Britton & Baker	44	45	2.74	2.16
Pachira aquatica Aublet	45	46	2.80	2.21
BORAGINACEAE				
Cordia dodecandra A. DC.	13	21	0.81	1.01
CAPPARACEAE				
Capparis sp.	1	1	0.06	0.05
Crataeva tapia L.	1	1	0.06	0.05
CARICACEAE				
Carica papaya L.	1	1	0.06	0.05
COCHLOSPERMACEAE				
Cochlospermun vitifolium (Willd.) Sprengel	2	2	0.12	0.10
COMPOSITAE				
No determinada	30	44	1.86	1.12
EBENACEAE				
Diospyros verae-crucis Standley.	23	27	1.43	1.30
Diospyros digyna Jacq.	5	5	0.31	0.24
ERYTHROXYLACEAE				
Erythroxylum areolatum L..	17	18	1.05	0.87
FLACOURTIACEAE	1	1	0.06	0.05
LAURACEAE				
Nectandra coriacea (Sw) Griseb	62	68	3.86	3.27
LEGUMINOSAE				
Andira galeottiana Standley.	52	56	3.23	2.69
Acacia cornigera (L.) Willd	1	6	0.06	0.29
Bauhinia divaricata L..	1	3	0.06	0.14
Dyphysa macrophylla Lundell.	21	24	1.31	1.15
Erythrina folkersii Krunkoff & Moldenke	5	5	0.31	0.24
Gliricidia sepium Steudel (Jacq.)	85	186	5.29	8.95
Inga vera Willd.	7	8	0.43	0.38
Lonchocarpus cruentus Lundell.	33	45	2.06	2.16
Leucaena leucocephala (Lam.) De wit.	2	2	0.12	0.10
MALPIGHIACEAE				
Bunchosia lanceolata Turcz	1	1	0.06	0.05

MELIACEAE				
Cedrela odorata L.	13	15	0.81	0.72
Guarea glabra Vahl	5	6	0.31	0.29
Trichilia havanensis Jacq.	18	18	1.12	0.87
MORACEAE				
Chlorophora tinctoria (L.) Gaudish.	23	25	1.43	1.20
Cecropia obtusifolia Bertol.	5	8	0.31	0.38
Ficus cotinifolia Kunth.	24	24	1.49	1.15
Ficus insipida Willd.	3	3	0.18	0.14
Ficus maxima Miller	12	14	0.74	0.67
Ficus obtusifolia Kunth.	15	19	0.93	0.91
Ficus pertusa L.	21	27	1.31	1.30
Ficus trigonata	4	6	0.24	0.29
Trophis racemosa (Liebm) Bur.	2	2	0.12	0.10
Brosimun alicastrum Sw.	17	17	1.05	0.82
MYRTACEAE				
Eugenia capuli (Cham. & Schldl.) Berg	2	2	0.12	0.10
NYCTAGINACEAE				
Pisonia aculeata L.	12	16	0.74	0.77
PALMAE				
Scheelea liebmannii Becc.	1	1	0.06	0.05
PIPERACEAE				
Piper amalago. L.	1	2	0.06	0.10
POLYGONACEAE				
Coccoloba barbadensis Jacq.	2	3	0.12	0.14
RUBIACEAE				
Randia monantha Berth.	49	50	3.05	2.41
RUTACEAE				
Zanthoxylum caribaeum Lambert	7	9	0.43	0.43
SAPINDACEAE				
Cupania dentata Poiret.	9	9	0.56	0.43
Sapindus saponaria L.	7	7	0.43	0.34
SAPOTACEAE				
Chrysophyllum mexicanum Brandeg. ex Standley	11	16	0.69	0.77
Manilkara zapota (L.) van Royen	1	1	0.06	0.05
Mastichodendron capiri (DC) Cronq.	20	24	1.24	1.15
Pouteria sp.	2	2	0.12	0.10
SOLANACEAE				
Cestrum nocturnum L.	7	17	0.43	0.82
STERCULIACEAE				
Guazuma ulmifolia Lambert	10	14	0.62	0.67
TILIACEAE				
Heliocarpus donnell-smithii Rosse	13	20	0.81	0.96
URTICACEAE				
Myriocarpa cordifolia Liebm.	37	69	2.31	3.31
TREES NOT IDENTIFIED				
Leportea mexicana	1	2	0.06	0.10
Not identified	54	68	3.36	3.27

DIETARY FLEXIBILITY, BEHAVIORAL PLASTICITY, AND SURVIVAL IN FRAGMENTS: LESSONS FROM TRANSLOCATED HOWLERS

Scott C. Silver and Laura K. Marsh[*]

1. INTRODUCTION

Examining the behavioral response of howler monkeys (*Alouatta pigra*) to translocation may lend some insight into their ability to repopulate and persist in tropical forest fragments. In fragmented landscapes, forest patches may be large enough to sustain small numbers of individuals and social groups, but not at levels high enough for populations to persist solely in these patches. In fragmented forest landscapes, the ability to colonize unfamiliar forest patches is necessary to maintain demographic and genetic variability. In effect, translocated individuals successfully migrate to another fragment and have to depend upon their ability to include novel food items and strata. Survival depends on the level of dietary flexibility and behavioral plasticity.

High levels of species rarity characterize the floral community of tropical forests, and both translocated primates and primates establishing themselves in previously unfamiliar forest fragments face an array of unfamiliar plant species and may need to overcome highly seasonal fluctuations in food abundance. Being able to document the development of a dietary strategy of howler monkeys translocated into an unfamiliar environment offers an opportunity to characterize the evolution and ecology of a foraging strategy that may help to explain the ability of howler monkeys to persist in fragmented forests.

2. FORAGING CHALLENGES IN UNFAMILIAR AREAS

Both howler monkeys translocated to unfamiliar environments and those monkeys colonizing new patches of tropical forest must be able to determine the suitability of unfamiliar potential food items for inclusion in the diet. Monkeys in new areas may

---

[*] Scott C. Silver, Queens Wildlife Center Wildlife Conservation Society 53-51 111th St. Flushing, NY 11368, USA. Laura K. Marsh, Los Alamos National Laboratory Ecology Group (RRES-ECO) Mail Stop M887 Los Alamos, NM 87545, USA. Correspondence to S. C. Silver (email: ssilver@wcs.org).

Primates in Fragments: Ecology and Conservation
Edited by L. K. Marsh, Kluwer Academic/Plenum Publishers, 2003

encounter a largely unfamiliar plant community and must decide whether novel items are suitable foods. Even among familiar species, intraspecific variability in toxin and nutrient levels (Glander, 1975, 1978, 1981; Ganzhorn and Wright, 1994) may present the monkeys with challenges assessing the suitability of food items. In Costa Rica, Glander found that the presence of plant defensive compounds varied intraspecifically in important howler food species, and some individual trees were regularly eaten from while other conspecifics were routinely ignored. If this is the case, even species familiar to the monkeys may be temporarily unavailable as food when they are first establishing a home range in an unfamiliar area.

The second challenge to new arrivals to a forest is an ignorance of the locations of high-quality food patches. Furthermore, the relative value of these food patches is constantly changing as seasonal abundance of food items change. For monkeys translocated to a contiguous forest, this problem may be greater than for the monkeys who find themselves in limited patches of forest with limited area, but in both cases, the monkeys need to be able to cope with variable resource abundance, which may result in periods of nutritional stress. This is something that howler monkeys may face routinely. Previous studies of *Alouatta* species have found the abundance of high-quality food sources to be highly variable both spatially and seasonally (Glander, 1978, 1981; Milton, 1980; Estrada, 1984; Silver, 1997; Marsh, 1999). This problem can be compounded for animals unfamiliar with their foraging area as they may be unaware of the location and phenological patterns of high-quality foods (Glander, 1992). Incomplete information about the abundance and distribution of food resources may affect foraging decisions such as where to search for food and when to leave a depleted food patch (Charnov, 1976; Pyke, 1984).

3. SPECIES COMPOSITION OF DIET

In a translocation study done in Belize, howler monkeys were translocated from the Community Baboon Sanctuary (CBS) in northern Belize and released in the Cockscomb Basin Wildlife Sanctuary (CBWS) approximately 100 km south of their natal home ranges (Figure 1) (see Koontz and Ostro [2001] for more details on the translocation project itself). Examining the plant species comprising their diet both before and after their translocation helps give us a picture of the howler monkeys' ability to add unfamiliar species to their diet. Determining how quickly, and to what degree, translocated howler monkeys rely upon unfamiliar species to fulfill their nutritional needs may lend some insight into why howler monkeys can survive well in fragmented landscapes when other primates may not. Finally, an increased ability to modify their behavior to cope with nutritional stress may increase the ability to persist in patches with limited food sources.

4. PLANT PART COMPOSITION OF DIET

Young and mature leaves are an integral component of the diet of howler monkeys (Milton, 1980; Julliot and Sabatier, 1993, Silver et al., 1998). Mature leaves are often a more abundant resource and are thought to be eaten when preferred items such as fruit and flowers are unavailable (c.f., Neville et al., 1988; Marsh, 1999). The diets of animals

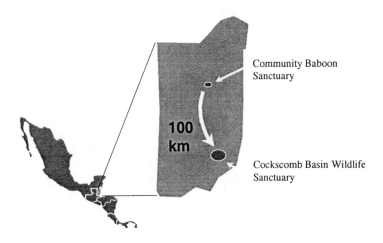

Community Baboon
Sanctuary

100
km

Cockscomb Basin Wildlife
Sanctuary

Figure 1. Location of the study areas, Belize.

that are unfamiliar with the locations of preferred, but clumped, food sources could be expected to be higher in those plant parts most easily located, so mature leaves may be expected to form a greater portion of the diet of animals newly arrived in an area.

5. STUDY SITES AND METHODS

5.1. Community Baboon Sanctuary

The study commenced in the village of St. Paul's Bank (17°33'N and 88°35'W) within CBS, between March 1994 and April 1995. The CBS consists of eight-member villages situated along the Belize River in northern Belize. The CBS is a subsistence agricultural community with a patchwork of pastures, gallery forest, and variable sized fragments of secondary forest. The area experiences a distinct dry season, usually lasting from January through May. The area also experiences seasonal flooding. Annual rainfall during the study was 1,955 mm (Silver, 1997). Mean annual temperatures during the study were 20°C minimum and 32°C maximum. Calculated densities of *A. pigra* within forest fragments throughout CBS range from 47 to 250 individuals per km^2 (Ostro et al., 2001). Within CBS, hunting pressure is high for most animals except the howler monkeys, and grazing pressure from livestock (horses and cattle) is also high.

5.2. Cockscomb Basin Wildlife Sanctuary

The CBWS is a 400-km^2 protected area along the eastern slope of the Maya Mountains in south-central Belize. The park is located about 15 km from the Caribbean coast at 16°49'N and 88°47'W. Elevations within the sanctuary range from 50 to 1,120 m above sea level (Kamstra, 1987). The vegetation in this area is a mixture of evergreen and semi-evergreen broadleaf forest. The terrain is comparatively flat, with 75% of the area lying below 200 m asl (Kamstra, 1987). The pattern of rainfall was similar to that of

CBS, with little rain between January and late May. Total annual rainfall during the study was 2,420 mm (Silver, 1997). The mean minimum temperature was 25°C, and the mean maximum temperature was 32°C. The only resident howler monkeys in CBWS before this study's translocation were 40 monkeys that had been similarly translocated in the two years preceding the 1994 release.

5.3. Vegetation Transects

Fourteen belt transects of 100 m by 10 m each were established within CBWS. The origins and direction of the transects were randomly assigned to coordinates within the home range of a howler troop. All stems within the transects ≥ 30 cm circumference at breast height were measured, tagged, and identified, usually to genus or species level. Transects were established until a plot of the species-area curve approached an asymptote. Vegetation was characterized with regard to overall stem density, species composition, and species richness. These measures were compared with the vegetation found on similar transects totaling one ha in CBS.

5.4. Study Troops and Behavioral Data Collection

Beginning in March 1994, two troops (T1 and T2) were followed in CBS for three days per month. In May 1994, the study troops were captured. The adults were fitted with radio transmitter collars and identifying bracelets and translocated to CBWS. After release, data were collected for 12 months from these troops and two troops (E1 and E2) whose members had been translocated into the basin in 1992 and 1993 (Koontz et al., 1994).

5.5. Dietary Comparisons

All foods eaten by the focal animals were categorized by species and plant part whenever possible. Foods were also categorized as belonging to species found within both CBS and CBWS (familiar) or exclusively in CBWS (novel). Information provided by field assistants, other researchers in CBS (Lyon personal communication; Marsh personal communication), our own observations, and published lists of vegetation (Horwich and Lyon, 1990) were used to determine if species in CBWS were also found in CBS. Time spent feeding on foods belonging to species not found in CBS by established and translocated troops were compared. Because foliage is more likely than fruits to be chemically defended, and thus may be avoided by translocated monkeys, time spent feeding on foliage of species not found in CBS was compared separately.

Dietary diversity for each troop was assessed on a monthly and yearly basis from May 1994 to May 1995. Dietary overlap between troops within CBWS and between groups in CBWS and CBS was determined by the formula:

$$Dietary\ overlap = \sum S_i , \qquad (1)$$

where dietary overlap is the sum of the percentages of diet shared by each of the troops (Si) during the period in question (Struhsaker, 1975).

5.6. Statistical Analyses

Diet composition data were not normally distributed. Square root transformations of these data yielded the lowest skewness and kurtosis coefficients. Monthly dietary diversity indices were calculated using the Shannon-Wiener diversity index (Pielou, 1966). Diversity indices were log transformed. We then compared differences between troops in monthly dietary diversity indices using a repeat measures analysis of variance.

A repeated measures analysis of variance was also used to analyze differences in the proportion of time each troop spent feeding on novel foods (i.e., from species not found in CBS). Pearson's product moment correlations were used to analyze the relationship between time spent feeding and time resting, as well as feeding and traveling.

6. RESULTS

6.1. Vegetation and Food Abundance

One-third (32.5%) of the relative density of the CBWS forest is comprised of trees also found in CBS. However, only 10.6% of the relative tree density is comprised of trees that the translocated monkeys had previously eaten in the CBS.

As expected, both mature leaves and young leaves eaten by the monkeys were available throughout the year. Not surprisingly, the abundance of fruit, flowers, and young leaves were highly seasonal (Figure 2). The months of April through August were the months of high fruit availability, and October, November, and December had low fruit availability. January through March had moderate fruit abundance.

6.2. Dietary Diversity and Overlap

Diversity indices for the entire study were higher for the two translocated troops (H' = 1.41 and 1.35) than for the established troops (H' = 1.26 and 1.23). There were no clear monthly differences in dietary diversity between troops ($p = 0.816$).

Dietary overlap between the two translocated troops in CBWS was low for the first six weeks after translocation was initially low, but reached 60% six weeks after their release (Figure 3). After that time, over the course of the study, overlap between T1 and T2 was 41%. There was little difference in the degree of dietary overlap between troops based upon their time of translocation. T1 and T2 overlapped E1 and E2 at most by 49% and at least by 42%. E1 and E2 shared 51% dietary overlap. The one important exception to this was the first two months following the translocation (May and June 1994). During this period dietary overlap between the two established troops remained high, but overlap between the two translocated troops, and between the established and translocated troops, was much lower (Figure 3). With only two dietary exceptions, all troops fed from the same major food items (Table 1). The mean annual dietary overlap between any two study troops at the same site was 61%. By comparison, the mean annual dietary overlap between the monkeys studied in CBWS and monkeys in CBS was 10%.

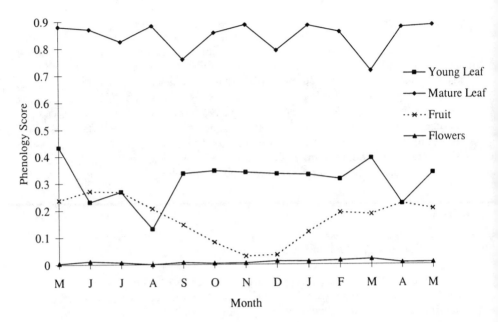

Figure 2. Food availability in the CBWS, May 1994 to May 1995. Monthly food abundance scores for young and mature leaves, fruits, and flowers.

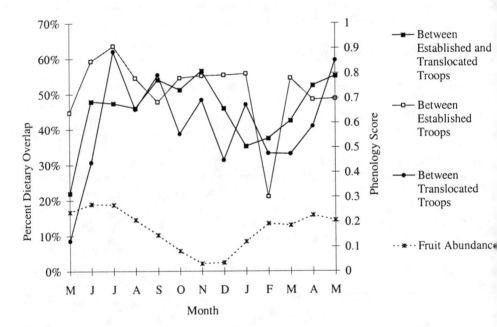

Figure 3. Monthly dietary overlap between two established troops; two translocated troops and the established and translocated troops. Also included is the monthly fruit abundance based upon the phenology score.

Table 1. Plant species eaten in the CBWS by recently translocated and established troops for >1% of the total feeding time for all troops. Data are presented in order of dietary frequency and categorized by troop use and familiarity. Values for all *Ficus* species are combined. Novel foods were species not found at the source site (CBS).

Species	Common name	% Diet	Parts eaten	Troops	Novel/ Familiar
Celtis schippei	Suc'luwiin	21.81	ML	ALL	N
Pouruma bicolor	Mt. Trumpet	10.58	FR	ALL	N
Cecropia spp.	Cecropia	10.58	ML, FR, PT	ALL	F
Cordia bicolor	Fiddlewood	5.99	FR	ALL	N
Arrabidae sp.*	Common vine	5.61	YL, ML, FL, TD	ALL	F
Brosimum guianense	Red Ramon	5.21	YL, ML, FR, FL	ALL	F
Ficus spp.	Ficus spp.	4.53	YL, ML, FR	ALL	F
Simarouba glauca	Negrito	4.29	FR	ALL	F
Combretum frucitosum*	Bottlebrush	3.41	YL, FR, FL, TD	ALL	F
Pterocarpus sp.	Kaway	3.21	YL, ML	ALL	N
Bellucia grossularides	Black broadleaf	3.03	FR	E2, T2	N
Daliocarpus sp.*	Sandpaper vine 2	2.61	YL, FR, TD	ALL	N
Schizolobium parahybum	Quamwood	1.95	YL, FR, FL	ALL	F
Acacia glomerosa	Wild Tamarind	1.93	YL, FR, FL	T1	N
Dialium guianensis	Ironwood	1.77	YL, FR	ALL	N
Brosimum sp. 2	Ramon	1.34	YL	ALL	N
Inga sp.	Tamatama	1.30	YL, ML, FR, FL	ALL	F
Davilla sp.*	Sandpaper vine-1	1.23	YL	ALL	F
Piptocarpha sp.*	White water vine	1.19	YL, ML, TD	ALL	F
unknown	Milk vine	1.18	YL, ML, TD	ALL	N

YL = young leaves; ML = mature leaves; FR = fruits; FL = flowers; PT = petioles;
TD = tendrils.
(*) indicates liana or vine species

6.3. Consumption of Novel Foods

Of the 75 species eaten in CBWS, 59% were novel species (Figure 4). Overall, focal animals spent 60% of their feeding time eating foods belonging to species not found in CBS. Of the 20 most frequently eaten food species, 10 were not found in CBS (Table 1). Translocated monkeys were consuming novel foods on the first day they were followed, post-release. The portion of time spent feeding on foods belonging to species not found in CBS was nearly identical between the translocated and established troops, with no statistical differences among troops when all foods are considered ($p = 0.795$, Figure 4).

6.4. Differences in Plant Part Composition of Diet

The study troops in CBWS spent 34% of their feeding time eating fruit, 6% eating flowers, 29% on mature foliage, 29% on immature foliage (or foliage of unknown maturity), and 3% on other items such as stems, petioles, and tendrils (Figure 5).

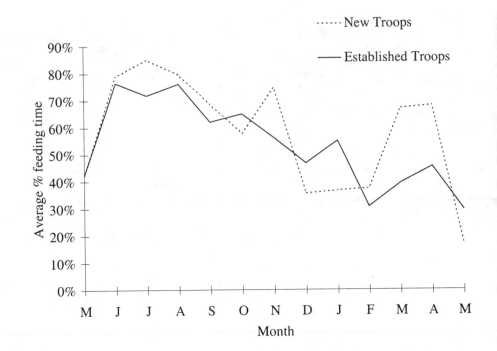

Figure 4. Consumption of foods belonging to species not found in CBS by new and established troops in the CBWS.

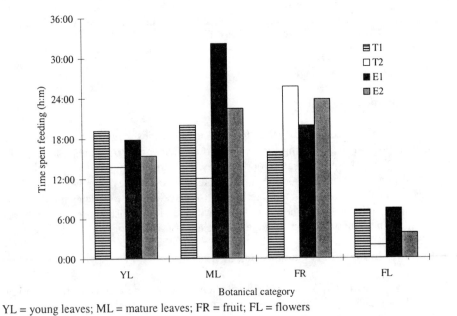

YL = young leaves; ML = mature leaves; FR = fruit; FL = flowers

Figure 5. Time engaged in eating different plant parts by all study troops.

There were no consistent differences in the overall time translocated troops spent eating fruits, flowers, or young leaves when compared to established troops (Figure 5). There does, however, appear to be a difference in the time spent eating mature leaves, with the total time of established troops much higher than that of translocated troops. The two translocated troops spent 28% and 20% of total time feeding upon mature leaves. This is compared with 30% and 37% for the two established troops.

6.5. Activity Budgets

A total of 1,540 hours of focal observations in CBWS were collected. Translocated animals traveled an average of 9.5% of the time, while the established troops traveled 10.1% of the time. The time spent feeding was less for both translocated troops (15.7% and 17.5% for T1 and T2, respectively) when compared to the established troops (20.9% and 17.9% for E1 and E2, respectively). Both translocated troops also rested more than established troops (73.3% and 71.6%, vs 65.7% and 69.4%, respectively) (Figure 6). Feeding and resting behavior were negatively correlated (p <0.001), but feeding and traveling were not ($p = 0.61$).

7. DISCUSSION

7.1. Novel Food Consumption

This study demonstrates how *A. pigra* has very plastic dietary and behavioral patterns. With only 10.6% of the trees in CBWS belonging to species previously eaten in

Figure 6. *A. pigra* resting in a *Ficus insipida* in the Community Baboon Sanctuary. Photo by Rob Horwich.

CBS, translocated monkeys encountered few familiar foods upon their release. New arrivals had little choice but to exploit unfamiliar food sources. Nonetheless, no preference for eating familiar food species was detected in the translocated monkeys' diet. They showed no increased reliance on familiar foods when their diet was compared with the established troops. The translocated monkeys did not exhibit any difficulty in recognizing suitable novel food species and including them in their diet. In fact, contrary to expectations based upon gradual acquisition of knowledge concerning the edibility of novel foods, time spent feeding upon novel foods decreased over time, rather than increased. This pattern reflects the seasonal availability of edible foods that happen to be found in CBS, as opposed to a reliance on the familiarity of the food. Thus, hypotheses that assign preeminence to the role of early-development experience learning mechanisms, such as food imprinting (Immelman, 1975) or social learning from adults during infancy (Glander, 1975) to explain diet selection were not supported by the observed pattern of novel food consumption. These findings may indicate that when howler monkeys living in fragmented landscapes need to colonize new patches of forest or are left in a remnant as the forest is destroyed, unfamiliar food sources are readily exploitable. This kind of flexibility allows howlers to persist in marginal habitats where other primate species fail.

7.2. Learning Processes and Seasonality

Patterns of dietary overlap in the translocated troops may be a reflection of acclimation processes by these troops. The two established troops generally had a high degree of dietary overlap, while the new arrivals in CBWS had less dietary overlap, either with the two resident troops or each other.

The measures of dietary overlap for the first two months of data collection in CBWS may reflect an adjustment period by translocated troops. Low levels of dietary overlap during this time may result from translocated monkeys initially selecting foods that were not the best nutritional choices, but adjusting their diet as they become more familiar with the relative nutritional value of food items and/or the locations of higher quality food sources. Three months after release, dietary overlap between the translocated troops was virtually as high as that between the two established troops, and the diet was very similar between translocated and established troops. This suggests translocated monkeys may have acquired enough information about the available foods to arrive at the same diet selection decisions as resident troops. As all CBWS howler troops maintained exclusive home ranges, all dietary decisions appeared to be reached independently by each troop, with no observational learning between troops likely.

Seasonal changes in resource abundance may force translocated troops to reshape their foraging strategies. Eight weeks after release, there was no discernible difference in dietary overlap between translocated and established troops until October, when diminished fruit abundance may force the monkeys to seek alternative diets. Then dietary overlap again drops between translocated troops, though not between established troops, and could reflect the same learning processes concerning the alternative food choices. Incremental increases in dietary overlap between translocated troops, and between translocated and established troops, support the idea that some gradual learning processes *are* shaping diet selection. At the beginning of each period where seasonal changes in food distribution occur, dietary overlap between translocated troops is lower than that between established troops. Overlap between the translocated troops then gradually approaches the level of established troops as the season progresses. Seasonal changes in the availability of foods present new challenges regarding the location and availability of food items. Thus, each change in season means meeting nutritional requirements with new food choices in new locations.

Sampling each troop for only four days per month makes it difficult to draw definitive conclusions about the role played by short-term learning processes such as one-trial avoidance learning (Garcia and Koelling, 1966), or the "sampling" behavior described by Glander (1975) and Coelho et al. (1976). These processes may be employed to ascertain a novel food item's palatability (as well as prepare the intestinal flora for processing of novel foods), but may occur outside of the data collection periods. Study methodology not withstanding, there were few instances when items were observed to be tasted and consequently ignored by adult monkeys, either within or outside of the data collection periods.

An alternative explanation for the observed diet selection patterns of the translocated animals may be "non-learned" palatabilities (see Chapman and Blaney, 1979; Cassini, 1994) where food palatability is determined by mechanisms of immediate sensory feedback. These can be employed in combination with generalized, learned associations to food item characteristics to formulate "rules-of-thumb," which determine dietary choices. Rules-of-thumb as applied to dietary choice (described by Cassini, 1994) does not require a demonstration of diet selection altered by experience, but regards diet selection as based upon simple rules concerning the characteristics of potential food choices. These rules would have been formed by the adults during immaturity previous to translocation, and because all monkeys came from the same source habitat, could result in all translocated monkeys in CBWS reaching the same decisions concerning the suitability of unfamiliar foods. This explains one of the most striking aspects of the consumption of unfamiliar foods, not only do new arrivals immediately begin eating foods belonging to species not found in CBS, but they begin eating the *same* items as the established monkeys.

For an animal such as a howler monkey, rules-of-thumb mechanisms for determining diet selection make ecological sense. Howler monkeys have the widest distribution of New World primates, inhabiting many different habitats with plant communities characterized by high degrees of endemism and low species abundances. Animals that follow generalized rules of diet selection that can be applied to a wide spectrum of potential foods would be at a great advantage over those that must rely on previous experience to determine a plant's edibility or nutritional value. This is particularly true when dispersal means encountering many new, or rare, plant species, and with rules-of-thumb, even areas with plant communities dominated by previously unknown species can be readily colonized.

7.3. Patterns of Food Abundance and Diet Composition

Although nutritionally desirable, ripe fruits often have a spatially and temporally heterogeneous distribution that makes them difficult to locate in continuous habitat. This is particularly true for new arrivals. Established troops have a great understanding of their range and may have "cognitive maps" of where fruiting trees are. Despite fewer trees used as food sources and less time feeding overall, the total time spent feeding on ripe fruits was similar between translocated and established troops. These results contrast with de Vries' study (1991) of translocated A. palliata in Costa Rica, who found that during their first four months of residency, translocated monkeys lost weight and fed significantly less on ripe fruits than control groups. In CBWS, translocated monkeys compensate for deficiencies in locating food sources by feeding for longer periods when fruit is found. Longer fruit-feeding sessions of translocated monkeys in comparison to established monkeys, particularly during the first three months in CBWS, may indicate a reluctance to leave a known, high-quality food patch in an unknown area (also see Rodriguez-Luna et al., this volume).

Howlers in the CBS fragments show a similar feeding pattern with scarce (sometimes single source) fruit resources (Marsh, 1999). It is possible that these individuals learned to exploit limited fruit sources from within their original fragments in the CBS, which enhanced their survival ability in a new forest. In Belize, translocated animals were intentionally released during the season estimated to be highest in fruit abundance, with the most common tree in the forest (*Pouruma bicolor*) bearing ripe fruit. Fruit remained widely distributed and abundant for the 10 weeks following translocation, then gradually decreased.

The prediction that new troops would exhibit a greater reliance on mature foliage than established troops was not supported. Established troops spent more time feeding on mature foliage than translocated troops. This difference is surprising if mature leaves are only eaten because they are an easy to locate, secondary alternative to fruits and young leaves. Yet in actuality, only a few specific mature leaf species are eaten. Established troops may be better able to exploit these mature leaves as a nutritional addition than recently translocated troops.

The most frequently eaten species, *Celtis schippii* (Ulmaceae), accounted for nearly 66% of the mature leaf feeding time and 22% of the time spent feeding overall. This species may be the one food item translocated troops gradually learned to eat. Unlike most species in which both young and mature leaves were eaten, the monkeys only ate mature foliage of *C. schippii*. Nutritional analysis of these leaves indicate a very high level of water soluble carbohydrates (Silver et al., 2000), a seasonally scarce nutrient. While the established troops were recorded as eating *C. schippii* from the first week of the study, it was four weeks before troop T2 was observed to feed upon this species, and 12 weeks before it was regularly represented in the diet. T1 was not observed to feed on *C. schippii* until 16 weeks from their release into CBWS, and while it eventually became a major dietary component during certain months, they consistently fed upon it less than the other three troops.

Tree distribution patterns may also help explain the differences in mature foliage consumption between new and established troops. The two most important mature leaf food sources, *C. schippii* and *Cecropia* spp., account for nearly 90% of the feeding time for mature foliage overall. The relative density of *C. schippii* is only 1.3%, and that of *Cecropia* spp. 1.8%, so the monkeys are being extremely selective in their mature leaf choices. Far from being abundant and evenly distributed, these species are uncommon and have a patchy distribution within the habitat. Translocated troops with less knowledge of the location of these trees may be at a disadvantage in finding mature leaves that they can exploit as a nutritional supplement, either because they cannot locate them effectively, or because their home ranges do not contain adequate numbers of these trees.

7.4. Differences in Activity

Howler monkeys faced with reduced nutritional intake may respond by reducing their energy expenditure. Initially, they appear to adopt a combination of increased rest and possibly, increased travel. This may be a short-term behavioral response to the acute stress of the translocation process and life in new surroundings. The 24- to 72-hour period following their release in CBWS showed the highest resting rates, combined with the lowest feeding rates recorded in the study. The increased time spent resting in response to reduced feeding supports the characterization of howler monkeys as energy minimizers (Neville et al., 1988). Excessive periods of rest (even for howler monkeys) immediately following release of translocated *A. seniculus* has also been reported (Vie and Richard-Hansen, personal communication). Bioenergetic constraints of a metabolism adapted to folivory may limit the howler monkeys' ability to increase travel time for extended periods.

The negative correlation between feeding and resting apparent throughout the study, and the lack of such a relationship between feeding and traveling, is the key to understanding the long-term behavioral response of howler monkeys to translocation. The long-term behavioral profiles of the translocated troops support the perception of howler monkeys as animals that behaviorally restrict their energy expenditure. The similar amounts of time spent traveling among all four troops, and the lack of a demonstrated correlation between travel time and feeding time, support Milton's (1978) contention that howler monkeys avoid extreme fluctuations in voluntary energy output. Less time spent feeding presents the howler monkeys with an energetic dilemma that can only be solved by reducing their voluntary energy use. During times of food abundance, high levels of rest may enable howler monkeys to maintain reserves in the face of seasonal fluctuations in readily available energy sources (Milton, 1978). In translocated animals this mechanism would also function as a cushion against foraging inefficiencies resulting from unfamiliar surroundings.

Behavioral modifications to reduce energy expenditure may allow howler monkeys to survive in habitats with a wide variety of food availability and abundance. Comparisons between monkeys translocated to CBWS and troops that remained in CBS show the mean portion of time spent feeding was much higher for monkeys studied at CBS (24.4%) than for all study troops in CBWS (18.2%). Not surprisingly, the portion of time spent resting was higher for the CBWS monkeys (70.6%) than those in CBS (61.9%). The portion of time spent traveling in both CBS and CBWS was 9.8%. These activity budgets are consistent with animals who maintain a constant travel time in response to lower food abundance.

The dietary overlap of approximately 10% between troops that remained in the source site (CBS) and the animals in CBWS illustrate the flexibility of the howler monkeys' diet. Monkeys living in CBWS spend much less time feeding on fruits and much more time feeding on foliage, particularly mature foliage. This flexibility in diet contrasts with the limited energy expended by traveling. The behavioral and dietary flexibility outlined here suggests howler monkeys may be both ideal candidates for translocation into suitable habitats, and well predisposed to colonize fragments, regardless of vegetative overlap with their site of origin. While the monkeys can adjust their diet composition in a new habitat, their short-term response of minimal activity reduces the chance of their immediate dispersal far from their release site.

8. SUMMARY

In this study, translocated howler monkeys exhibited few detectable signs of difficulty adjusting to unfamiliar surroundings. This dietary and behavioral plasticity emerged as integral components for survival in novel habitats. Two main points emerge that may suggest reasons why they can readily colonize forest fragments when other species cannot: 1) Although the patterns of use for a few food species may reflect some gradual learning mechanisms in diet selection, howler monkeys appeared able to exploit novel dietary items immediately upon introduction to a new area. The howler monkeys' diet appears flexible enough that vegetative overlap between forest fragments may not be important in colonizing a new forest area, and 2) by making behavioral adjustments that presumably minimize voluntary energy expenditure, howler monkeys can survive in diverse habitats with large differences in food abundance. By increasing the relative time resting in response to reduced feeding (while maintaining a steady travel time), howler monkeys may be well suited to overcome the nutritional stresses imposed by colonizing new areas or areas with lower nutritional sources. The same bioenergetic strategy that enables howler monkeys to follow a folivorous diet may also make them well suited to survival in fragments with nutritionally poor diet choices. Other folivorous primates who modify their activity patterns in a similar way may also be particularly good at persisting in forest fragments.

9. ACKNOWLEDGMENTS

The howler translocation study was supported by the Wildlife Conservation Society. Linde Ostro was a principal investigator of the study, and Fred Koontz and Rob Horwich contributed logistically and intellectually as well. We would like to thank the Belize Audubon Society and the people of St. Paul's Bank and Maya Centre for their assistance throughout the project.

10. REFERENCES

Cassini, M. H., 1994, Behavioral mechanisms of selection of diet components and their ecological implications in herbivorous mammals, *J. of Mammal.* **75**:733–740.
Chapman, R. F., and Blaney, W. M., 1979, How animals perceive secondary plant compounds, in: *Herbivores*, G. A. Rosenthal and D. H. Janzen, eds., Academic Press, New York, pp. 161–199.
Charnov, E. L., 1976, Optimal foraging: The marginal value theory, *Theoretical Population Biology* **9**:129–136.
Coelho, A. M., Jr., Bramblett, C. A., Quick, L. B., and Bramblett, S., 1976, Resource availability and population density in primates: A socio-bioenergetic analysis of the energy budgets of Guatemalan howler and spider monkeys, *Primates* **17**:63–80.
de Vries, A., 1991, Translocation of Mantled Howling Monkeys (*Alouatta palliata*) in Guanacaste, Costa Rica, M. A. Thesis, University of Calgary.
Estrada, A., 1984, Resource use by howler monkeys (*Alouatta palliata*) in the rainforest of Los Tuxtlas, Veracruz, Mexico, *Int. J. Primatol.* **5**:105–131.
Ganzhorn, J. U., and Wright, P. C., 1994, Temporal patterns in primate leaf eating: The possible role of leaf chemistry, *Folia Primatol.* **63**:203–208.
Garcia, J., and Koelling, R. A., 1966, Relation of cue to consequence in avoidance learning, *Science* **4**:23–124.
Glander, K. E., 1975, Habitat and Resource Utilization: An Ecological View of Social Organization in Mantled Howling Monkeys, Ph.D. dissertation, University of Chicago.

Glander, K. E., 1978, Howling monkey feeding behavior and plant secondary compounds: A study of strategies, in: *The Ecology of Arboreal Folivores*, G. G. Montgomery, ed., Smithsonian Institution Press, Washington D.C., pp. 561–574.

Glander, K. E., 1981, Feeding patterns in mantled howling monkeys, in: *Foraging Behavior*, A. C. Kamil and T. D. Sargen, eds., Garland series in ethology, New York, Garland STPM Press, pp. 231–257.

Glander, K. E., 1992, Dispersal patterns in Costa Rican mantled howling monkeys, *Internat. J. of Primatol.* **13**:415–436.

Horwich, R. H., and Lyon, J. A., 1990, *A Belizean Rainforest*, Orang-Utan Press, Gays Mills, WI.

Immelman, K., 1975, Ecological significance of imprinting and early learning, *Ann. Rev. Ecol. System.* **6**:15–37.

Julliot, C., and Sabatier, D., 1993, Diet of the red howler monkey (*Alouatta seniculus*) in French Guiana, *Int. J. Primatol.* **14**:527–550.

Kamstra, J., 1987, An Ecological Survey of the Cockscomb Basin, Belize, Masters Thesis, York University, Ontario, Canada.

Koontz, F. W., Horwich, R., Saqui, E., Saqui, H., Glander, K., Koontz, C., and Westrom, W., 1994, Reintroduction of black howler monkeys (*Alouatta pigra*) into the Cockscomb Basin Wildlife Sanctuary, Belize, in: *American Zoo and Aquarium Association Annual Conference Proceedings*, AZA, Bethesda Maryland, pp. 104–111.

Koontz, F., and Ostro, L. E. T., 2001, Translocation of black howler monkeys (*Alouatta pigra*) in Belize: A multidisciplinary team approach for effective wildlife conservation, *AZA Field Conservation Manual*, Bethesda, Maryland.

Marsh, L. K., 1999, Ecological effect of the black howler monkey (*Alouatta pigra*) on fragmented forests in the Community Baboon Sanctuary, Belize, PhD thesis, Washington University, St. Louis, MO.

Milton, K., 1978, Behavioral adaptation to leaf-eating by the mantled howler monkey (*Alouatta palliata*), in: *The Ecology of Arboreal Folivores,* G. G. Montgomery ed., Smithsonian Institution Press, Washington D.C., pp. 535–551.

Milton, K., 1980, *The Foraging Strategy of Howler Monkeys*, Columbia University Press, New York.

Neville, M. K., Glander, K. E., Braza, F., and Rylands, A. B., 1988, The howling monkeys, Genus *Alouatta*. in: *Ecology and Behavior of Neotropical Primates*, R. A. Mittermeier, A. B. Rylands, A. F. Coimbra-Filho, and G. A. B. de Fonseco, eds., World Wildlife Fund, pp. 349–454.

Ostro, L. E. T., Silver, S. C., Koontz, F. W., Horwich, R. H., and Brockett, R., 2001, Shifts in Social Structure of Black Howler (*Alouatta pigra*) Groups Associated with Natural and Experimental Variation on Population Density. *International Journal of Primatology* **22(5)**:733–748.

Pielou, E. C., 1966, Shannon's formula as a measure of specific diversity: Its uses and misuses, *Am. Nat.* **100**:463–465.

Pyke, G. H., 1984, Optimal foraging theory: A critical review, *Ann. Rev. Ecol. System.* **15**:523–576.

Silver, S. C., 1997, The feeding ecology of translocated howler monkeys in Belize, Ph.D. Dissertation, Fordham University.

Silver, S. C., Ostro, L. E. T., Yeager, C. P., and Horwich, R., 1998, The feeding ecology of the black howler monkey (*Alouatta pigra*) in northern Belize, *Am. J. Primatol.* **44**:263–279.

Silver, S. C., Ostro, L. E. T., Yeager, C. P., and Dierenfeld, E., 20002, Phytochemical and mineral components of foods consumed by black howler monkeys (*Alouatta pigra*) at two sites in Belize, *Zoo Biology* **19**:95–109.

Strushaker, T. T., 1975, Feeding habits of five monkey species in the Kibale Forest, Uganda, in: *Recent Advances in Primatology*, D. J. Chivers and J. Herbert, eds., Academic Press, New York.

HOWLER MONKEYS (*ALOUATTA PALLIATA MEXICANA*) AS SEED DISPERSERS OF STRANGLER FIGS IN DISTURBED AND PRESERVED HABITAT IN SOUTHERN VERACRUZ, MÉXICO

Juan Carlos Serio-Silva and Victor Rico-Gray[*]

1. INTRODUCTION

Recent studies suggest that the identification of interactions among keystone species in ecosystems is one of the most important aspects to be considered in conservation programs (Thompson, 1994; Howe and Miriti, 2000). We also need to evaluate how interspecific interactions are modified by habitat fragmentation and destruction (Nathan and Muller-Landau, 2000), which will provide information required to make informed policy decisions regarding conservation and forest management (Garber and Lambert, 1998). In this chapter, we address the effects of habitat fragmentation on the regeneration dynamics of tropical rain forests, particularly the ecological interactions among strangler figs (*Ficus* spp.: Moraceae) and Mexican howler monkeys (*Alouatta palliata mexicana*) in southeastern Mexico.

1.1. Figs and Primates

Figs are distributed in tropical and subtropical areas worldwide, and the genus is considered one of the most numerous in species (ca. 800) among woody plants. Many of these species (ca. 280) are hemiepiphytic "stranglers," especially within the subgenus *Urostigma* (Putz and Holbrook, 1986). Figs have been regarded as keystone species for conservation in many tropical rain forests (Terborgh, 1986; McKey, 1989) and have been suggested as an important food source for invertebrates and vertebrates, including birds, bats, and monkeys, who may also disperse their seeds (Banaccorso, 1979; Janzen, 1979; Gautier-Hion et al., 1985; de Figueiredo, 1993; Marsh, 1999). However, the latter may not be true under all circumstances (Gauthier-Hion and Michaloud, 1989).

[*] Serio-Silva and Rico-Gray, Divisiûn Acadèmica de Ciencias Biologicas, Universidad Juarez Autûnoma de Tabasco, km 0.5 carretera Villahermosa-Cardenas, entrada por Bosques de Saloya, CP 86039, Villahermosa, Tabasco, México. Correspondence to J. C. Serio-Silva (email: juancarlosserio@hotmail.com).

Primates in Fragments: Ecology and Conservation
Edited by L. K. Marsh, Kluwer Academic/Plenum Publishers, 2003

Frugivore strategies and the interactions between fruiting plants (including figs) and primate dispersers have been documented for *Pongo pygmaeus* in Asia (Galdikas, 1982), for *Papio anubis* and *Gorilla gorilla gorilla* in Africa (Lieberman et al., 1979; Tutin et al., 1991), and for *Alouatta palliata mexicana, A. palliata palliata, Cebus capucinus, Saguinus mystax,* and *Saguinus fuscicollis* in the Neotropics (Howe, 1980; Estrada and Coates-Estrada, 1984, 1986, 1991; Garber, 1986). Some authors have suggested that as a consequence of their flowering asynchrony, figs should be particularly vulnerable to forest fragmentation and area reduction (McKey, 1989; Janzen, 1986). Other observations suggest that due to forest fragmentation, figs, and specially hemiepiphytic figs, may be more vulnerable to extinction than other tropical forest trees, because their minimum viable population size has to be large but adults often occur at low densities (McKey, 1989). However in some cases, we can find a disproportionately high density of figs in fragments with howlers (*A. pigra*) (Marsh, 1999). In these cases the history of each fragment, age, and causes of isolation all influence the floristic composition at each site. Figs in fragmented forests have strong dependence on the behavior of seed dispersers, who may alter the shape of the seed shadow (i.e., the spatial distribution of seeds following dispersal). For example, if monkeys restricted to matured forest are important dispersal agents for particular fig species, the seed shadow may be truncated by fragmentation, since these animals will remain in the forest fragment, and the density of dispersed seeds may increase only within the fragment. Birds or bats, in contrast, may carry fig seeds from one forest fragment to another, with few seeds dropped in suitable habitats between fragments (Galindo-González, 1998). The direction and magnitude of the effect of forest fragmentation on fig density and seed dispersal will surely vary from case to case, and fig species may thus pose difficult but crucially important problems for conservation biology. Therefore, how important are some primate species as dispersal agents of fig seeds?

1.2. Figs and Mexican Howler Monkeys

Figs have been considered one of the most important food sources for primates when they are available (Serio-Silva, 1996, Serio-Silva et al., 2002; Silver et al., 1998; Marsh, 1999), particularly during general food scarcity periods (Wrangham et al., 1993). *Ficus* trees contribute young and mature leaves, and particularly fruits, to the diet of all howler monkeys (*Alouatta* spp.) (Milton et al., 1980; Serio-Silva, 1996, 1997; Silver et al., 1998, Marsh, 1999). Furthermore, recent studies suggest that howler monkeys are particularly important seed dispersers of many tree species including *Ficus* (Coates-Estrada and Estrada, 1986; Estrada and Coates-Estrada, 1991; Marsh, 1999). The main results suggest that *Ficus* seeds consumed by these primates and deposited with their feces attain a higher germination success (de Figueiredo, 1993; Serio-Silva and Rico-Gray, 2002a, b). However, the specific association between these primates and *Ficus* species, in the subgenus *Urostigma*, has not been studied in detail (but see, de Figueiredo, 1993; Wrangham et al., 1993).

This ecological process can be very important in the regeneration dynamic of Neotropical rain forests, especially among howler monkeys and *Ficus* species. Species in this subgenus are hemiparasitic and need a host tree. Howler monkeys use the canopy for their normal activities, and, after a long digestive process ($\pm$ 20 hours), they move to different places where seeds can be deposited with the feces on host trees (e.g., branches, holes, bark [Milton, 1980]). Seeds will then face microclimatic factors that influence their germination and later establishment. However, these microclimatic factors are not uniform in all the forested areas (Laman, 1995), which will be evident when we compare undisturbed and disturbed habitats.

Many frugivores allow the passage of fig seeds undamaged through their digestive tracts (Medellin and Gaona, 1999). However, few quantitative studies exist on the role of howler monkeys as dispersers and their effect on the germination of fig seeds (Estrada and Coates-Estrada, 1986; de Figueiredo, 1993). Specifically, we do not know how seed dispersal and germination are altered by changes produced by forest fragmentation. Thus, we focused on the dispersal distances where figs seeds were deposited by Mexican howler monkeys in preserved and disturbed habitats, and the effect of howlers on germination of ingested (versus noningested) *Ficus perforata* and *Ficus lundelli* seeds. We addressed the following questions: 1) Do dispersal distances differ between disturbed and preserved habitats? 2) Does ingestion by Mexican howler monkeys affect germination success under controlled conditions? 3) Does germination success vary as an effect of microclimatic conditions (light, temperature, and humidity) on the canopy in preserved and disturbed habitat? and, 4) Do microclimatic factors vary between arboreal strata on disturbed and preserved habitat, and how does this influence germination success of ingested and noningested *F. perforata* and *F. lundelli* seeds?

2. MATERIALS AND METHODS

2.1. Description of Study Sites

Field observations, seed sampling, and planter box experiments were done between 1998 and 1999 in two contrasting habitats, both of them inhabited by howler monkey troops, in southern Veracruz, México.

The disturbed site (DIS) was a relatively small isolated forest patch (40 ha) modified by humans (e.g., extraction of species, artificial fires, agriculture, cattle raising) located in Playa Escondida, San Andrés Tuxtla, Veracruz, México (18°35' N, 95°03' W; altitude 70 m) (Figure 1). The climate is warm and humid, total annual precipitation is approximately 2,000 mm, the highest precipitation occurs between June and February and the lowest in May. Maximum and minimum temperatures are 39.5°C and 9°C, respectively, with a mean of 25.1°C. The main vegetation type is tropical rain forest, intermingled with patches of secondary vegetation. The most common plant families are Moraceae, Cecropiaceae, and Leguminosae. The density of the *Ficus* species selected was *F. perforata* = 1.2 individuals per ha and *F. lundelli* = 0.9 individuals per ha (Table 1).

The preserved site (PRE) is a relatively large forest patch (600 ha), which is connected with other forested areas and protected from human activities, located in the Carolino Anaya Reserve, near Coatzacoalcos, Veracruz, México (18°07' N, 94°18' W; altitude 50 m) (Figure 1). The climate is warm and humid, total annual precipitation is approximately 2,500 mm. Maximum and minimum temperatures are 43.5°C and 18°C, respectively, with a mean of 25.5°C. The main vegetation type is tropical rain forest. The most common plant families are Moraceae, Rubiaceae, Orchidaceae, and Leguminosae. The density of the *Ficus* species selected was *F. perforata* = 0.2 individuals per ha, and *F. lundelli* = 0.1 individuals per ha (Table 1).

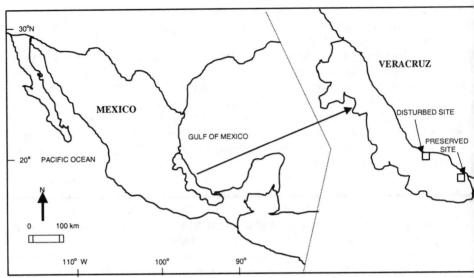

Figure 1. Location of the two study sites in southern Veracruz, Mexico: the disturbed site at Playa Escondida and the preserved site in Carolina Anaya reserve.

Table 1. General characteristics of the study sites.

Parameter	Disturbed	Preserved
Size	40 ha	600 ha
Location	18°35' N, 95°03' W	18°07' N, 94°18' W
Climate	Warm, subhumid	Warm, humid
Annual precipitation	2,000 mm	2,500 mm
Temperature (max-mean-min)	39.5 – 25.1 – 9°C	43.5 – 25.5 – 18 °C
Altitude	70 m	50 m
Vegetation type	Mostly patches of secondary vegetation derived from tropical rain forest	Tropical rain forest
Main plant families	Moraceae, Cecropiaceae, Fabaceae	Moraceae, Rubiaceae, Orchidaceae, Fabaceae
Density of *Ficus perforata*	1.2 per ha	0.2 per ha
Density of *Ficus lundelli*	0.9 per ha	0.1 per ha
Number of howler troops	4	8
Howler density	1.05 per ha (4 troops, 42 ind.)	0.12 per ha (8 troops, 74 ind.)
Home range per troop	10 ha	75 ha

For both study sites we estimated the distribution pattern for *F. perforata* and *F. lundelli*, using the Morisita Index of Dispersion (I_d, Brower et al., 1989),

$$I_d = N \frac{\Sigma X^2 - N}{N\,(N-1)}, \tag{1}$$

where N is the number of plots (0.5 ha in this case; DIS N = 4, PRE N = 20), N is the total number of individuals counted on all N plots, and ΣX^2 is the square of the number of individuals per plot, summed over all plots (I_d = 1.0 for random dispersal, I_d = 0 for uniform dispersal, and I_d = N for maximally aggregated). F statistic was used to obtain significance (Poole, 1974).

2.2. Study Troops

We studied two howler monkey troops in each study site. Total troops in DIS were four (ca. 10 ha home range) and in PRE, eight (ca. 75-ha home range). The troops selected in DIS had eight to nine individuals, and those in PRE had eight to ten individuals.

2.3. Seed Collection and Selection

Mature fruits of *F. perforata* and *F. lundelli* and monkey feces were collected until enough seeds were obtained for germination trials. Fruits were weighed and dried and seeds were randomly selected until enough seeds for germination trials were obtained. Monkey fecal samples were collected from the forest floor, mostly under the trees where the monkeys defecated, particularly under those trees where they begin their locomotion (Serio-Silva, personal observation). Feces were washed to separate leaf fiber from seeds, which were then randomly selected for the germination trials. Seeds from both feces and syconia were observed under the microscope to confirm the presence of an embryo, and those used by fig wasps (Agaonidae) were discarded. (Serio-Silva and Rico-Gray, 2002a, b.)

2.4. Controlled Germination Trials

Seeds were placed in each 9-cm Petri dish and covered with Whatman 42 filter paper. Forty Petri dishes (N = 50 seeds each) were allocated per *Ficus* species and per study site (DIS, PRE). Twenty dishes were used for feced seeds and 20 dishes for seeds from fruits (control). Petri dishes were placed in a thermostatically controlled chamber and exposed during 30 days to different temperatures based on field data (recorded in an unpublished previous study) that alternated within a 24-hour period: 16 hours at 27°C and 8 hours at 30°C. Lids were kept in place to minimize evaporation and distilled water was added to maintain humidity. Evidence of germination (appearance of the radicle) was noted every two days until no further germination was observed in seven successive counts. The position of Petri dishes was modified during each daily survey (one row from the front to the back of the chamber). We calculated the proportion of seeds germinating per treatment, per species.

2.5. Planter Box Experiment

In each site one *Diospyros dygina* tree (a typical *Ficus* host) was selected for the germination planter boxes experiments. In each selected tree, located in the most central and representative area of each DIS and PRE sites, were placed 10 germination planter boxes on each strata (high and low), all of them separated by at least 5 m between them (Figure 2). In each box we placed 200 fig seeds distributed in a random position, obtained

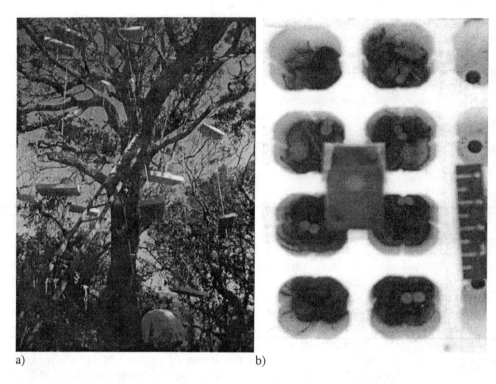

a)   b)

Figure 2. Images showing planter boxes on the tree selected (a) and fig seeds after germination (b). Photos by Juan Carlos Serio-Silva

from the following treatments: a) 50 *F. perforata* seeds (fruits), b) 50 *F. perforata* (feces), c) 50 *F. lundelli* seeds (fruits), d) 50 *F. lundelli* seeds (feces). On each germination box and arboreal stratum we placed calibrated light-sensitive *Diazo* paper for measuring integrated light ($\log_{10}$PAR) (Bardon et al., 1995). At the same time, for each arboreal strata in both sites we placed a hygrometer with a daily sampling of temperature in °C and humidity (%). Germination (appearance of the radicle) was checked every two days for 40 days.

2.6. Dispersal Distances

To evaluate seed dispersal distance (i.e., from site of fruit ingestion to site of defecation), we selected individual trees of *F. perforata* and *F. lundelli* where a monkey troop was feeding. We then followed the troop during several days. To be sure of the dispersal distance of the particular *Ficus* tree we had observed, we only obtained the distances after the second day of *Ficus* fruit ingestion and defecation (Katherine Milton, personal communication). Finally, for each study site we estimated the linear distance (9 distances for *F. perforata* and 10 distances for *F. lundelli* per site, respectively) from fruit tree to defecation site using the howler digestive process estimation of 20 hours (Milton et al., 1980).

3. STATISTICAL ANALYSES

Logit link function was used in the analysis of germination proportions using the GLIM-4 statistical package (Francis et al., 1993) to test the hypothesis that differences in germination are related to the factors considered for a) controlled germination [species *(F. perforata, F. lundelli)*, site (PRE, DIS), and seed source (seeds from fruits, seeds from feces) and their interactions] and b) field germination in planter boxes [species *(F. perforata, F. lundelli)*, site (PRE, DIS), arboreal strata (high, low), and seed source (seeds from fruits, seeds from feces)]. As a result of data overdispersion, the goodness-of-fit was evaluated with an *F*-test, using a binomial error distribution over the proportions, the scaled deviance is calculated and the change in variance can be compared directly with *F* tables to assess its significance (Crawley, 1993). To compare microclimatic factors [(light $(\log_{10}PAR)$, temperature ($^{\circ}$C), and humidity (%)], we used RM-ANOVA's per site, and per arboreal strata. For each factor we used a Tukey test to evaluate all pairwise multiple comparison for significant differences (Zar, 1996). Finally, we used *t*-tests (Zar, 1996) to compare the seed dispersal distance per each *Ficus* species, per site and their interactions.

4. RESULTS

4.1. Controlled Experiment

Seeds of both *Ficus* species germinated regularly between days 9 and 11 after the beginning of the trials, peaking around day 16, and fully developed after 28 to 30 days. For both sites and species, a higher germination percentage was obtained for seeds taken from feces than for seeds taken directly from fruits (Table 2). The log-linear model was significant ($F_{1,152} = 3.84$, $p<0.0001$) in predicting variance in germination: seed source (feces, fruit) and the interaction 'species × source.' These variables accounted for 87.9% and 5.06%, respectively, of the variation explained by the model (Table 3).

4.2. Field Experiment (Planter Boxes)

For both sites and species, a higher germination percentage was obtained for seeds taken from feces and germinated in high and low arboreal strata than seeds taken directly from fruits in high and low arboreal strata. The log-linear model showed a significant difference ($p<0.0001$) in germination between site, species, and seed source and for the interaction site × species. These variables accounted for 16.3%, 10.4%, 40.7%, and 1.1%, respectively, of the total variation explained by the model (Table 4).

4.3. Light, Temperature, and Humidity

We obtained different values for integrated light (PAR) between sites and arboreal strata. In general, the test showed significant differences ($p<0.001$) with higher PAR values for the DIS in both arboreal strata than in the PRE (Figure 3a). Our results show significant differences ($p<0.001$) in temperature, with higher temperature levels on both arboreal strata for DIS than for PRE (Figure 3b). Finally, the humidity was significantly higher ($p<0.001$) on both arboreal strata of PRE than in DIS (Figure 3c).

Table 2. Mean ($\pm$ SE) number of seeds per fruit, mean number of viable and nonviable seeds, mean fresh and dry weight of feces of Mexican howler monkeys, mean number of seeds of *Ficus perforata* and *Ficus lundelli* in feces, germination percentage of seeds from feces and fruit, and seed dispersal distances in a disturbed and a preserved site.

Parameter	F. perforata	F. lundelli
Seeds per fruit (N = 100)	123.6 $\pm$ 2.6	118.8 $\pm$ 2.2
Viable seeds per fruit (N = 100)	56.9 $\pm$ 2.0	54.2 $\pm$ 2.0
Nonviable seeds per fruit (N = 100)	66.6 $\pm$ 2.5	64.6 $\pm$ 1.6
Fresh weight of feces (g) (N = 30)	41.7 $\pm$ 3.8	32.7 $\pm$ 1.7
Dry weight of feces (g) (N = 30)	20.4 $\pm$ 1.6	17.0 $\pm$ 0.7
Mean number of seeds in feces (both sites)	2132 $\pm$ 145.8	1897 $\pm$ 36.5
Mean number of seeds in feces (DIS)	2224.8 $\pm$ 111.5	1917 $\pm$ 110.3
Mean number of seeds in feces (PRE)	2040.1 $\pm$ 111.5	1877.4 $\pm$ 49.3
Germination of seeds from feces (DIS)	78%	59.5%
Germination of seeds from feces (PRE)	72%	68.1%
Germination of seeds from fruits (DIS)	0%	9.2%
Germination of seeds from fruits (PRE)	0.3%	6.0%
Dispersal distance (DIS)	72.71 $\pm$ 21.42 m	57.48 $\pm$ 7.9 m
Dispersal distance (PRE)	172.25 $\pm$ 15.64 m	248.3 $\pm$ 15.74 m

Table 3. Analyses of proportions using a *logit* model and a binomial error to check for differences in germination of *Ficus* as a function of the following factors: species (*F. lundelli, F. perforata*), site (disturbed, preserved), and source (fruits, faces). $F_{1,152} = 3.84$, $p<0.0001$, df = degrees of freedom, p = probability, VEM = variation explained by model. Missing value are not significant.

Source of variation	Variance	df	Mean squares	F	p	VEM (%)
Species	1.545	1	1.545	0.75	0.387	–
Site	0.0002	1	0.0002	0.0009	0.9761	–
Source	2,072	1	2,072	1,005.8	<0.0001	87.9
Species × source	119.3	1	119.3	57.9	0.0001	5.06
Error	147.14	152	0.96	–	–	–
Total	2,357	159	–	–	–	All others 7.04

All other interactions were nonsignificant.

Table 4. Analyses of proportions using a *logit* model and a binomial error to check for differences in germination of *Ficus* as a function of the following factors: species (*F. lundelli, F. perforata*), site (disturbed, preserved), and source (fruits, feces) and arboreal strata (high, low). $F_{1,144} = 78.69$, $p<0.0001$, df = degrees of freedom, p = probability, VEM = variation explained by model.

Source of variation	Sum of squares	df	F	p	VEM (%)
Sites	78.69	1	78.69	0.0001	16.32
Species	50.36	1	50.36	0.0001	10.44
Source	196.1	1	196.1	0.0001	40.68
Arboreal strata	1.649	1	1.649	0.1991	–
Sites x Species	5.317	1	5.317	0.0211	1.10
Error	141.97	144	–	–	–
Total	482	159	–	–	All others 29.46

All other interactions were nonsignificant.

a)

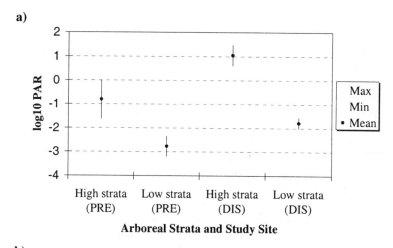

b)

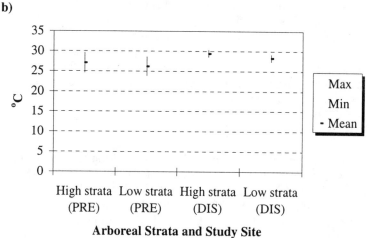

c)

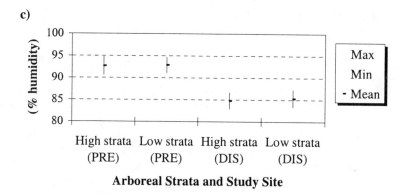

Figure 3 (a–c). Mean (± s.d.) of light (a), temperature (b), and humidity (c) for each arboreal strata for DIS and PRE sites with planter boxes containing *Ficus perforata* and *Ficus lundelli* seeds consumed and not consumed by howler monkeys (see text).

4.4. Dispersal Distance

Our observations show that seeds of both *Ficus* species pass through the digestive system of howler monkeys in about the same time (from time of ingestion to time of defecation): *F. perforata*, 21:58 ± 0:32 h, and *F. lundelli*, 20:54 ± 0:45 h. Howlers exhibited different home range size (DIS = 10 ha, PRE = 75 ha) and density (DIS = 1.05 individuals per ha, PRE = 0.12 individuals per ha) between habitats (Table 1). Seeds from each species were deposited by monkeys farther away from the parent tree in the PRE rather than in the DIS site (*F. perforata*, $t = 3.374$, df = 8, $p<0.01$; *F. lundelli*, $t = 10.138$, df = 9, $p<0.01$). We found no significant differences in the number of *Ficus* seeds in feces between sites (*F. perforata*, $t = 0.615$, df = 10, $p = 0.552$; *F. lundelli*, $t = 0.519$, df = 8, $p = 0.618$) (Table 2). In general, seed dispersal distance was significantly higher in the preserved site (X = 217.8 ± 16.4 m) than in the disturbed site (X = 65.7 ± 11.9 m) ($t = 7.594$, df = 18, $p<0.001$) (Table 2).

Our analyses show a particular distribution pattern of the *Ficus* population for each site. In this sense, while in the disturbed site there were 84 individuals of both *Ficus* species exhibiting a clumped distribution pattern (I_d, *F. perforata* = 1.25, *F. lundelli* = 1.18), in the PRE there were 180 individuals uniformly dispersed (I_d, *F. perforata* = 0.79, *F. lundelli* = 0.86). After testing each value for the significance of the deviation from 1 by the statistic F (Poole, 1974), we did not find significant differences; however, the clumped and uniformly dispersed assumptions should be considered with caution.

5. DISCUSSION

Howler monkeys had a positive and significant effect on the germination of *Ficus* seeds, similar to *Papio anubis* (Lieberman et al., 1979), *Saguinus mystax*, and *S. fuscicollins* (Garber, 1986), *A. palliata* (Estrada and Coates-Estrada, 1991), and *Pan troglodytes* (Wrangham et al., 1993, 1994). Although in the controlled experiment, seed source (fruit vs feces) explained most of the variation in seed germination, the significant 'source × species' interaction shows, however, that aside from the effect of passage through the digestive tract of the monkeys, there is a combined effect in germination that involves the species of *Ficus* consumed. Even though the germination of *Ficus* seeds evidently benefits from the interaction with howlers, the latter suggests that not all *Ficus* species exhibit the same response, and that each *Ficus* species has its own response to this process. Our field experiments suggest that establishment success could be affected by habitat fragmentation and the overall alteration of microclimatic factors, which have an influence on seed germination. Although Titus et al. (1990) suggest that consumption of *Ficus* by animals is not an obligatory condition for seed germination, we can assume that the optimal conditions for *Ficus perforata* and *Ficus lundelli* seed germination include feces as a seed source, as well as a preserved site and interactions with the levels of light, temperature, and humidity. Laman (1995) in his study of germination in Bornean tropical rain forest using canopy planter boxes showed that, given adequate water and nutrients, fig seedlings will grow faster in a higher light environment. However, here we show that *Ficus* species can exhibit different germination responses depending on seed source, and that habitat conditions and canopy microclimate affect germination response in these species.

The hemiepiphytic habit has been considered an adaptation to exploit the high-light canopy environment (Laman, 1994; Dobzhansky and Murca-Pires, 1954; Ramirez, 1977; Putz and Holbrook, 1986; Daniels and Lawton, 1991). Even though we agree with the latter, we should also consider host density (Athreya, 1999) and, particularly, relative humidity and temperature of the germination site and its surroundings. Sites in the canopy with these combination of conditions of light, humidity, and temperature for fig seed germination appear to be very scarce, particularly in the disturbed sites. Under these circumstances, areas with preserved forest, which maintain optimal conditions and the presence of substrates with good moisture retention, should highly influence seed germination success (Putz and Holbrook, 1989).

Since *Ficus* trees lose approximately 50% of their seeds to agaonid wasps (Ramírez, 1976), the efficient dispersal (sensu Fleming and Sosa, 1994) of the remaining seeds should be of the utmost ecological importance and for conservation purposes. Primates should be considered an important seed dispersal vector, since they make up 25% to 40% of the animal biomass of tropical forests. In a typical day howlers in particular can ingest a much higher volume of *Ficus* fruits (and seeds) than other important frugivores (e.g., birds or bats) (Coates-Estrada and Estrada, 1986). However, the effect of differences in digestion time and patterns of defecation among these vertebrates (e.g., Morrison, 1980; Fleming, 1988; Lambert, 1989a, b; Charles-Dominique, 1991; Midya and Brahmachary, 1991; Medellín and Gaona, 1999) on the dispersal and germination of strangler fig seeds has rarely been considered. Howler monkeys ingest and disperse more *Ficus* seeds (e.g., 2,015 seeds per individual, this study) than other sympatric vertebrate frugivores (e.g., birds 294 seeds per individual, and bats 154 seeds per individual; Medellín and Gaona, 1999), suggesting that howlers may have a profound impact not only on *Ficus*, but on the structure of the forest as well. Moreover, monkeys confer a second advantage by limiting the effect of predators and fungi and allowing seeds a better chance for germination and establishment (Martínez-Mota and Serio-Silva, unpublished data).

The *Alouatta palliata mexicana – Ficus* (*Urostigma*) interaction is beneficial for both interacting species and has important implications for their conservation. Howlers gain from the ingestion of an important and key food source (Milton et al., 1980; Serio-Silva, 1996), and the plants could be increasing their germination rate and obtaining efficient dispersal. Moreover, this interaction is also beneficial for the general dynamic of tropical rain forests in southeastern Mexico, because a number of other invertebrate and vertebrate species are involved. The effective dispersal of key species (e.g., *Ficus* spp.) will contribute to decrease problems derived from low minimum viable population sizes (McKey, 1989; Zuidema et al., 1996) or a decrease in genetic fitness (Turner, 1996). Finally, it is increasingly important to study how disperser - plant interactions change with habitat fragmentation and a decrease in forested area.

5.1. Dispersal Distances

Our results show that dispersal distances differed between sites, with the greatest in the PRE. However, this leads us to a circular argument because PRE was 15 times larger than DIS. To determine if the distances traveled were proportional, we related dispersal distance in meters (DD) to the number of *Ficus* individuals (NF), and area in hectares (A) per site. When dispersal distance was related to number of *Ficus* individuals (DD/NF), it was relatively larger in the preserved site (1.17 [PRE] vs 0.78 [DIS]). However, when dispersal distance was related to area (DD/A), it was relatively shorter in the preserved site (0.35 [PRE] vs 1.63 [DIS]). We suggest that the above differences in dispersal distance between habitats are the result of two factors acting simultaneously: (1) the social structure of a troop, and (2) plant phenology, abundance, and distribution. Troop displacements depend on its social structure within a home range of flexible size (Zucker et al., 1996). In DIS (40 ha) the four monkey troops must divide the fragment into small home ranges (ca. 10 ha), and troops moving through neighboring home ranges exhibit constant agonistic encounters (vocalizations). The eight troops in PRE (600 ha) can occupy home ranges of up to 75 ha, where physical encounters among troops are quite rare. Moreover, monkeys must move constantly among *Ficus* trees to monitor and cope with their asynchronic fruiting pattern (Milton et al., 1980; Serio-Silva, 1996). The latter is further complicated by the distribution pattern of the *Ficus* population in PRE. For example, while in DIS there were 84 individuals of both *Ficus* species exhibiting a clumped distribution pattern, in PRE there were 180 individuals dispersed throughout the site, and thus, the spatial distribution of the *Ficus* population should exert a significant influence on troop movements. Even though howlers move relatively longer distances in DIS, they are moving within a clump of *Ficus* individuals. However, monkeys in PRE, which moved relatively shorter distances, may be more effective in placing seeds on suitable sites for germination and future establishment, i.e., furthest away from their parent tree and on a tree different than *Ficus* (Chapman, 1995).

5.1.1. Directions for Future Research

Howler monkeys in natural habitats (disturbed or preserved) performed an important role in depositing seeds in optimal or safe sites for germination and establishment, although host microclimatic conditions are also of the utmost importance. In this sense, although figs have commonly been regarded as a keystone species for conservation in many tropical rainforests (Terborgh, 1986; McKey, 1989; Leighton and Leighton, 1983), habitat fragmentation can decrease the ecological interactions between howlers and figs and thus the possibility of seed dispersal.

In summary, the biology of forest regeneration is very complex and can easily be altered due to forest transformation (Howe and Miriti, 2000; Zuidema et al., 1996). It is clear that the range of herbivore behavior and their interactions with plants should be studied in detail, especially under the environmental conditions resulting from an increase in habitat fragmentation. The increasing habitat loss, hunting, and other forms of human ecosystem interference suggest that it is necessary to increase the different efforts on studies of seed dispersal, which also can contribute directly to primate conservation and their ecological interactions.

6. SUMMARY

Mexican howler monkeys (*Alouatta palliata mexicana*) consume large amounts of fig fruits (*Ficus* spp.), and some reports suggest that seeds expelled in primate's feces have higher germination success. We studied how germination rate and dispersal distance of seeds of *Ficus perforata* and *F. lundelli* dispersed by howler monkeys, changed between a disturbed (Playa escondida (PE), Veracruz, Mexico – 40 ha-) and a preserved ("Carolino Anaya" reserve (COA), Coatzacoalcos, Veracruz, Mexico – 600 ha -) tropical rain forest site in southern Veracruz, Mexico. Passage though the digestive tract of howler monkeys had a significant positive effect on germination of seeds recovered from feces relative to germination of seeds from fruits. Dispersal distances differed between sites, being significantly larger in the preserved site. However, when seed dispersal distance was related to number of *Ficus* individuals located in each site, it was larger in the preserved site (1.17 vs 0.78), whereas when related to area of each site it was shorter in the preserved site (0.35 vs 1.63). On other hand, although the howlers show an important role in increasing velocity and total amount of *Ficus* seed germination and optimal dispersal, we found that diverse microclimatic factors of each study site can influence the success of seed germination in *F. perforata* and *F. lundelli*. Our experimental data show that differences in germination were the result of the arboreal strata where the seeds were deposited, as well as the quality (degree of conservation) of the habitat where the monkeys are found (either preserved or disturbed).

We discussed the possibility that changes in landscape structure influence disperser behavior and thus the dispersal patterns in the forest dynamic. We suggest the results of these ecological interactions are the consequence of simultaneously-acting factors: the social structure of a troop, and plant phenology, abundance and distribution. Howlers move longer distances in the disturbed site, but they are moving within a clump of *Ficus*. Howlers in the preserved site moved shorter distances, but may be more effective in placing seeds on suitable sites for germination and future establishment. Previous studies have shown that ingestion of strangler fig seeds *(Ficus Urostigma)* by *Alouatta palliata mexicana* increases their germination. However, the results show greater germination in the tree canopy for seeds of both species; moreover, seed germination was significantly higher in the preserved site for seeds from feces. The latter is probably the effect on seed germination of the higher relative humidity and low light and temperature levels in the host trees for hemiepiphytic *Ficus*, to which seeds should adjust their germination response.

7. ACKNOWLEDGMENTS

We thank Albright and Wilson Troy of Mexico for permission to work in "Carolino Anaya Reserve," and V. J. Sosa-Fernández, V. Parra-Tabla, and R. Manson for their comments and suggestions to earlier versions of this manuscript, and J. Bello-Gutiérrez for his help with the statistical analyses. Research was supported by Instituto de Ecologia, A.C (902-16), equipment donation by *Idea Wild* and a CONACyT scholarship to JCSS.

8. REFERENCES

Athreya, V. R., 1999, Light or presence of host-trees: Which is more important for the strangler fig? *J. Trop. Ecol.* **15**:589–603.

Banaccorso, F. J., 1979, Foraging and reproductive ecology in a Panamanian bat community, *Bull. Florida St. Mus. Biol. Sci.* **24**:59–408.

Bardon, R. E., Countryman, D. W., and Hall, R. B., 1995, A reassessment of using light-sensitive *Diazo* paper for measuring integrated light in the field, *Ecology* **76**:1,013–1,016

Brower, E. J., Zar, H. J., and von Ende, C. N., 1989, *Field and Laboratory Methods for General Ecology,* W. C. Brown Company Publishers, USA.

Chapman, C. A., 1995, Primate seed dispersal: Coevolution and conservation implications, *Evol. Anth* **4**:74–82.

Charles-Dominique, P., 1991, Feeding strategy and activity budgets of the frugivorous bat *Carollia perpiscillata* (Chiroptera:Phyllostomidae) in French Guyana, *J. Trop. Ecol.* **7**:243–256.

Coates-Estrada, R., and Estrada, A., 1986, Fruiting and frugivores at a strangler fig in the tropical rain forest of Los Tuxtlas, México, *J. Trop. Ecol.* **2**:349–357.

Crawley, M., 1993, *GLIM for Ecologists*, Blackwell Scientific Publications, Oxford.

Daniels, J. D., and Lawton, R. O., 1991, Habitat and host preferences of *Ficus crassiuscula*, a Neotropical strangling fig of the lower-montane rain forest, *J. Ecol.* **79**:129–141.

de Figueiredo, R. A., 1993, Ingestion of *Ficus enormis* seeds by howler monkeys (*Alouatta fusca*) in Brazil: Effects on seed germination, *J. Trop. Ecol.* **9**:541–543.

Dobzhansky, T., and Murca-Pires, J., 1954, Strangler trees, *Sci. Am.* January 78-80.

Estrada, A., and Coates-Estrada, R., 1984, Fruit eating and seed dispersal by howling monkeys (*Alouatta palliata*) in the tropical rain forest of Los Tuxtlas, Mexico, *Int. J. Primatol.* **5**:105–131.

Estrada, A., and Coates-Estrada, R., 1986, Frugivory in howling monkeys (*Alouatta palliata*) at Los Tuxtlas: Seed dispersal and fate of seeds, in: *Frugivores and Seed Dispersal*, A. Estrada and T. H. Fleming, eds., Dr W. Junk Publishers, The Hague, pp. 93–104.

Estrada, A., and Coates-Estrada, R., 1991, Howling monkeys (*Alouatta palliata*), dung beetles (Scarabaeidae), and seed dispersal: Ecological interactions in the tropical rain forest of Los Tuxtlas, Mexico, *J. Trop. Ecol.* **7**:459–474.

Fleming, T. H., 1988, *The Short-tailed Fruit Bat: A Study in Plant-Animal Interactions,* University of Chicago Press, Chicago.

Fleming, T. H., and Sosa, V. J., 1994, Effects of nectarivorous and frugivorous mammals on reproductive success of plants, *J. Mamm.* **75**:845–851.

Francis, B., Green, M., and Payne, C., 1993, *The GLIM System: Release 4 Manual,* Clarendon Press, Oxford.

Galdikas, B. M. F., 1982, Orangutans as seeds dispersers in Tanjung Puting, central Kalimatan: Implications for conservation, in: *The Orangutan. Its Biology and Conservation,* L. E. M. de Boer, ed., Junk Publishers, Dordrecht, The Netherlands, pp. 285–299.

Galindo-González, J., 1998, Dispersión de semillas por murciélagos: su importancia en la conservación y regeneración del bosque tropical, *Acta Zool. Mex. (nueva serie)* **73**:57–74.

Garber, P. A., 1986, The ecology of seed dispersal in two species of callitrichid primates (*Saguinus mystax* and *Saguinus fuscicollis*), *Am. J. Primatol.* **10**:155–170.

Garber, P. A., and Lambert, J. E., 1998, Primates as seed dispersers: Ecological process and directions for future research, *Am. J. Primatol.* **45**:3–8.

Gauthier-Hion, A., Duplantier, J. M., Feer, F., Sourd, C., Decoux, J. P., Dubost. J. G., Emmons, L., Erard, C., Hecketsweiler, P., Moungazi, A., Roussilhon, C., and Thiollay, J. M., 1985, Fruit characteristics as a basis for fruit choice and seed dispersal in a tropical forest vertebrate community, *Oecologia* **65**:324–337.

Gauthier-Hion, A., and Michaloud, G., 1989, Are the figs always keystone resources for tropical frugivores vertebrates? A test in Gabon, *Ecology* **70**:1,826–1,833

Howe, H. F., 1980, Monkey dispersal and waste of a Neotropical fruit, *Ecology* **61**:944–959.

Howe, H. F., and Miriti, M. N., 2000, No question: Seed dispersal matters, *TREE* **15**:434–436.

Janzen, D. H., 1979, How to be a fig, *Ann. Rev. Ecol. System.* **10**:13–51.

Janzen, D. H., 1986, Mice, big mammals, and seeds: It matters who defecates what where, in: *Frugivores and Seed Dispersal*, A. Estrada and T. H. Fleming, eds., Dr. W. Junk, publisher, The Hague, pp. 251–271.

Laman, T. G., 1994, The ecology of strangler figs (hemiepiphytic *Ficus* spp.) in the rain forest canopy of Borneo, Ph.D. dissertation, Harvard University, Cambridge, MA.

Laman, T. G., 1995, *Ficus stupenda* germination and seedling establishment in a Bornean rain forest canopy, *Ecology* **76**:2,617–2,626.

Lambert, F., 1989a, Fig eating by birds in a Malaysian lowland rainforest, *J. Trop. Ecol.* **5**:410–412.

Lambert, F., 1989b, Pigeons as seed predators and dispersers of a fig in a Malaysian lowland forest, *J. Trop. Ecol.* **131**:512–527.

Leighton, M., and Leighton, D. R., 1983, Vertebrate responses to fruiting seasonality within a Bornean rain forest, in: *Tropical Rain Forest: Ecology and Management*, S. L. Sutton, T. C. Whitemore, and A. C. Chadwick, eds., Blackwell, London, England, pp. 181–196.

Lieberman, D., Hall, J.B., Swaine, M. D., and Lieberman, M., 1979, Seed dispersal by baboons in the Shai Hills, Ghana, *Ecology* **60**:65–75.

Marsh, L. K., 1999, Ecological effect of the black howler monkey (*Alouatta pigra*) on fragmented forest in the Community Baboon Sanctuary, Belize, Ph.D. dissertation, Washington University, WA, USA.

McKey, D. B., 1989, Population biology of figs: Applications for conservation, *Experientia* **45**:661–673.

Medellín, R. A., and Gaona, O., 1999, Seed dispersal by bats and birds in forest and disturbed habitats of Chiapas, México, *Biotropica* **31**:478–485.

Midya, S. and Brahmachary, R. L., 1991, The effect of birds upon germination of banyan (*Ficus bengalensis*) seeds, *J. Trop. Ecol.* **7**:537–538.

Milton, K., van Soest, P. J., and Robertson, J. B., 1980, Digestive efficiencies of wild howler monkeys, *Physiol. Zool.* **53**:402–409.

Morrison, D. W., 1980, Foraging and day-roosting dynamics of the canopy fruits bats in Panama, *J. Mamm.* **61**:20–29.

Nathan, R., and Muller-Landau, H. C., 2000, Spatial patterns of seed dispersal, their determinants and consequences for recruitment, *TREE* **15**:278–285.

Poole, R. W., 1974, *An Introduction to Quantitative Ecology,* McGraw-Hill series in population biology, New York, USA.

Putz, F. E., and Holbrook, N. M., 1986, Notes on the natural history of hemiepiphytes, *Selbyana* **9**:61–69.

Putz, F. E., and Holbrook, N. M., 1989, Strangler fig rooting habits and nutrient relations in the Llanos of Venezuela, *Am. J. Bot.* **76**:781–788.

Ramirez, B. W., 1976, Germination of seeds of new world *Urostigma (Ficus)* and *Morus rubra* L. *(Moraceae),* *Revista Biol. Trop.* **24**:1–6.

Ramírez, W. B., 1977, Evolution of the strangling habit in *Ficus* subgenus *Urostigma* (Moraceae), *Brenesia* **12/13**:11–19.

Serio-Silva, J. C., 1996, Calidad del alimento consumido por Alouatta palliata mexicana en condiciones de semilibertad, MSc thesis, Instituto de Neuroetología, Universidad Veracruzana, Xalapa, México, pp. 66.

Serio-Silva, J. C., 1997, Studies of howler monkeys (*Alouatta palliata*) translocated to a Neotropical rain forest fragment. I. Social distance in translocated howler monkeys, II. Activity patterns and feeding habits of translocated howler monkeys, *Lab. Primate News.* **36**:11–14.

Serio-Silva, J. C., and Rico-Gray, V., 2002a, Influence of microclimate at different canopy heights on the germination of *Ficus (Urostigma)* seeds dispersed by Mexican howler monkey (*Alouatta palliata mexicana*), *Interciencia* **27**:186–190.

Serio-Silva, J. C., and Rico-Gray, V., 2002b, Interacting effects of forest fragmentation and howler monkey foraging on germination and dispersal of fig seeds, *Oryx* **36**:266–271.

Serio-Silva, J. C., Rico-Gray, V., Hernández-Salazar, L. T., and Espinosa-Gomez, R., 2002, The role of *Ficus (Moraceae)* in the diet of a troop of Mexican howler monkeys, *J. Trop. Ecol.* In press.

Silver, S. C., Ostro, L. E. T., Yeager, C. P., and Horwich, R., 1998, Feeding ecology of the black howler monkey (*Alouatta pigra*) in northern Belize, *Am. J. Primatol.* **45**:263–270.

Terborgh, J., 1986, Keystone plant resources in the tropical rain forest, in: *Conservation Biology,* M. Soulé, ed., Sinauer Associates, Sunderland, MA, pp. 330–344.

Titus, J. H., Holbrook, N. M., and Putz, F. E., 1990, Seed germination and seedling distribution of *Ficus pertusa* and *F. tuerckheimii*: Are strangler figs autotoxic? *Biotropica* **22**:425–428.

Thompson, J. N., 1994, *The Coevolutionary Process,* The University of Chicago Press, Chicago, USA.

Turner, I. M., 1996, Species loss in a fragment of tropical rain forest: A review of the evidence, *J. App. Ecol.* **33**:200–204.

Tutin, C. E. G., Williamson, E. A., Rogers, M. E., and Fernández, M., 1991, A case study of a plant animal relationship: *Cola lizae* and lowland gorillas in the Lope Reserve, Gabon, *J. Trop. Ecol.* **7**:181–199.

Wrangham, R. W., Conklin, N. L., Etot, G., Obua, J., Hunt, K. D., Hauser, M. D., and Clark, A. P., 1993, The value of figs to chimpanzees, *Int. J. Primatol.* **14**:243–256.

Wrangham, R. W., Chapman, C. A., and Chapman, L. J., 1994, Seed dispersal of forest chimpanzees, *J. Trop. Ecol.* **10**:355–368.

Zar, J., 1996, *Biostatistical Analysis.* Prentice Hall, Englewood Cliffs, NJ.

Zucker, E. L., Clarke, M. R., Glander, K. E., and Scott, N. J., Jr., 1996, Sizes of home ranges and howling monkey groups at hacienda La Pacífica, Costa Rica: 1972–1991, *Brenesia* **45-46**:153–156.

Zuidema, P. A., Sayer, J. A., and Dijkman, W., 1996, Forest fragmentation and biodiversity: The case for intermediate-sized conservation areas, *Env. Cons.* **23**:290–297.

HOW DO HOWLER MONKEYS COPE WITH HABITAT FRAGMENTATION?

Júlio César Bicca-Marques[*]

1. INTRODUCTION

Howler monkeys (genus *Alouatta*) represent a successful radiation of at least nine species[†] (Rylands et al., 2000). Among all Neotropical primate genera, *Alouatta* inhabits the widest variety of forested habitats and presents the largest geographic distribution, occurring from Mexico to Argentina and South Brazil (Crockett and Eisenberg, 1987; Neville et al., 1988). Howlers are known for their ability to survive in intact or disturbed anthropogenic ecosystems, such as forest fragments as small as a few hectares (Schwarzkopf and Rylands, 1989; Bicca-Marques, 1994; Chiarello and Galetti, 1994; Estrada and Coates-Estrada, 1996; Crockett, 1998; Marsh, 1999; Juan et al., 2000; Gilbert and Setz, 2001).

Howler monkeys have a folivorous-frugivorous diet (Crockett and Eisenberg, 1987). Their success in coping with habitat fragmentation has been related to their capacity to include a great amount of leaves in their diet, while living in relatively small home ranges (Neves and Rylands, 1991; Estrada and Coates-Estrada, 1996; Crockett, 1998; Estrada et al., 1999; Marsh, 1999; Juan et al., 2000). According to Marsh (1999), howlers are capable of increasing the amount (both in species and in quantity) of leaves eaten during unusually lean fruit and flower times, and this ability helps them to survive in small fragments. In addition, howlers are reported to be able to adjust their choice of food plant species to survive in forests with varying floristic compositions (Bicca-Marques and Calegaro-Marques, 1994a, b; DeLuycker, 1995; Crockett, 1998; Kowalewski and Zunino, 1999; Marsh, 1999).

The ecology and behavior of howler monkeys have been extensively recorded in the wild (for reviews, see Crockett and Eisenberg, 1987; Neville et al., 1988; Kinzey, 1997). Researches analyzing the effect of habitat fragmentation on their habits, however, are scarce (Estrada et al., 1999; Juan et al., 2000). Marsh (1999), for example, compared the ecology and behavior of six groups of *A. pigra* living in fragments differing in size from

[*] Pontifícia Universidade Católica do Rio Grande do Sul, Faculdade de Biociências, Av. Ipiranga 6681 Pd 12A, Porto Alegre, RS 90619-900, Brazil. Email: jcbicca@terra.com.br.
[†] In spite of the recent revision by Rylands et al. (2000), in this paper I follow the systematic arrangement used by Crockett and Eisenberg (1987), with the single substitution of the specific name *fusca* for *guariba*.

Primates in Fragments: Ecology and Conservation
Edited by L. K. Marsh, Kluwer Academic/Plenum Publishers, 2003

1.3 to 80 ha in the Community Baboon Sanctuary, Belize. She observed that, independent of the size and floristic composition of the forest patch, all study groups behaved in a species-specific way (sensu Sussman, 1987) by feeding on similar proportions of plant items from a similar number of species, although the species exploited could be different (Marsh, 1999). In only one fragment did a group adjust their diet toward a greater amount of fruit resources than the others based on continuous overlap of *Ficus* fruiting. On the other hand, during a six-month study Juan et al. (2000) observed different trends in diet diversity, the contribution of leaves and fruits to the diet, and in time spent traveling among three groups of *A. palliata* living in fragments of 3, 35, and 250 ha at Los Tuxtlas, Mexico. Specifically, they found that in the largest fragment howlers used a greater number of plant species as food sources, fed more on fruits and less on leaves, and spent more time traveling than in the smallest fragment. Recently, studies analyzing the effect fragmentation and habitat quality have on howler population density have also been published (Estrada and Coates-Estrada, 1996; Peres, 1997; Chapman and Balcomb, 1998; Sorensen and Fedigan, 2000; Zunino et al., in press; see Onderdonk and Chapman, 2000, for similar data on a folivorous Old World monkey, *Colobus guereza*).

In this paper, I perform a cross-study comparison of the way howler monkeys cope with varying degrees of habitat fragmentation. I analyze the effect of fragmentation on the use of space, diet composition, and activity patterns. Specifically, I test whether forest size can predict the (1) size of the home range, (2) distance traveled daily, (3) contribution of food items to the diet, (4) number of plant species used as food sources, (5) number of species responsible for the bulk of the diet, (6) number of plant species consumed on a daily basis, and (7) activity budget.

2. METHODOLOGICAL APPROACH

I limited the analysis to studies lasting nine or more months to reduce the influence of seasonality on the results. Data from more than 42 howler groups living in 27 study sites were used in this analysis (Table 1, Figure 1). The forests where the studies were conducted varied from fragments of 1.3 ha (*A. pigra*; Marsh, 1999) to continuous forests of 1,240,000 ha in size (*A. seniculus*; Queiroz, 1995), with a median size of 82 ha (N = 35; Appendix I).

The size of the study site (forest), referred to in this report as fragment size, was logged for all analyses. For each variable analyzed, I first tested for the occurrence of species-specific differences by using a Kruskal-Wallis one-way analysis of variance (Sokal and Rohlf, 1995). This was done to avoid taking species-specific differences in behavior as a consequence of habitat fragmentation. In the absence of such differences I analyzed the information from all species as a single data set. When there are interspecific differences that allow clear separation of groups of species, I treated these groups as separate batches of data. Regression analyses were performed to determine the strength of the relationship between forest size and each of the variables discussed; whereas, the correlation between two variables was tested by the Spearman rank correlation coefficient (Sokal and Rohlf, 1995). Differences in sample size among tests are attributed to the fact that most studies do not provide all of the information discussed in this paper (Appendices I to III). All tests were performed using Systat™ (Wilkinson, 1990).

Table 1. List of studies included in the analysis on habitat fragmentation and howler monkey ecology and behavior.

Species	Study site (coordinates)	References
A. belzebul	Fazenda Pacatuba (7°03'S, 35°09'W)	Bonvicino, 1989
	Fazenda Universal (9°34'S, 56°19'W)	Pinto, 2002
A. caraya	Bosque Municipal Dr. Fábio de Sá Barreto (23°10'S, 47°48'W)	Alves, 1983
	River Riachuelo (27°30'S, 58°41'W)	Rumiz et al., 1986; Zunino, 1986
	Estância Casa Branca (29°37'S, 56°17'W)	Bicca-Marques, 1993, 1994; Bicca-Marques and Calegaro-Marques, 1994b
A. guariba	Estação Biológica de Caratinga (19°50'S, 41°50'W)	Mendes, 1989
	Mata Boa Vista (22°02'S, 43°11'W)	Limeira, 1996
	Ribeirão Cachoeira (22°45'S, 46°52'W)	Gaspar, 1997
	Fazenda Rio Claro (22°48'S, 48°55'W)	Martins, 1997
	Reserva Santa Genebra (22°49'S, 47°06'W)	Chiarello, 1993a, b, 1994
	Parque Estadual da Cantareira (23°22'S, 46°26'W)	Lunardelli, 2000
	Parque Estadual Intervales (24°12'S, 48°03'W)	Steinmetz, 2000
	Floresta Nacional de Três Barras (26°12'S, 50°17'W)	Perez, 1997
	Estação Ecológica de Aracuri (28°13'S, 51°10'W)	Marques, 1996
	Campo de Instrução de Santa Maria (29°43'S, 53°42'W)	Fortes, 1999
	Parque Estadual de Itapuã (30°21'S, 51°01'W)	Cunha, 1994
A. palliata	Barro Colorado Island (9°10'N, 79°50'W)	Milton, 1980
	Santa Rosa National Park (10° N, 85° W)	Chapman, 1987, 1988; Larose, 1996
	La Selva Biological Reserve (10°26'N, 83°59'W)	Stoner, 1996
	Hacienda La Pacifica (10°28'N, 85°07'W)	Glander, 1978
	Agaltepec Island (18°27'N, 95°02'W)	Serio-Silva, 1997
	Station "Los Tuxtlas" (18°34'N, 95°09'W)	Estrada, 1984; Estrada et al., 1999
A. pigra	Community Baboon Sanctuary (17°30'N, 88°25'W)	Marsh, 1999; Silver et al., 1998; Ostro et al., 1999
A. seniculus	Estação Ecológica Mamirauá (2°58'S, 64°55'W)	Queiroz, 1995
	Finca Merenberg (2°14'N, 76°08'W)	Gaulin and Gaulin, 1982
	Nourague Station (4°05'N, 52°41'W)	Julliot, 1994
	Hato Masaguaral (8°34'N, 67°35'W)	Neville, 1972

Figure 1. Map showing the location of the study sites discussed in this paper. *A. belzebul*: 1-Fazenda Pacatuba, 2-Fazenda Universal; *A. caraya*: 3-Bosque Municipal Dr. Fábio de Sá Barreto, 4-River Riachuelo, 5-Estância Casa Branca; *A. guariba*: 6-Estação Biológica de Caratinga, 7-Mata Boa Vista, 8-Ribeirão Cachoeira, 9-Fazenda Rio Claro, 10-Reserva Santa Genebra, 11-Parque Estadual da Cantareira, 12-Parque Estadual Intervales, 13-Floresta Nacional de Três Barras, 14-Estação Ecológica de Aracuri, 15-Campo de Instrução de Santa Maria, 16-Parque Estadual de Itapuã; *A. palliata*: 17-Barro Colorado Island, 18-Santa Rosa National Park, 19-La Selva Biological Reserve, 20-Hacienda La Pacifica, 21-Agaltepec Island, 22-Station "Los Tuxtlas," *A. pigra*: 23-Community Baboon Sanctuary; *A. seniculus*: 24-Estação Ecológica Mamirauá, 25-Finca Merenberg, 26-Nourague Station, and 27-Hato Masaguaral.

3. HABITAT FRAGMENTATION AND HOWLER MONKEYS

3.1. Use of Space

Habitat fragmentation will result in a decrease in home range size, especially in small fragments where the animals are unable to increase their area of activity beyond the limits of the forest. Interspecific differences in home range size were observed (Table 2). This result contradicts contention by Crockett and Eisenberg (1987) that, among howlers, home range varies more within than between species. In *A. palliata*, the howler that has larger home ranges than the other species, home range size was directly related to group size (F-ratio = 15.533, N = 11, p = 0.003); whereas, with *A. guariba* (F-ratio = 1.142, N = 11, p = 0.313), it was not. Considering all species together, except *A. palliata*, group size cannot predict home range size (F-ratio = 0.550, N = 28, p = 0.465).

Fragment size did not influence home range size in *A. palliata* (F-ratio = 2.854, N = 11, p = 0.125). However, fragment size is a good predictor of home range in the other species (F-ratio = 10.025, N = 28, p = 0.004). A similar result was observed when all species were analyzed together (F-ratio = 13.262, N = 39, p = 0.001; Figure 2a). The howler groups traveled from 11 to 1,564 m each day, with an average of 497 m (Appendix I).

Day range measurements were similar among all species (Table 2) and were not influenced by fragment size (average: F-ratio = 0.458, N = 24, p = 0.505, Figure 2b; minimum: F-ratio = 0.462, N = 13, p = 0.511; maximum: F-ratio = 0.356, N = 13, p = 0.563). As described by Crockett and Eisenberg (1987), mean day range and home range size were not correlated (r_s = 0.015, N = 24, p>0.50). Mean day range, however, was positively correlated with the average number of species utilized as food sources per day (r_s = 0.664, N = 12, p<0.05).

3.2. Diet

In this paper, howler diet is divided into three categories. The category "leaves" includes leaf buds, young and mature leaves, and petioles. "Fruits" refers to ripe and unripe pulp and seeds. Seeds also includes those of the coniferous *Araucaria angustifolia*. "Flowers" comprises all floral reproductive parts, regardless of stage of development. A cross-species comparison indicated differences in the amount of leaves and fruits in the diet of howlers (Table 2). These differences were driven by a greater use of fruits and a lower use of leaves by *A. belzebul*. All other howler monkey species included similar amounts of these food items in their diet. The use of leaves (excluding data on *A. belzebul*: F-ratio = 0.116, N = 34, p = 0.736; Figure 3a), mature leaves (F-ratio = 0.200, N = 24, p = 0.659), immature leaves (F-ratio = 0.021, N = 22, p = 0.885), fruits (excluding data on *A. belzebul*: F-ratio = 0.956, N = 35, p = 0.335; Figure 3b), and flowers (F-ratio = 0.753, N = 36, p = 0.392) were not altered by the size of the fragment.

Howler species utilized similar numbers of plant species as food sources on an annual basis (Table 2). This variable can be predicted by fragment size (F-ratio = 6.392, N = 36, p = 0.016; Figure 4). The number of sources of leaves was different among species (Table 2), especially because of a lower variety eaten by *A. caraya* and *A. belzebul*. On the other hand, the diversity of sources of fruits and flowers was similar among species (Table 2). When all species are analyzed together, fragment size predicts

Table 2. Cross-species comparisons of the behavioral and ecological variables discussed in the text. Results in bold are significant. Raw data used in these analyses are shown in Appendices I to III.

Variable	Kruskal-Wallis one-way analysis of variance
Home range size	**K-W = 21.004, N = 39, d.f. = 5, p = 0.001**
Day range: mean	K-W = 9.440, N = 24, d.f. = 5, p = 0.093
Day range: minimum	K-W = 3.297, N = 13, d.f. = 3, p = 0.348
Day range: maximum	K-W = 1.813, N = 13, d.f. = 3, p = 0.612
Diet: % leaves	**K-W = 13.109, N = 36, d.f. = 5, p = 0.022**
Diet: % leaf buds + young leaves	K-W = 7.832, N = 22, d.f. = 5, p = 0.166
Diet: % mature leaves	K-W = 8.792, N = 24, d.f. = 5, p = 0.118
Diet: % fruits	**K-W = 13.146, N = 37, d.f. = 5, p = 0.022**
Diet: % flowers	K-W = 3.595, N = 36, d.f. = 5, p = 0.609
# plant species used as food sources	K-W = 7.788, N = 36, d.f. = 5, p = 0.168
# sources of leaves	**K-W = 12.162, N = 25, d.f. = 5, p = 0.033**
# sources of fruits	K-W = 8.240, N = 27, d.f. = 5, p = 0.144
# sources of flowers	K-W = 7.204, N = 27, d.f. = 5, p = 0.206
# food species eaten per day	K-W = 3.277, N = 12, d.f. = 3, p = 0.351
# food species making up 50% of diet	K-W = 7.635, N = 33, d.f. = 5, p = 0.178
# food species making up 80% of diet	**K-W = 12.403, N = 31, d.f. = 4, p = 0.015**
Activity budget: % resting	K-W = 5.753, N = 24, d.f. = 5, p = 0.331
Activity budget: % feeding	K-W = 9.412, N = 24, d.f. = 5, p = 0.094
Activity budget: % moving	K-W = 8.580, N = 24, d.f. = 5, p = 0.127

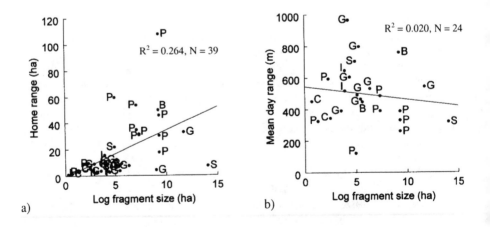

Figure 2. Relationship between fragment size and (a) home range size and (b) mean day range. B = *A. belzebul*, C = *A. caraya*, G = *A. guariba*, I = *A. pigra*, P = *A. palliata*, S = *A. seniculus*.

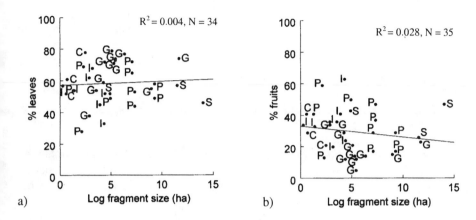

Figure 3. Relationship between fragment size and the contribution of (a) leaves and (b) fruits to the diet (both excluding data on *A. belzebul*). C = *A. caraya*, G = *A. guariba*, I = *A. pigra*, P = *A. palliata*, S = *A. seniculus*.

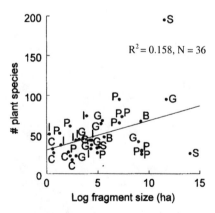

Figure 4. Relationship between fragment size and the number of plant species used as food sources. B = *A. belzebul*, C = *A. caraya*, G = *A. guariba*, I = *A. pigra*, P = *A. palliata*, S = *A. seniculus*.

the diversity of leaves (F-ratio = 5.610, N = 25, p = 0.027; Figure 5a) and fruits (F-ratio = 7.197, N = 27, p = 0.013; Figure 5b) consumed, but not of flowers (F-ratio = 1.723, N = 27, p = 0.201). Excluding *A. caraya* and *A. belzebul* from the first analysis, the significant relationship between fragment size and the diversity of leaves eaten disappears (F-ratio = 3.942, N = 20, p = 0.063).

The average number of food species exploited per day varied from 3.5 to 10.8 (mean = 6.7 ± 2.3, N = 12; Appendix III) and did not differ among howler species (Table 2). This variable cannot be predicted by fragment size (F-ratio = 0.237, N = 12, p = 0.637).

The number of food species contributing with 50% and 80% of the diet is also a measurement of diet diversity that may be affected by a decrease in plant species richness in forest fragments. From one to as much as 40 plant species (mean = 5.4 ± 6.5, N = 33; Appendix III) made up 50% of howlers' diet. No differences were found among howler species (Table 2) and fragment size did not predict this number (F-ratio = 2.945, N = 33, p = 0.096; Figure 6). In addition, four to more than 40 food species accounted for 80% of the diet of the study groups (mean = 11.7 ± 6.6, N = 31; Appendix III). Interspecific differences in the latter variable were observed especially because of the high importance of a more limited number of species to *A. caraya*'s diet (Table 2). Independent of the exclusion of *A. caraya* from this analysis, the number of plant species responsible for 80% of the diet was not altered as a consequence of habitat fragmentation (including *A. caraya*: F-ratio = 0.604, N = 31, p = 0.443; excluding *A. caraya*: F-ratio = 0.005, N = 28, p = 0.942).

Data on the top ranking food species are available for 34 study groups. In 20 cases, these species belonged to the genus *Ficus* (Table 3). The top ranking species belonged to families Moraceae (N = 20), Leguminosae (N = 10), Araucariaceae (N = 2), Cecropiaceae (N = 1), and Tiliaceae (N = 1). The second most important food species belonged to a larger array of genera (Table 3). Again, Leguminosae (N = 10) and Moraceae (N = 7) were the most important families, followed by Burseraceae (N = 3), Apocynaceae (N = 2), and Bignoniaceae, Cecropiaceae, Combretaceae, Lauraceae, Lecythidaceae, Melastomataceae, Nyctaginaceae, Phytolaccaceae, Rutaceae, Sapindaceae, and Ulmaceae (N = 1 each).

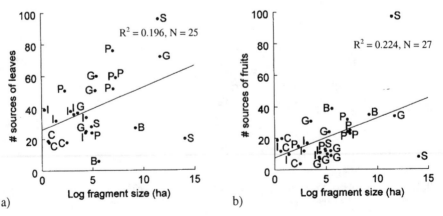

Figure 5. Relationship between fragment size and the number of plant species used as sources of (a) fruits and (b) leaves. B = *A. belzebul*, C = *A. caraya*, G = *A. guariba*, I = *A. pigra*, P = *A. palliata*, S = *A. seniculus*.

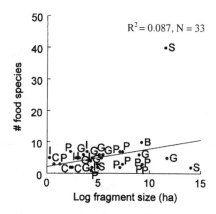

Figure 6. Relationship of fragment size and the number of plant species making up 50% of howler monkeys' diet. B = *A. belzebul*, C = *A. caraya*, G = *A. guariba*, I = *A. pigra*, P = *A. palliata*, S = *A. seniculus*.

Table 3. Genera of the first and second most important plant species used as food sources by howlers. The numbers indicate the frequency of study groups of each species using a plant species of each genus as the top (or second) species in the diet. B = *A. belzebul*, C = *A. caraya*, G = *A. guariba*, I = *A. pigra*, P = *A. palliata*, S = *A. seniculus*.

Plant genus (family)	B	C	G	I	P	S
Top ranking						
Ficus (Moraceae)	-	2	2	6	9	1
Araucaria (Araucariaceae)	-	-	2	-	-	-
Andira (Leguminosae)	-	-	-	-	1	-
Apuleia (Leguminosae)	-	-	1	-	-	-
Cecropia (Cecropiaceae)	-	-	-	-	-	1
Dialium (Leguminosae)	1	-	-	-	-	-
Enterolobium (Leguminosae)	-	-	-	1	-	-
Inga (Leguminosae)	-	-	1	-	-	-
Luehea (Tiliaceae)	-	-	1	-	-	-
Lutzelburgia (Leguminosae)	-	-	1	-	-	-
Myrocarpus (Leguminosae)	-	-	1	-	-	-
Parapiptadenia (Leguminosae)	-	1	-	-	-	-
Pterocarpus (Leguminosae)	-	-	1	-	1	-
Second most important plant						
Brosimum (Moraceae)	-	-	1	-	3	-
Inga (Leguminosae)	-	-	-	3	1	-
Bursera (Burseraceae)	-	-	-	-	3	-
Forsteronia (Apocynaceae)	-	2	-	-	-	-
Acacia (Leguminosae)	1	-	-	-	-	-
Andira (Leguminosae)	-	-	-	1	-	-
Cariniana (Lecythidaceae)	-	-	1	-	-	-
Castilla (Moraceae)	-	-	-	-	1	-
Cecropia (Cecropiaceae)	-	-	1	-	-	-
Celtis (Ulmaceae)	-	-	1	-	-	-

Table 3. (continued).

Plant genus (family)	B	C	G	I	P	S
Combretum (Combretaceae)	-	-	-	1	-	-
Coussapoa (Moraceae)	-	-	-	-	-	1
Cupania (Sapindaceae)	-	-	1	-	-	-
Enterolobium (Leguminosae)	-	-	1	-	-	-
Erythrina (Leguminosae)	-	-	-	1	-	-
Guapira (Nyctaginaceae)	-	-	1	-	-	-
Miconia (Melastomataceae)	-	-	-	1	-	-
Morus (Moraceae)	-	-	-	-	-	1
Nectandra (Lauraceae)	-	-	-	-	1	-
Phytolacca (Phytolaccaceae)	-	1	-	-	-	-
Pithecellobium (Leguminosae)	-	-	-	-	1	-
Platypodium (Leguminosae)	-	-	-	-	1	-
Pyrostegia (Bignoniaceae)	-	-	1	-	-	-
Zanthoxylum (Rutaceae)	-	-	1	-	-	-

3.3. Activity Budget

Resting accounted for an average of 65.5 ± 7.8% (range: 53% to 80%, N = 24) of the daily activities. Feeding was the second most important behavior, occurring from 6% to 24% of the time (mean = 17.3 ± 4.5%, N = 24), and, on average, during 12.8 ± 4.8% of the time the howlers were moving (range: 2 to 19, N = 24; Appendix II). Resting, feeding, and moving were consistent among species (Table 2). Fragment size cannot predict time spent in any of these activities (resting: F-ratio = 0.172, N = 24, p = 0.682, Figure 7a; feeding: F-ratio = 1.357, N = 24, p = 0.257, Figure 7b; moving: F-ratio = 2.680, N = 24, p = 0.116).

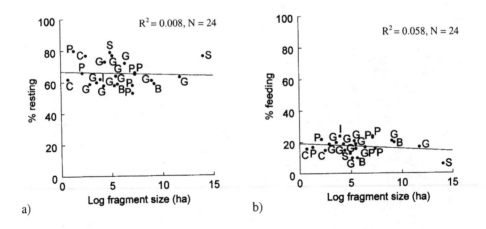

Figure 7. Relationship of fragment size and time spent (a) resting and (b) feeding by howler monkeys. B = *A. belzebul*, C = *A. caraya*, G = *A. guariba*, I = *A. pigra*, P = *A. palliata*, S = *A. seniculus*.

4. DISCUSSION

Most of the aspects of *Alouatta* ecology and behavior analyzed in this paper do not vary among species and are not predicted by fragment size. Although the size of the fragment may have a direct influence on home range size, the number of plant species used as food sources, and the diversity of fruits eaten, it does not affect day range, the contribution of food items to the diet, the diversity of sources of leaves and flowers, the number of species making up most of the diet, the number of plant species consumed on a daily basis, and the activity budget. This supports Jones' (1995) contention that howlers appear to be preadapted to cope with habitat fragmentation. Onderdonk and Chapman (2000) found that the folivorous black-and-white colobus (*Colobus guereza*) living in forest fragments near Kibale National Park, Uganda, presented changes in home range size and diet composition in terms of plant species. The activity budget and contribution of food items to the howler diet, on the contrary, were very similar to those presented by groups observed in Kibale. These results, however, contradict observations made by Juan et al. (2000) on howler monkey groups living in fragments of different sizes in Mexico.

Because howlers can live in very small forests, the observation that fragment size is a good predictor of home range size is not a surprise. On the other hand, the lack of relationship between fragment size and day range is explained by the energetic constraints derived from their behavioral adaptation to a highly folivorous diet (Milton, 1978; Crockett and Eisenberg, 1987). Day range similarity among howler populations exploiting different-sized home ranges was described by Bicca-Marques (1994).

Fragment size was not a good predictor of the amount of leaves, fruits, or flowers consumed. These findings contradict Johns' (1986) and Juan et al.'s (2000) observations on *Presbytis melalophos*, *Hylobates lar*, and *Alouatta palliata* living in altered habitats. However, the present study indicates groups living in smaller fragments tend to feed on a lower number of species and to exploit fewer sources of fruits than groups living in larger patches. This pattern may be related to the fact that many tropical tree species are clumped or randomly distributed over the landscape. As a consequence, habitat fragmentation and isolation tend to alter plant diversity in forest patches. Most likely, the smaller the fragment the lower its species richness. This in turn affects the array of food species available to howlers.

One genus (*Ficus*) and two families (Moraceae and Leguminosae) appear to be important components of the howler diet whenever present in a forest fragment. According to Julliot (1994), this preference may derive from the fact that *Ficus* spp. as well as other Moraceae species are particularly abundant in secondary environments, where most of the research on *Alouatta* has been conducted. Data from Queiroz' (1995) study group of *A. seniculus* in Mamirauá do not support Julliot's contention. In this relatively undisturbed site, *Ficus* was also the main howler food source. The fact that *Ficus* is characterized by intraspecific fruiting asynchrony with fruits being produced throughout the year may help explain its prevalence in howler diet.

Whatever their food preferences howlers can adapt to changes in floristic composition through flexible diet selection (Bicca-Marques and Calegaro-Marques, 1994a, 1994b; DeLuycker, 1995; Crockett, 1998; Marsh, 1999; but see Estrada et al., 1999). The *A. caraya* study group of Bicca-Marques and Calegaro-Marques (1994a) is perhaps the most striking case of this flexibility. This group survives in an extremely small forest patch by feeding on a diet whose contribution of exotic foods (especially fruits of *Citrus sinensis*, leaves of *Chorisia speciosa*, *Pereskia aculeata*, and *Melia azedarach*, and seeds and bark of *Eucalyptus* spp. among others) represented 38% of the annual feeding records. According to these authors, the fruits of exotic species were essential during a six-month period in which native fruits were extremely scarce (Figure 8). Howlers' ability to adapt their diet to the plant species available (including exotics) and the high contribution reached by some of them are aspects of their foraging strategy that are crucial for their success in surviving in fragments (see Onderdonk and Chapman, 2000, for similar observations on *Colobus guereza*).

Time spent resting, feeding, and moving, like most aspects of howlers' diet, are not predicted by fragment size. Howlers' very conservative activity budget is also probably a consequence of their energy-saving behavioral adaptation to a folivorous diet. Studies with other primates both agree (Onderdonk and Chapman, 2000) and disagree (Johns, 1986) with these results.

We urgently need comparative studies analyzing the behavior of particular species under varying degrees of habitat fragmentation. Preferably, this must be done with populations living in different-sized fragments that previously belonged to the same continuous forest (Estrada et al., 1999). This is important to reduce the influence that different floristic compositions may have on the resulting patterns when comparing fragments distributed over a wide latitudinal range as was done in this study. In addition, very few data are available for some species, especially *A. belzebul*. Different species of howler monkeys possibly respond differently to habitat fragmentation and that making genus-based generalizations is not the best approach. While such information is not available, the conclusions of this paper need to be taken as preliminary.

Are howlers safe in fragments? The answer is no. Despite their ability to survive in forest patches with few changes in behavior, howlers are more vulnerable to hunting, diseases, and predation in these habitats (Chiarello and Galetti, 1994; Cruz et al., 2000). Cruz et al. (2000), for example, found that black-and-gold howler monkeys living in a small and degraded fragment showed a higher diversity and prevalence of parasites than groups living in less disturbed habitats. In addition, fragmentation facilitates the occurrence of inbreeding depression in isolated populations. It may also compromise resource supply as a consequence of changes in interspecific relationships (Estrada and Coates-Estrada, 1996; Crockett, 1998). According to Onderdonk and Chapman (2000), long-term studies are necessary to determine the stability of a species' presence in fragments. In the case of howlers, we also need information on the biotic, physical, and historical characteristics that determine their presence in or absence from fragments (Estrada and Coates-Estrada, 1996).

Regardless of the long-term viability of fragmented populations, fragments may play an important role in howler monkey conservation. Since howlers are poor breeders in captivity (Crockett, 1998), zoological gardens should establish small forest patches composed of plant food species to serve as "maternity" for endangered howler species and, consequently, as sources of individuals for management programs. Human-made patches should also be established to serve as corridors between isolated forests (Bicca-Marques and Calegaro-Marques, 1994a; Horwich, 1998).

Figure 8. A juvenile female black-and-gold howler monkey visiting a fruiting *Citrus sinensis* tree in a winter day, a time of lean native fruit availability. Photo by Julio Cesar Bicca-Marques.

5. SUMMARY

Howler monkeys are known for their ability to survive in both intact and disturbed habitats of varying size. Their broad tolerance to changes in habitat quality has been related to a high degree of folivory, a diverse and flexible diet, and the use of small home ranges. However, whether their behavior and ecological relationships vary accordingly to the size of the available habitat (forest fragment) is unknown. This paper sheds light on this question through genus-based comparison of studies lasting nine months or more conducted at 27 Neotropical forests ranging from 1.3 to 1,240,000 ha in size. The effect of the size of the fragment on the use of space, diet composition, and activity patterns was determined through regression analysis. The variables analyzed were home range size, daily path length, contribution of food items to the diet, number of plant species used as food sources, number of species responsible for the bulk of diet, number of species consumed on a daily basis, two most important genera in the diet, and percentage of time spent resting, feeding, and moving.

Fragment size predicts home range size, the number of plant species used as food sources, and the diversity of leaf and fruit sources. On the other hand, it does not predict day range, the contribution of food items to the diet, the diversity of sources of flowers, the number of species making up most of the diet, the number of plant species consumed on a daily basis, the main food genera selected, and the activity budget. Therefore, howlers cope with habitat fragmentation without showing directional changes in most aspects of their feeding ecology and behavior. Their ability to contend with varying floristic compositions by feeding on a flexible diet composed of a large array of plants, including exotic species, is of special relevance to their success in forest fragments.

In spite of their ability to survive in forest patches, howlers are not safe in fragments on a long-term basis. Under these conditions, they are reported to be more vulnerable to hunting, diseases, predation, food shortages, and inbreeding depression. After all, small fragments may play an important role in conservation by serving as naturalistic environments for breeding howler monkeys in zoological gardens. Further research is needed to corroborate the results of this study.

6. ACKNOWLEDGMENTS

I thank Dr. Laura K. Marsh for inviting me to contribute to this volume and for critical suggestions to improve the final version. I also thank my wife Cláudia, my son Gabriel, and my daughter Ana Beatriz for giving me family support and inspiration.

7. REFERENCES

Alves, I. M. S. C., 1983, Comportamento e hábito alimentar de um grupo de bugios (Alouatta caraya Humboldt, 1812) em ambiente semi-natural (Primates, Cebidae), Bachelor's dissertation, Universidade de São Paulo, Ribeirão Preto.

Bicca-Marques, J. C., 1993, Padrão de atividades diárias do bugio-preto Alouatta caraya (Primates, Cebidae): Uma análise temporal e bioenergética, in: A Primatologia no Brasil - 4, M. E. Yamamoto and M. B. C. Sousa, eds., Editora Universitária-UFRN, Natal, pp. 35–49.

Bicca-Marques, J. C., 1994, Padrão de utilização de uma ilha de mata por Alouatta caraya (Primates: Cebidae), Rev. Brasil. Biol. 54:161–171.

Bicca-Marques, J. C., and Calegaro-Marques, C., 1994a, Exotic plant species can serve as staple food sources for wild howler populations, Folia Primatol. 63:209–211.

Bicca-Marques, J. C., and Calegaro-Marques, C., 1994b, Feeding behavior of the black howler monkey (Alouatta caraya) in a seminatural forest, Acta Biol. Leopoldensia 16:69–84.

Bonvicino, C. R., 1989, Ecologia e comportamento de Alouatta belzebul (Primates: Cebidae) na Mata Atlântica, Rev. Nordestina Biol. 6:149–179.

Chapman, C., 1987, Flexibility in diets of three species of Costa Rican primates, Folia Primatol. 49:90–105.

Chapman, C., 1988, Patterns of foraging and range use by three species of Neotropical primates, Primates 29:177–194.

Chapman, C. A., and Balcomb, S. R., 1998, Population characteristics of howlers: Ecological conditions or group history, Int. J. Primatol. 19:385–403.

Chiarello, A. G., 1993a, Activity pattern of the brown howler monkey Alouatta fusca, Geoffroy 1812, in a forest fragment of southeastern Brazil, Primates 34:289–293.

Chiarello, A. G., 1993b, Home range of the brown howler monkey, Alouatta fusca, in a forest fragment of southeastern Brazil, Folia Primatol. 60:173–175.

Chiarello, A. G., 1994, Diet of the brown howler monkey Alouatta fusca in a semi-deciduous forest fragment of southeastern Brazil, Primates 35:25–34.

Chiarello, A. G., and Galetti, M., 1994, Conservation of the brown howler monkey in southeast Brazil, Oryx 28:37–42.

Crockett, C. M., 1998, Conservation biology of the genus Alouatta, Int. J. Primatol. 19:549–578.

Crockett, C. M., and Eisenberg, J. F., 1987, Howlers: Variations in group size and demography, in: Primate Societies, B. B. Smuts, D. L. Cheney, R. M. Seyfarth, R. W. Wrangham, and T. T. Struhsaker, eds., The University of Chicago Press, Chicago, pp. 54–68.

Cruz, A. C. M. S., Borda, J. T., Patiño, E. M., Gómez, L., and Zunino, G. E., 2000, Habitat fragmentation and parasitism in howler monkeys (*Alouatta caraya*), *Neotrop. Primates* **8**:146–148.

Cunha, A. S., 1994, Aspectos sócio-ecológicos de um grupo de bugios (*Alouatta fusca clamitans*) do Parque Estadual de Itapuã, RS, M.Sc. dissertation, Universidade Federal do Rio Grande do Sul, Porto Alegre.

DeLuycker, A., 1995, Deforestation, selective cutting, and habitat fragmentation: The impact on a black howler monkey (*Alouatta caraya*) population in northern Argentina, *Bol. Primatol. Lat.* **5**:17–24.

Estrada, A., 1984, Resource use by howler monkeys (*Alouatta palliata*) in the rain forest of Los Tuxtlas, Veracruz, Mexico, *Int. J. Primatol.* **5**:105–131.

Estrada, A., and Coates-Estrada, R., 1996, Tropical rain forest fragmentation and wild populations of primates at Los Tuxtlas, Mexico, *Int. J. Primatol.* **17**:759–783.

Estrada, A., Juan-Solano, S., Martinez, T. O., and Coates-Estrada, R., 1999, Feeding and general activity patterns of a howler monkey (*Alouatta palliata*) troop living in a forest fragment at Los Tuxtlas, Mexico, *Am. J. Primatol.* **48**:167–183.

Fortes, V. B., 1999, Dieta, atividades e uso do espaço por *Alouatta fusca clamitans* (Cabrera, 1940) (Primates: Cebidae) na Depressão Central do Rio Grande do Sul, M.Sc. dissertation, Universidade Federal do Rio Grande do Sul, Porto Alegre.

Gaspar, D. A., 1997, Ecologia e comportamento do bugio ruivo, *Alouatta fusca* (Geoffroy, 1812, Primates: Cebidae), em fragmento de mata de Campinas, SP, M.Sc. dissertation, Universidade Estadual Paulista, Rio Claro.

Gaulin, S. J. C., and Gaulin, C. K., 1982, Behavioral ecology of *Alouatta seniculus* in Andean cloud forest, *Int. J. Primatol.* **3**:1–32.

Gilbert, K. A., and Setz, E. Z. F., 2001, Primates in a fragmented landscape: Six species in Central Amazonia, in: *Lessons from Amazonia: The Ecology and Conservation of a Fragmented Forest*, R. O. Bierregaard, C. Gascon, T. E. Lovejoy, and R. Mesquita, eds., Yale University Press, New Haven, pp. 262–270.

Glander, K. E., 1978, Howling monkey feeding behavior and plant secondary compounds: A study of strategies, in: *The Ecology of Arboreal Folivores*, G. G. Montgomery, ed., Smithsonian Institution Press, Washington, D.C., pp. 561–574.

Horwich, R. H., 1998, Effective solutions for howler conservation, *Int. J. Primatol.* **19**:579–598.

Johns, A. D., 1986, Effects of selective logging on the behavioral ecology of west Malaysian primates, *Ecology* **67**:684–694.

Jones, C. B., 1995, Howler monkeys appear to be preadapted to cope with habitat fragmentation, *Endangered Species UPDATE* **12**:9–10.

Juan, S., Estrada, A., and Coates-Estrada, R., 2000, Contrastes y similitudes en el uso de recursos y patrón general de actividades en tropas de monos aulladores (*Alouatta palliata*) en fragmentos de selva en Los Tuxtlas, México, *Neotrop. Primates* **8**:131–135.

Julliot, C., 1994, Diet diversity and habitat of howler monkeys, in: *Current Primatology - Vol. I: Ecology and Evolution*, B. Thierry, J. R. Anderson, J. J. Roeder, and N. Herrenschmidt, eds., ULP, Strasbourg, pp. 67–71.

Kinzey, W. G., 1997, *Alouatta*, in: *New World Primates: Ecology, Evolution, and Behavior*, W. G. Kinzey, ed., Aldine de Gruyter, New York, pp. 174–185.

Kowalewski, M. M., and Zunino, G. E., 1999, Impact of deforestation on a population of *Alouatta caraya* in northern Argentina, *Folia Primatol.* **70**:163–166.

Larose, F., 1996, Foraging strategies, group size, and food competition in the mantled howler monkey, *Alouatta palliata*, Ph.D. thesis, University of Alberta, Edmonton.

Limeira, V. L. A. G., 1996, Comportamento alimentar, padrão de atividades e uso do espaço por *Alouatta fusca* (Primates, Platyrrhini) em um fragmento degradado de Floresta Atlântica no estado do Rio de Janeiro, M.Sc. dissertation, Universidade Federal do Rio de Janeiro, Rio de Janeiro.

Lunardelli, M. C., 2000, Padrões de atividade e efeitos de compostos fenólicos na ecologia alimentar de um grupo de bugios-ruivos (*Alouatta fusca*) no sudeste brasileiro, M.Sc. dissertation, Universidade de São Paulo, São Paulo.

Marques, A. A. B., 1996, O bugio ruivo *Alouatta fusca clamitans* (Cabrera, 1940) (Primates, Cebidae) na Estação Ecológica de Aracuri, RS: Variações sazonais de forrageamento, M.Sc. dissertation, Pontifícia Universidade Católica do Rio Grande do Sul, Porto Alegre.

Marsh, L. K., 1999, Ecological effect of the black howler monkey (*Alouatta pigra*) on fragmented forests in the Community Baboon Sanctuary, Belize, Ph.D. thesis, Washington University, St. Louis.

Martins, C. S., 1997, Uso de habitat pelo bugio, *Alouatta fusca clamitans*, em um fragmento florestal em Lençóis Paulista – SP, M.Sc. dissertation, Universidade Estadual de Campinas, Campinas.

Mendes, S. L., 1989, Estudo ecológico de *Alouatta fusca* (Primates: Cebidae) na Estação Biológica de Caratinga, MG, *Rev. Nordestina Biol.* **6**:71–104.

Milton, K., 1978, Behavioral adaptations of leaf eating by the mantled howler monkey (*Alouatta palliata*), in: *The Ecology of Arboreal Folivores*, G. G. Montgomery, ed., Smithsonian Institution Press, Washington, D.C., pp. 535–549.

Milton, K., 1980, *The Foraging Strategy of Howler Monkeys: A Study in Primate Economics*, Columbia University Press, New York.

Neves, A. M. S., and Rylands, A. B., 1991, Diet of a group of howling monkeys, *Alouatta seniculus*, in an isolated forest patch in Central Amazonia, in: *A Primatologia no Brasil - 3*, A. B. Rylands, and A. T. Bernardes, eds., Fundação Biodiversitas para a Conservação da Diversidade Biológica, Belo Horizonte, pp. 263–274.

Neville, M. K., 1972, The population structure of red howler monkeys (*Alouatta seniculus*) in Trinidad and Venezuela, *Folia Primatol.* **17**:56–86.

Neville, M. K., Glander, K. E., Braza, F., and Rylands, A. B., 1988, The howling monkeys, genus *Alouatta*, in: *Ecology and Behavior of Neotropical Primates - Vol. 2*, R. A. Mittermeier, A. B. Rylands, A. F. Coimbra-Filho, and G. A. B. Fonseca, eds., World Wildlife Fund, Washington, D.C., pp. 349–453.

Onderdonk, D. A., and Chapman, C. A., 2000, Coping with forest fragmentation: The primates of Kibale National Park, Uganda, *Int. J. Primatol.* **21**:587–611.

Ostro, L. E. T., Silver, S. C., Koontz, F. W., Young, T. D., and Horwich, R. H., 1999, Ranging behavior of translocated and established groups of black howler monkeys *Alouatta pigra* in Belize, Central America, *Biol. Conserv.* **87**:181–190.

Peres, C. A., 1997, Effects of habitat quality and hunting pressure on arboreal folivore densities in Neotropical forests: A case study of howler monkeys (*Alouatta* spp.), *Folia Primatol.* **68**:199–222.

Perez, D. M., 1997, Estudo ecológico do bugio-ruivo em uma floresta com araucária do sul do Brasil (*Alouatta fusca*, IHERING 1914 - PRIMATES, ATELIDAE), M.Sc. dissertation, Universidade de São Paulo, São Paulo.

Pinto, L. P., 2002, Dieta, padrão de atividades e área de vida de *Alouatta belzebul discolor* (Primates, Atelidae) no sul da Amazônia, M.Sc. dissertation, Universidade Estadual de Campinas, Campinas.

Queiroz, H. L., 1995, *Preguiças e Guaribas: Os Mamíferos Folívoros Arborícolas do Mamirauá*, CNPq/Sociedade Civil Mamirauá, Brasília.

Rumiz, D. I., Zunino, G. E., Obregozo, M. L., and Ruiz, J. C., 1986, *Alouatta caraya*: Habitat and resource utilization in northern Argentina, in: *Current Perspectives in Primate Social Dynamics*, D. M. Taub and F. A. King, eds., Van Nostrand Reinhold, New York, pp. 175–193.

Rylands, A. B., Schneider, H., Langguth, A., Mittermeier, R. A., Groves, C. P., and Rodríguez-Luna, E., 2000, An assessment of the diversity of New World primates, *Neotrop. Primates* **8**:61–93.

Schwarzkopf, L., and Rylands, A. B., 1989, Primate species richness in relation to habitat structure in Amazonian rain forest fragments, *Biol. Cons.* **48**:1–12.

Serio-Silva, J. C., 1997, Studies of howler monkeys (*Alouatta palliata*) translocated to a Neotropical rain forest fragment, *Lab. Primate Newsl.* **36**:11–14.

Silver, S. C., Ostro, L. E. T., Yeager, C. P., and Horwich, R., 1998, Feeding ecology of the black howler monkey (*Alouatta pigra*) in northern Belize, *Am. J. Primatol.* **45**: 263–279.

Sokal, R. R., and Rohlf, F. J., 1995, *Biometry*, 3rd ed. W. H. Freeman, New York.

Sorensen, T. C., and Fedigan, L. M., 2000, Distribution of three monkey species along a gradient of regenerating tropical dry forest, *Biol. Cons.* **92**:227–240.

Steinmetz, S., 2000, Ecologia e comportamento do bugio (*Alouatta guariba clamitans*, Atelidae-Primates) no Parque Estadual Intervales, SP, M.Sc. dissertation, Universidade de São Paulo, São Paulo.

Stoner, K. E., 1996, Habitat selection and seasonal patterns of activity and foraging of mantled howling monkeys (*Alouatta palliata*) in northeastern Costa Rica, *Int. J. Primatol.* **17**:1–30.

Sussman, R. W., 1987, Species-specific dietary patterns in primates and human dietary adaptations, in: *The Evolution of Human Behavior: Primate Models*, W. G. Kinzey, ed., State University of New York Press, Albany, pp. 151–179.

Wilkinson, L. E., 1990, *Systat*, Systat Inc., Evanston.

Zunino, G. E., 1986, Algunos aspectos de la ecología y etología del mono aullador negro (*Alouatta caraya*) en habitat fragmentados, Ph.D. thesis, Universidad de Buenos Aires, Buenos Aires.

Zunino, G. E., González, V., Kowalewski, M. M., and Bravo, S. P., In press, *Alouatta caraya*: Relations among habitat, density, and social organization, *Primate Rep.*

Appendix I

Study site number (see Figure 1), forest size (in hectares), howler group name, group size range, home range size (in hectares), and minimum-maximum day range (mean).

Site #	Forest size	Group ID	Group size	Home range	Day range	Reference
A. belzebul						
1	271	Group A	6 to 8	9.5	350-650 (450)	Bonvicino, 1989
2	10,000		7 to 9	50.1	167-1425 (761)	Pinto, 2002
A. caraya						
3	1.8		13 to 20	1.8	-	Alves, 1983
4	10	[1]	6-8	-	-	Rumiz et al., 1986
	12	G1	6	5.5	130-1200 (345)	Zunino, 1986
		G2	6	6.3	-	Zunino, 1986
5	2		15 to 17	2.0	156-893 (454)	Bicca-Marques, 1994
A. guariba						
6	570	IP	7	7.9	197-1010 (534)	Mendes, 1989
7	80	G1	3 to 4	11.6	235-1527 (607)	Limeira, 1996
8	234		7 to 9	8.5	-	Gaspar, 1997
9	165		3 to 4	12.5	(494)	Martins, 1997
10	230		6	4.1	241-808 (467)	Chiarello, 1993b
11	7,900		8 to 10	4.5	-	Lunardelli, 2000
12	120,000		5 to 6	33.0	50-1280 (546)	Steinmetz, 2000
13	17.7		7	9.3	-	Perez, 1997
14	70		10 to 13	9.2	(968)	Marques, 1996
15	200		7 to 9	7.2	376-1564 (799)	Fortes, 1999
16	35		7 to 10	3.9	155-741 (393)	Cunha, 1994
A. palliata						
17	1,550	Scarface	17	31.1	104-792[2] (488)	Milton, 1980
		Old Forest	17	31.7	(392)	Milton, 1980
18	10,800		40	108.0	-	Chapman, 1988
		Sendero	20 to 28	30.5	(390)	Larose, 1996
		Exclosure	10 to 14	46.0	(331)	Larose, 1996
		San Emílio	4 to 7	17.8	(262)	Larose, 1996
19	1,200	Troop 1	20 to 23	54.0	-	Stoner, 1996
		Troop 2	11 to 14	35.0	-	Stoner, 1996
20	10	Group 1	13	9.9	207-1261 (596)	Glander, 1978
21	8.4		-	-	-	Serio-Silva, 1997
22	140	Troop S	14	60	11-503 (123)	Estrada, 1984
	3.6		7	3.6	(326)	Estrada et al., 1999
A. pigra						
23	50	C1	7	10.4	(520)	Ostro et al., 1999
		C2	10	15.8	(648)	Ostro et al., 1999
	1.3	School	5	1.3	-	Marsh, 1999
	4	Baptist	5	3.1	-	Marsh, 1999
	17.5	Ruben's II	8	3.0	-	Marsh, 1999
	24	Fig	7	3.0	-	Marsh, 1999
	75	OTC	6	3.5	-	Marsh, 1999
	82	West Dellas	5	3.0	-	Marsh, 1999
A. seniculus						
24	1,240,000	G2	8	7.5	(320)	Queiroz, 1995
25	135	Group 1	9 to 10	22.0	(706)	Gaulin and Gaulin, 1982
26	100,000		6	-	-	Julliot, 1994
27	93.4		Troop 1	6 to 9	7.1	-	Neville, 1972

[1] Data compiled from four groups.
[2] Does not inform to which group minimum and maximum day range values refer to.

Appendix II

Study site number (see Figure 1), contribution of plant items to the diet (in %; LV = leaves, YL = leaf buds + young leaves, ML = mature leaves, FR = fruits, and FL = flowers), and % time spent resting, feeding, and moving.

Site #	% diet composition					% time			Reference
	LV	YL	ML	FR	FL	Rest	Feed	Move	
A. belzebul									
1	13	6	3	59	28	59	10	19	Bonvicino, 1989
2	25	20	5	56	6	59	20	18	Pinto, 2002
A. caraya									
3	52	-	-	41	8	-	-	-	Alves, 1983
4	78	35	19	21	1	77	15	2	Zunino, 1986
5	61	19	34	29	3	62	16	18	Bicca-Marques, 1993; Bicca-Marques and Calegaro-Marques, 1994b
A. guariba									
6	77	-	-	14	9	72	17	11	Mendes, 1989
7	72	21	43	12	10	73	15	11	Limeira, 1996
8	74	-	-	12	9	64	16	16	Gaspar, 1997
9	79	10	66	21	0	77	10	12	Martins, 1997
10	73	43	23	5	12	64	19	13	Chiarello, 1993a, 1994
11	55	50	5	15	29	61	21	14	Lunardelli, 2000
12	74	37	25	23	1	63	17	16	Steinmetz, 2000
13	38	-	-	34	18	59	19	15	Perez, 1997
14	59	26	24	29	12	58	19	19	Marques, 1996
15	71	47	14	14	12	58	21	17	Fortes, 1999
16	54	27	22	36	10	60	20	13	Cunha, 1994
A. palliata									
17	44	-	<1	47	10	66	16	10	Milton, 1980
	53	-	3	37	9	65	16	10	Milton, 1980
18	49	21	28	29	23	-	-	-	Chapman, 1987
	58	37	21	17	-	-	-	-	Larose, 1996
	-	-	-	17	20	-	-	-	Larose, 1996
19	72	65	6	17	11	53	23	17	Stoner, 1996
	65	62	2	29	6	58	24	10	Stoner, 1996
20	69	44	19	13	18	-	-	-	Glander, 1978
21	28	-	-	59	<1	66	22	12	Serio-Silva, 1997
22	49	39	10	50	<1	-	-	-	Estrada, 1984
	54	47	8	41	1	80	17	2	Estrada et al., 1999
A. pigra									
23	45	37	8	41	11	62	24	10	Silver et al., 1998[1]
	57	-	-	34	9	-	-	-	Marsh, 1999
	54	-	-	33	13	-	-	-	Marsh, 1999
	62	-	-	36	2	-	-	-	Marsh, 1999
	68	-	-	20	11	-	-	-	Marsh, 1999
	33	-	-	63	4	-	-	-	Marsh, 1999
	52	-	-	24	24	-	-	-	Marsh, 1999
A. seniculus									
24	46	20	26	47	2	76	6	16	Queiroz, 1995
25	52	45	7	43	5	79	13	6	Gaulin and Gaulin, 1982
26	57	-	-	26	13	-	-	-	Julliot, 1994

[1] Data compiled from two groups (C1 & C2).

Appendix III

Study site number (see Figure 1), number of plant species used as food sources (PS), as sources of leaves (SL), fruits (SF), and flowers (SW), mean number of species eaten per day (DA), and number of food species making up 50% (50) and 80% (80) of diet.

Site #	PS	SL	SF	SW	DA	50	80	Reference
A. belzebul								
1	47	6	39	5	-	-	-	Bonvicino, 1989
2	67	27	35	-	6.5	10	-	Pinto, 2002
A. caraya								
3	32	19	12	6	-	-	-	Alves, 1983
4	23	-	-	-	-	2	4	Rumiz et al., 1986[1]
	18	18	4	1	-	2	5	Zunino, 1986
5	27	18	20	8	7.7	3	6	Bicca-Marques and Calegaro-Marques, 1994b
A. guariba								
7	37	25	7	7	5.6	2	7	Limeira, 1996
8	-	-	9	11	-	6	24	Gaspar, 1997
9	34	-	10	0	-	6	15	Martins, 1997
10	68	60	12	12	-	6	24	Chiarello, 1994
11	41	-	-	-	-	6	17	Lunardelli, 2000
12	95	72	34	9	10.8	5	14	Steinmetz, 2000
13	23	-	-	-	-	-	-	Perez, 1997
14	43	-	-	20	8.8	4	12	Marques, 1996
15	64	51	24	14	9.2	5	14	Fortes, 1999
16	45	37	31	19	6.0	6	12	Cunha, 1994
A. palliata								
17	73	59	23	16	7.4	3	17	Milton, 1980
	73	59	25	13	8.0	7	32	Milton, 1980
18	-	-	-	-	-	3	9	Chapman, 1988
	30	-	-	-	3.5	3	7	Larose, 1996
	29	-	-	-	3.8	3	8	Larose, 1996
	25	-	-	-	3.6	3	7	Larose, 1996
19	95	76	32	25	-	7	14	Stoner, 1996
	65	52	24	17	-	2	12	Stoner, 1996
20	61	51	15	31	-	7	17	Glander, 1978
21	28	-	-	-	-	-	-	Serio-Silva, 1997
22	27	24	12	1	-	2	6	Estrada, 1984
	52	-	-	-	-	3	8	Estrada et al., 1999
A. pigra								
23	74	-	-	-	-	7	18	Silver et al., 1998[2]
	51	39	19	5	-	5	-	Marsh, 1999
	37	32	10	3	-	-	5	Marsh, 1999
	43	38	12	4	-	5	10	Marsh, 1999
	44	36	17	7	-	5	10	Marsh, 1999
	32	24	8	5	-	1	5	Marsh, 1999
	43	34	11	5	-	5	10	Marsh, 1999
A. seniculus								
24	26	20	8	2	-	2	6	Queiroz, 1995
25	33	28	13	4	-	3	7	Gaulin and Gaulin, 1982
26	195	96	97	36	-	40	-	Julliot, 1994

[1] Data compiled from four groups.
[2] Data compiled from two groups (C1 & C2).

SECTION III: CONSERVATION AND MANAGEMENT

Laura K. Marsh

1. INTRODUCTION

Linking population dynamics, genetics, and behavioral ecology allows for a discussion of conservation and management. Inherent in creating systems for maintaining primate populations in fragments is the need for creating generalizations (c.f., Marsh et al., this volume). Authors in this section discuss needs for each of the species they study specific to their region. Every situation is slightly different, and this difference slows the pace of conservation on a regional scale. The final chapter in this section attempts to summarize an action plan for all primates in fragments.

Once again, the matrix plays a key role for the primates, especially concerning their conservation status. Laurance (1994) noted three key advantages of matrix-tolerant species: 1) they can disperse between fragments, or between continuous forest and fragments, and therefore increase genetic viability of the population; 2) they can recolonize fragments following local extinctions; and 3) since they tend to be generalists, they often exploit ecological changes in fragments, like edge effects. While these traits in general allow some species to move between fragments or forest, they can also have grave consequences for those moving though human dominated landscapes. Species who use a matrix dominated by human agriculture or other human use areas may be walking out into a hostile land, one that is not necessarily safe even if it is ecologically useful to the primates. Many communities see primates outside of forested areas as pests. In every case in this section primates move through or use human created matrix. For some it is positive (McCann et al., this volume) and for others it is not (Reynolds et al., Eniang, this volume).

2. AGRICULTURAL MATRIX

Reynolds et al. in Chapter 17 describe the combination of fragments, sugar and chimpanzees in the Masindi District, western Uganda. There are about 600 chimps in the study region. A group of 12 chimps living in a riverine forest fragment outside the Budongo Forest Reserve were the focus of this chapter. Sugar cane growers have surrounded this fragment, nearly isolating it from the reserve. Chimps are at risk of being killed if they enter into the cane fields to disperse or to raid the crops. Local human population pressures and the hostile attitude toward the chimps makes conservation difficult. A study is being conducted to determine long-term viability, and the presence of local researchers is beginning to reduce some pressure.

McCann et al. discuss in Chapter 18 shade coffee plantations as wildlife refuges for *Alouatta palliata* in Nicaragua. Mombacho Reserve is one of the few remaining cloud forests in the southwest of the country, which makes the region attractive to coffee growers. Out of 97 howler troops located in this survey, only 13 were recorded within the Reserve itself, representing 13% of the population. The buffer area surrounding the reserve is important to the howlers in the region. The authors argue that shade grown coffee on the plantations and lands surrounding Mombacho Reserve can have a positive effect on the howlers if the coffee owners plant fruit species that support monkeys and other local wildlife. Establishing and managing agricultural lands and buffer zones is critical for the long-term protection of the area's biodiversity.

3. MUTIPLE ENCROACHMENTS

Eniang presents in Chapter 19 conservation considerations for the Cross River gorilla (*Gorilla gorilla dielhi*) in fragmented forest of southeastern Nigeria. The current population status of the gorillas is unknown but estimated at 100 to 250 individuals in the Cross River region. Enclave communities, highways, bushmeat hunting, logging, non-forest timber harvesting and agriculture all lead to fragmentation, isolation and disturbance of remaining groups. The author suggests many conservation measures, such as adapting traditional conservation strategies, enforcing wildlife laws, closing the main highway, and providing alternatives to bushmeat that may be the ultimate answer to reducing the gorilla's state of decline.

4. COMPILATION

Marsh in Chapter 20 summarizes conservation methods for primates in fragments with the "wild zoo" concept. Traditionally, strategies for managing primates in the wild include preservation, reintroduction, captive breeding, and translocation. The "wild zoos" methods are a combination of established methodology for primates, methods for managing fragments, and additional techniques specific to a 'hands on' approach for managing primates in fragments. This chapter brings together many ideas that can be used in various combinations to maintain the viability of primates in threatened habitats.

5. SUMMARY

Much of the conservation that needs to be done in tropical forests has not been tested or proven effective without some basic truths. If a region has protected status then there must be enforcement of wildlife laws governing hunting and use of habitat. Reserves have a better chance of protecting than private lands. If there is no protection, then local communities must be called upon to be stewards of the land. Only if local people can have all of their needs met will they protect habitat and species. When humans and primates are in agreement with how to use the matrix or buffer area surrounding fragments, then there are few conflicts. This was the case in Nicaragua where McCann et al. found the howlers and coffee growers "getting along." On the other hand, in countries where land is a premium and the population needs more of it to sustain themselves, like in Uganda and Nigeria, there is less tolerance for primates who invade crops or are simply "in the way." The challenge is to develop plans specific to nations and to regions that promote the needs of people while being sensitive to the needs of local primate populations. These matters are complex and we need clarity of action and cohesive thought to drive conservation in the future. We further develop goals for the conservation of primates in fragments in the next section.

6. REFERENCES

Laurance, W. F., 1994, Rainforest fragmentation and the structure of small mammal communities in tropical Queensland, *Biol. Cons.* **69**:23–32.

FRAGMENTS, SUGAR, AND CHIMPANZEES IN MASINDI DISTRICT, WESTERN UGANDA

Vernon Reynolds, Janette Wallis, and Richard Kyamanywa[*]

1. INTRODUCTION

Masindi District in western Uganda contains the Budongo Forest, with a population of approximately 600 chimpanzees (*Pan troglodytes schweinfurthii*). In 1999, a group of chimpanzees was discovered living in semi-isolation in the Kasokwa riverine forest fragment outside the main forest block (Figure 1). Although there is growing local support among village residents for saving the chimpanzees, immigrant sugar farmers pose a problem as they have been removing forest cover at an alarming rate.

This chapter details the history of deforestation in Masindi District and the problem of the sugarcane industry. In addition, we provide some preliminary information from study of the Kasokwa chimpanzees and describe the development of a conservation program.

2. FRAGMENTATION OF THE BUDONGO FOREST

The Budongo Forest Reserve is the largest mahogany forest in East Africa and the largest remaining forest block in Uganda. As stated, it is home to an estimated 600 chimpanzees, 20% of the approximately 3,000 chimpanzees remaining in Uganda today (Plumptre and Reynolds, 1996). Budongo was one of the first sites for wild chimpanzee research (Reynolds and Reynolds, 1965) and is now home base of the Budongo Forest Project (BFP). The Budongo chimpanzees probably form 10 to 15 communities, one of which—the Sonso Community comprising 46 to 50 chimpanzees—has been studied for over a decade (Reynolds, 1998).

As with many forests in Africa, human encroachment has occurred in the region surrounding the Budongo Forest Reserve over the last several decades. Local human

[*] Vernon Reynolds, Institute for Biological Anthropology, Oxford University, U.K., and Budongo Forest Project, Masindi, Uganda. Janette Wallis, Department of Psychiatry and Behavioral Sciences, University of Oklahoma, U.S.A. Richard Kyamanywa, Budongo Forest Project, Masindi, Uganda. Correspondence to V. Reynolds (email: vernon.reynolds@bioanthropology.oxford.ac.uk).

Primates in Fragments: Ecology and Conservation
Edited by L. K. Marsh, Kluwer Academic/Plenum Publishers, 2003

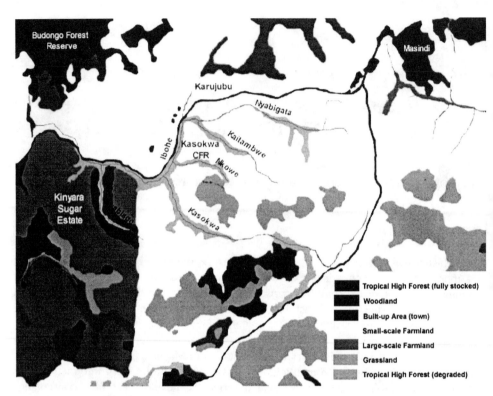

Figure 1. The Kasokwa Central Forest Reserve is a small riverine forest fragment located southeast of the Budongo Forest Reserve (upper left).

populations have deforested some areas for farming purposes, and a large sugar plantation has grown unrestricted within recent years.

For 40 years population expansion has occurred in western Uganda, establishing a trend of destruction (Hamilton, 1984) of unprotected trees in riverine forests and wood lots. Wood is needed for cooking fires, for building poles, for charcoal (especially for curing tobacco, a prominent local industry), and for other purposes. Trees are therefore very much at risk unless they are in a protected area. This destruction in and around the periphery of the Budongo Forest Reserve has ultimately led to a number of forest fragments scattered in the region. There is little doubt that, in the recent past, the whole course of the river was forested. Today the forest cover has been removed outside the Kasokwa Central Forest Reserve (CFR), and the line of the river between Kasokwa CFR and the main forest block is without trees.

The Kasokwa CFR is 73 ha in extent and is part of a complex series of riverine forests that are not CFRs but can be called community riverine forests. Some of these riverine forests occur along tributaries of the Kasokwa River and other nearby rivers. The names and estimated sizes of these other forests are as follows: Myabigata, 80 ha; Nkohe, 50 ha; Kinyara, 36 ha; Kasokwa, 146 ha; and Kaifambwe, 60 ha. Between these forest fragments are a variety of land-use types, including villages with houses and gardens;

fields of cassava, maize, beans, tobacco, and other crops; sugarcane grown on farmers' fields; and huge swathes of land owned by Kinyara Sugarworks, Ltd. (KSW).

In the case of Kasokwa CFR, because the riverine forest to the south of the Masindi-Butiaba Road was designated as a CFR (and thus under Forest Department control), much of the tree cover has been left intact, though some encroachment has occurred and is still occurring. The result is that parts of this forest strip are degraded and contain both farmers' fields and their houses. The study and restoration of such remnant forests are important and result in conservation activities with strong possibilities for success (Lamb et al., 1997; Tutin, 1999). The presence of researchers in these areas helps protect against further encroachment, and community-based education programs focus on reforestation and creation of forest corridors to reunite fragmented populations of various wildlife species.

At a meeting in Masindi with local officials we learned that the people who live between the Kasokwa CFR and the main forest block are running short of water in the dry season. This was formerly not the case. By removing the tree cover from the Kasokwa River, it appears that they have reduced the water-holding capacity of the soil and now their river runs dry seasonally. For this reason, a number of local leaders are eager to plant trees along the river in the hope that this will improve their water supply in future years.

Tree planting is a feature of an education program currently being run by the BFP. The loss of trees both inside and around Budongo Forest Reserve began to assume serious proportions in the 1970s, after the rise of Idi Amin, although selective logging had been a feature of Budongo forestry since around 1930. With the loss of control of logging by the Forest Department, pit-sawing became very widespread and almost all large mahogany trees were felled. This generated an attitude about trees that is only now being reversed, namely that all trees could be felled with impunity. Today we are seeing signs that this attitude has been counter-productive; and there are signs of a return to controlled, selective logging.

3. ADD SUGAR: THE DEVELOPMENT OF AN INDUSTRY

In recent years, there has been expansion of a large sugar factory near the Kasokwa forest strip, the KSW. This has led to employment opportunities and an unprecedented influx of migrant workers from various parts of Uganda into the region. As a result, many of the fields to the south of the main Budongo Forest block have been planted with sugarcane, and there has been much removal of tree cover in village wood lots and other unprotected areas.

To increase production, the Sugarworks also has an outgrowers' scheme whereby privately owned fields are put under sugar and the owners are paid by KSW for the sugar they grow. This provides a very lucrative source of income for owners of fields with few or no sources of income. As a result, much land to the north of the Masindi-Butiaba Road adjoining the Sugarworks has been planted with sugar. This private land extends right up to the edge of the main forest block and it covers the area between the main forest block and the Kasokwa riverine forest to the south of the Masindi Road.

4. THE CHIMPANZEES OF KASOKWA FOREST

During 1999, a small community of chimpanzees was discovered inhabiting at least one forest fragment in Masindi District, southeast of the Budongo Forest main block. This community ranges to the southeast of Budongo, into a riverine forest strip large enough to have been designated as Kasokwa CFR. The strip is approximately 7 km in length and is approximately 100 to 200 m wide along the Kasokwa River. According to members of the local village of Karajubu, this strip has been utilized by chimpanzees over a long period. At the present time, a number of chimpanzees appear to be living either permanently or for considerable periods of time in the Kasokwa Forest Reserve.

For chimpanzees to reach the Kasokwa CFR to find natural foods, they have to cross an area of at least 2 km where the tree cover along the Kasokwa River has been removed. In this area they must pass through villages, traditional gardens, or *"shambas,"* and newly planted fields of sugarcane grown by outgrowers for the KSW. The chimpanzees must then cross the main Masindi-Butiaba Road. When they reach the tree cover of Kasokwa Forest they are close to human habitations, newly made fields established by squatters who are occupying Forest Department land illegally, and the borders of a vast area of sugarcane belonging to KSW (Figure 2).

Figure 2. This photograph shows clearly the plight of the Kasokwa chimpanzees. In the foreground are houses alongside cleared farmland by the forest strip. In the background are fields of sugarcane. Between the cane and the houses runs a small forest strip–home to at least 12 chimpanzees. Photo by Richard Kyamanywa.

4.1. The Kasokwa Chimpanzee Project

In October of 1999, personnel of the BFP dispatched a locally trained field assistant to investigate the behavioral ecology of these chimpanzees. One aim was to determine if the forest patch was suitable for their long-term survival—or whether plans should be made to translocate the chimpanzees to a more suitable environment.

With proper information on the subjects' behavioral ecology, we will be able to determine whether the chimpanzees can survive in the forest fragment for the next several years. In addition to this forest fragment at Kasokwa, several other forest fragments in the district will be surveyed during the coming years as part of a general assessment of the region. By monitoring the behavioral ecology of the Kasokwa chimpanzees, we will determine their population density, home range, activity patterns, behavior, and diet. We will learn the extent of their raiding of nearby crops and whether they are able to make contact with other chimpanzee communities.

Thus, the ongoing study of the Kasokwa chimpanzees aims to determine the following:

- the size and demographic profile of the community of chimpanzees living in this forest strip and whether it is permanent or temporary;
- whether the reason for the long-term presence of these chimpanzees is because this habitat is viable in the long term or whether it is due to the difficulty of reaching and leaving this riverine forest from the main Budongo Forest block in the modern situation of human population expansion and sugarcane growing;
- the movement patterns of chimpanzees in this community, in particular, whether they move between the Budongo Forest Reserve and the Kasokwa Forest Reserve, and whether they also move into other riverine forest strips farther south;
- whether the individuals studied in Kasokwa are a sub-community and form part of a wider community within which there are movements of individuals or parties, and, if so, how often such movements occur; and
- to what extent the food supply in the Kasokwa riverine forest strips are sufficient for the chimpanzees' needs and to what extent they depend on crop raiding, in particular for sugarcane, to survive.

4.2. Data Collection Techniques

The primary method used to obtain information about the Kasokwa chimpanzees has been to locate and follow them in the forest strip. One field assistant (R. Kyamanywa) who lives locally was aware of the existence of these chimpanzees and was employed by the BFP in September 1999. He was trained by BFP chimpanzee field assistants with 10 years' experience to observe the Kasokwa chimpanzees. Although this forest does not have a regular trail system, there are human paths in the forest along the course of the river so that movement is possible without noise or disturbance. In September 2000, a second field assistant was employed.

The field assistants' duties are to enter the forest each morning (at approximately 0600) and locate chimpanzees. Due to the nature of the fission-fusion social organization of chimpanzees, one rarely finds all community members together–even for this small community. Instead, the community will routinely break into small groups that may or may not meet during the course of the day. The Field Assistants use prepared data sheets and collect pertinent information (date, time of observation, individual chimpanzee identities, and location when observed). Detailed behavioral data are recorded that include social behaviors such as sex, play, grooming, aggression, hunting, and solitary behaviors such as resting and feeding. In addition, information on diet and vocalization is recorded. Attempts are made to follow the chimpanzees until the time they make a nest for sleeping at night, thus providing their location for finding them the next day.

4.3. Subjects and Community Composition

By April 2000, the composition of the community had been established. Preliminary data indicated that at least 12 chimpanzees lived in the small strip of riverine forest along the Kasokwa River: one adult male, four adult females, three subadult males (Figure 3), two juvenile males, one juvenile female, and one infant.

The chimpanzees were effectively cut off from the main forest block, although continued study will determine if inter-group transfer may still occur naturally. Whereas the forest appears to support an adequate food supply for much of the year, during the dry season the chimpanzees were observed to leave the forest and eat sugarcane and other crops raised by local people. This encroachment by the chimpanzees (but, more often, baboons) has led some local farmers to set traps. Consequently, at least 50% of the

Figure 3. This young male is one of at least 12 chimpanzees living in the small forest fragment along the Kasokwa River, Masindi District, Uganda. Photo by Janette Wallis.

chimpanzees first identified (N = 7) had snare wounds (missing or deformed appendages) (Kyamanywa, 2000).

The group of chimpanzees in Kasokwa Forest Reserve possibly lived there permanently and without immigration or emigration, having been cut off from the main forest block by human settlements and tree clearance. This was a particularly troubling prospect since only one adult male, *Kigere*, was in the group. In fact, this male died during June 2000 (for details see below). The loss of the only adult male was a great blow to this small population; the next oldest male in the group was estimated to be a young adolescent. Remarkably, however, within a couple of months of *Kigere's* death, a new adult male and two adult female chimpanzees (one with an infant) arrived in the Kasokwa forest. We do not know from which direction these newcomers arrived. They possibly came from another of the forest fragments to the south, rather than from the Budongo Forest Reserve proper. The transfer into a community by an adult male chimpanzee is a very unusual event. Thus, further study of this new male and his behavior is warranted, as is the continued survey of the area for additional forest fragments containing small isolated groups of chimpanzees.

4.4. Preliminary Information of the Kasokwa Chimpanzees' Behavior and Diet

Study of the Kasokwa chimpanzees has revealed a number of things about their daily activities and travel patterns. In general, the chimpanzees form parties to search for food. They spend 45% to 55% of the day feeding, with morning and afternoon peaks. They feed between 0630 or 0700 and 1130. They tend to rest during midday and resume feeding at approximately 1500 until approximately 1800. Beginning at 1800, they start nesting. This pattern resembles the feeding and nesting schedules of chimpanzees living in less disturbed habitats (Wrangham, 1977; Brownlow et al., 2001).

During the dry season (December–February), the chimpanzees of Kasokwa depend on 40% natural foods (e.g., fruits, flowers, leaves, and shoots) and 55% human foods (e.g., sugarcane near the forest, paw-paws, and mangoes), and 5% meat (e.g., colobus monkey). During the wet season (March–November), when fruit availability is enhanced, the chimpanzees' diet relies much less on human foods; they raid human crops less often at that time (R. Kyamanywa, personal observation; Figure 4).

As the forest corridor is predominantly riverine forest, the most common tree species include *Pseudospondias microcarpa* and *Phoenix reclinata*. Both are typically found in wet areas. The most common fig tree is *Ficus vallis choudae*.

4.5. Chimpanzee-Human Interactions

On 6 June 2000 the dominant male chimpanzee of the Kasokwa community, *Kigere* (a name meaning 'foot' because he had lost his right foot in a snare or trap), was caught by the right hand in a large iron 'man trap' 40 cm in length and weighing 10 kg. Having two major trap injuries on the same side of the body would have been a difficult, if not impossible, burden to overcome. The trap snapped shut over his right fingers but did not sever them. Instead, this individual dragged and carried the trap around with him for 11 days before he finally succumbed to gangrene and septicemia. He was found dead on 17 June 2000 (Munn and Kalema, 1999–2000).

Figure 4. (Top) A female chimpanzee walks from the sugarcane field to the forest strip on the right. (Bottom) Later examination of the sugarcane reveals clear evidence of "crop raiding" by the chimpanzee. Photos by Richard Kyamanywa.

On 1 July 2000, a human boy aged 6 to 8 months was seized by a young female chimpanzee of the Kasokwa community and taken into the forest. A search was mounted for him and after some hours he was found dead and already cold, lying on the forest floor. He had received bites in the nasal region, a laceration on the back of the neck, a laceration on the scalp, and multiple puncture wounds to the body. The index and smallest digit of the right hand and the left ear were missing. He was taken to Masindi Hospital where he was certified dead on arrival.

As the above events portray, relations between the humans and chimpanzees at Kasokwa have been antagonistic at times. However, the situation is complex. The mother whose child was taken was one of a family of squatters living illegally on forest edge land. After an initial outcry against the chimpanzees, it later emerged that the local residents who have lived in the area for a long time did not sympathize with the family whose child was lost. They do not want the squatters in their area and they made this clear at a public meeting (F. Babweteera, personal communication). Indeed, there is widespread feeling among the local traditional residents that the migrant workers who have come into the area seeking work at the Sugarworks are spoiling the environment. Trees are being cut down, forest destroyed, and the 2 km course of the Kasokwa River where it runs from the remaining riverine forest into the main forest block, now denuded of trees, dries up in the dry season, which never happened when the river was under trees. This level of environmental degradation is beginning to make itself felt and it may be that, in the future, steps can be taken to safeguard the area from the new immigrants.

5. SOLUTIONS: PLANNING FOR THE FUTURE

Since BFP took an interest in this chimpanzee community, the Jane Goodall Institute-Uganda has also become involved (Cox et al., 2000). Funding has been received from a number of agencies, and several meetings have been held with local officials and others with an interest in chimpanzee conservation. Chimpanzees are an endangered species, protected by international laws, but these laws are not enforced in the situation at Kasokwa. The BFP is running an educational outreach program to try and sensitize people to the plight of the chimpanzees, but this is not likely to have an impact on the sugar outgrowers because their livelihoods are at stake. In the case of sugar outgrowers, paid as they are by the weight of the sugar they produce, any crop raiding must be prevented and for this they use children, dogs, snares, and traps. As this activity occurs on private land, there is no law prohibiting it.

The human population seems to be divided between the older traditional residents who want the chimpanzees to be protected and, in addition, hope to benefit from ecotourism and the younger immigrant sugar workers (and would-be workers) whose interests are primarily in financial gain (Lauridsen, 1999) and to whom all forms of wildlife constitute a nuisance. Still, there has been a surge of interest in the Kasokwa chimpanzees, and this may materialize in a concerted plan to save them and their forest from further damage. Some of the older, traditional residents clearly do not object to having chimpanzees in the area even if they do some limited raiding of mango, paw-paw, banana, and maize crops. A common belief is that chimpanzees are more selective than baboons, which constitute a major threat to the crops because they do more damage than chimpanzees (Hill, 2000).

When people living adjacent to Parks and Reserves are surveyed, most reveal that they are in favor of protecting forests. A study by Newmark et al. (1993) found that people living near Tanzanian National Parks supported abolition of the parks only if they had experienced past problems with wildlife or shortage of farming land. Similarly, people in and around Budongo Forest Reserve reveal positive attitudes about conservation (BFP, 2000a), though many remain ignorant about their rights in the forest (BFP, 2000b). Because a positive conservation attitude was to a large extent related to the level of education, it appears there is a good basis for improvement through a proper education campaign in the Masindi District (BFP, 2000a).

A community-based organization, Nature Conservation and Promotion Association (NACOPRA), was formed in 2000 to protect and preserve the remaining forest and wildlife living within the area. This grassroots organization functions through voluntary activity of local people, with a small amount of financial aid from the Environmental Protection and Economic Development project and BFP. Because community support is necessary for any conservation program to succeed (McNeely, 1992), on 13 April 2000, representatives of the Uganda Wildlife Authority, JGI-Uganda, and BFP organized a meeting in Masindi Town. The gathering included the organizers, NACOPRA, local community leaders, a senior KSW representative, officials from the Forest Department, and the authors. After much discussion, it was agreed by all that the chimpanzees should remain under protective study, and conservation efforts should be heightened to preserve the forest and eventually reconnect it to the main forest block. Eventual translocation of the chimpanzees was not ruled out if forest depletion were to continue. However, too little is presently known about the chances for long-term survival in the Kasokwa forest fragment and it is known that similar remnant communities of chimpanzees exist nearby.

With the continuation of the project, the BFP will supplement NACOPRA, which has implemented a local conservation education program. NACOPRA has enlisted the aid of local community leaders in protecting the forest and its inhabitants. In addition to generally helping the local people learn about sustainable use of the forest, we are obtaining the support of the District Forest Department authorities to begin reforestation plans that will produce firewood and building poles as well as create corridors to allow easy passage of chimpanzees between the forest fragment and the main forest of Budongo.

The future for the Kasokwa chimpanzees and others living in forest outliers and riverine strips near the Budongo Forest lies in the success of conservation agencies in enabling local people to reverse current trends. This can only be done with substantial amounts of money and by careful planning. Tree corridors, buffer zones, compensation schemes, and other such measures will be needed if these forest fragment chimpanzee populations are to be saved.

6. SUMMARY

Masindi District in western Uganda contains the Budongo Forest with its population of about 600 chimpanzees. In 1999, a group of 12 chimpanzees was discovered living in semi-isolation in the Kasokwa Forest Reserve, a riverine forest fragment outside the main Budongo Forest Reserve. This riverine forest was formerly joined to Budongo Forest along the line of a river flowing into and through Budongo. However, sugarcane outgrowers have isolated the Kasokwa Forest so now it has become a fragment separated by a minimum distance of 2 km from the main Budongo forest block.

The chimpanzees are now at risk of being killed by surrounding farmers when they raid the sugar. There is, however, support for saving them among residents of local villages, but immigrant sugar farmers are hostile to both chimpanzees and baboons, and are removing forest cover along this riverine fragment and setting snares and traps. A study of the chimpanzees is underway to determine their feeding habits, ranging patterns, and long-term viability. The presence of two local observers is reducing encroachment and damage to the apes and their habitat, and, recently, public pressure was brought on the Forest Department, forcing them to demarcate the borders of the Kasokwa Forest Reserve and remove squatters who were building houses and cultivating in this protected area.

7. ACKNOWLEDGMENTS

We are grateful to the National Geographic Society for its support of RK and of the chimpanzee research of the BFP; NORAD for core support to BFP; the staff of BFP for training assistance; Jane Goodall Institute (in particular, Sam Mugume and Julia Lloyd) for their valuable survey of the Kasokwa community in early 2000; and the American Society of Primatologists for supplemental support. In addition, we gratefully acknowledge a large number of local residents and officials in Masindi Town and surrounding villages who have been willing to give time to attend meetings and discuss the current range of problems discussed in this paper.

8. REFERENCES

Brownlow, A., Plumptre, A. J., Reynolds, V., and Ward, R., 2001, Sources of variation in the nesting behavior of chimpanzees (*Pan troglodytes schweinfurthii*) in the Budongo Forest, Uganda, *Am. J. Primatol.* **55**(1):49–55.

Budongo Forest Project, 2000a, Policy brief: Local attitudes towards the Budongo Forest, adapted from the research conducted by Kirstin Johnson.

Budongo Forest Project, 2000b, Policy brief: Understanding of national and local laws among villagers living to the southwest of Budongo Forest Reserve, Uganda, adapted from the research conducted by Lucy Bannon.

Cox, D., Mugume, S., and Lloyd, J., 2000, Investigation into the survival of a small chimpanzee community under immediate threat, JGI Report.

Hamilton, A. C., 1984, *Deforestation in Uganda*, Oxford University Press, Nairobi.

Hill, C. M., 2000, Conflict of interest between people and baboons: Crop raiding in Uganda, *Int. J. Primatol.* **21**(1):77–84.

Kyamanywa, R., 2000, Masindi chimpanzees, progress report for Budongo Forest Project.

Lamb, D., Parrotta, J., Keenan, R., and Tucker, N., 1997, Rejoining habitat remnants: Restoring degraded rainforest lands, in: *Tropical Forest Remnants: Ecology, Management, and Conservation of Fragmented Communities*, W. F. Laurence and R. O. Bierregaard, Jr., eds., The University of Chicago Press, Chicago, pp. 366–385.

Lauridsen, M., 1999, Workers in a forest: Understanding the complexity of incorporating local people in modern management, M.S. thesis, Institute of Anthropology, University of Copenhagen.

McNeely, J. A., 1992, Protected areas in a changing world: The management approaches that will be required to enable primates to survive into the 21st century, in: *Topics in Primatology, Volume 2, Behavior, Ecology, and Conservation*, N. Itoigawa, Y. Sugiyama, G. P. Sackett, and R. K. R. Thompson, eds., University of Tokyo Press, Tokyo, pp. 373–384.

Munn, J., and Kalema, G., 1999–2000, Death of a chimpanzee (*Pan troglodytes schweinfurthii*) in a trap in Kasokwa Forest Reserve, Uganda, *Af. Prim.* **4(1 and 2)**:58–62.

Newmark, W. D., Leonard, N. L., Sariko, H. I., and Gamassa, D. M., 1993, Conservation attitudes of local people living adjacent to five protected areas in Tanzania, *Biol. Cons.* **63**:177–183.

Plumptre, A. J., and Reynolds, V., 1996, Censusing chimpanzees in the Budongo Forest, Uganda, *Int. J. Primatol.* **17(1)**:85–99.

Reynolds, V., 1998, Demography of chimpanzees *Pan troglodytes schweinfurthii* in Budongo Forest, Uganda, *Af. Prim.* **3(1-2)**:25–28.

Reynolds, V., and Reynolds, F., 1965, Chimpanzees of the Budongo Forest, in: *Primate Behavior: Field Studies of Monkeys and Apes*, I. DeVore, ed., Holt, Rinehart & Winston, New York, pp. 368–424.

Tutin, C. E. G., 1999, Fragmented living: Behavioral ecology of primates in a forest fragment in the Lope Reserve, Gabon, *Primates* **40(1)**:249–265.

Wrangham, R. W., 1977, Feeding behaviour of chimpanzees in Gombe National Park, Tanzania, in: *Primate Ecology: Studies of Feeding and Ranging Behavior in Lemurs, Monkeys, and Apes*, T. H. Clutton-Brock, ed., Academic Press, New York, pp. 503–538.

SHADE COFFEE PLANTATIONS AS WILDLIFE REFUGE FOR MANTLED HOWLER MONKEYS (*ALOUATTA PALLIATA*) IN NICARAGUA

Colleen McCann, Kimberly Williams-Guillén, Fred Koontz, Alba Alejandra Roque Espinoza, Juan Carlos Martínez Sánchez, and Charles Koontz[*]

1. INTRODUCTION

1.1. Forest Fragmentation and Ecological Corridors in Central America

Flanked on either coast by contrasting ocean bodies and north and south by two continents, Central America is a unique and intriguing account of natural and cultural history (Coates, 1997). The Central American isthmus is a corridor of biological and cultural networks both between and within countries making it a diverse and complex region. Once a vast mosaic of diverse ecosystems, including coral reefs, savannas, semi-arid lowlands, rain-forested foothills, cloud forest, and pine-forested volcanoes (Wallace, 1997), the ever-increasing pressures of human densities and economic practices of the 20th century threatened the survival of these fragile systems and the natural resources they supported (Heckadon-Moreno, 1997). The colonization into forested areas and their subsequent transformation into agro-forested lands resulted in a corridor of fragmented forest patches, often associated with continued land modification, human poverty, and political turmoil (Illueca, 1997).

The history and pattern of land modification in Nicaragua has a similar theme of that of Central America as a whole. Since 1950, Nicaragua's forests, which once covered 8 million ha, have been reduced in size by 50% (Heckadon-Moreno, 1997). Moreover, the dry forests of the Pacific lowlands have nearly disappeared and the pine savannas of the Caribbean's Miskito Coast have been severely degraded. The first centers of human colonization and land use modification were the Pacific and central mountainous zones,

---

[*] Colleen McCann, Wildlife Conservation Society, Bronx, New York 10460, USA. Kimberly Williams-Guillén, NYCEP/New York University, New York, NY 10021, USA. Fred Koontz, Wildlife Trust, Pallisades, NY 10964, USA. Alba Alejandra Roque Espinoza, Fundación Cocibolca, Managua, Nicaragua. Juan Carlos Martínez Sánchez, Fundación Cocibolca, Managua, Nicaragua. Charles Koontz, Wildlife Trust, Pallisades, NY 10964, USA. Correspondence to C. McCann (email: cmccann@wcs.org).

resulting in a landscape of human habitation and forest fragmentation along the Pacific coastline. However, as farming and cattle pressures for land expansion intensified, modern colonization centers have begun to spread eastward into the large forested areas of BOSAWAS (area named for the abbreviation of the Bocay, Sang Sang, and Waspuk rivers) Biosphere Reserve and the Indio-Maiz Biological Reserve (Heckadon-Moreno, 1997). Despite the serious threat of continued colonization in these critical areas, Nicaragua retains some of the largest tracks of forested area in Central America, and efforts to safeguard them will contribute to the preservation of biodiversity for the region.

1.2. The Agroforestry of Coffee

Coffee is the second largest source of export earnings for the majority of the world and the most valuable tropical agricultural commodity (Talbot, 1995). In Latin America alone, nearly 50% of its perennial croplands are devoted to coffee, covering approximately 2.7 million ha and resulting in one-third of global coffee crop production (Perfecto et al., 1996). Traditionally, coffee has been grown under the shade of forest trees protecting the coffee plants from the harsh sun and rain of the tropics. In some areas of the world (e.g., the Pacific corridor of northern Latin America) where both deforestation rates and human densities are high, "shade" coffee plantations represent a significant percentage of the remaining forest cover in these areas (Pimentel et al., 1992; Toledo et al., 1994; Wille, 1994).

Coffee production and consumption has been on the rise since 1950 with an even greater demand in recent years for gourmet coffees (World Resources Institute, 1998). In the 1970s, partly due to protection against the coffee leaf rust fungus and in part due to the desire to increase production yields, farmers began converting their shade coffee plantations to "technified" or "sun" coffee production systems (Rice, 1997; Toledo and Moguel, 1997; Pendergrast, 1999) (Figure 1a). Although sun coffee produces higher yields, there are many associated direct and indirect costs ranging from increased agrochemical use to the loss of watersheds and rapid soil erosion. In the last 10 years, 40% of shade grown plantations in Mexico, Central America, Colombia, and the Caribbean were converted to technified coffee production systems, directly contributing to the loss of biodiversity in these areas (Perfecto et al., 1996).

1.2.1. Coffee Production Systems

Moguel and Toledo (1999) describe five distinct categories of coffee production systems, ranging from the least to the greatest modification of the original landscape. On the one end of the spectrum are unshaded monoculture plantations (i.e., sun coffee), which are the least floristically complex and most intense level of agroindustry. In sun coffee cultivation, the forest is cleared of trees to allow coffee bushes to grow in the direct sun. While this system typically produces a greater yield of coffee crop per acre, it is highly labor intensive and requires chemical fertilizers and pesticides to protect the plants from the loss of natural defenses. At the other end of the spectrum is the traditional rustic or mountain coffee cultivation (i.e., traditional shade coffee), which minimizes the alteration of the forest while maintaining the greatest structural complexity. In this cultivation system only the understory plants are cleared for the coffee bushes while the forest cover is kept intact, thus having a minimal impact on the original ecosystem. Because there are more natural defenses in place for shade coffee, many farmers reduce

Figure 1. Monocultural sun coffee plantation (a) in comparison to a traditional rustic shade coffee plantation (b) in Nicaragua. Photos by Colleen McCann.

or eliminate chemical treatments providing an organic form of agricultural management (Figure 1b).

The differences between sun and shade grown coffee systems are far reaching and extend beyond the structure of the forest to a variety of environmental parameters. The outcome of these agricultural management systems is a cumulative impact on forest structure, species richness and diversity, temperature and humidity fluctuations, watershed quality, soil erosion rates, carbon dioxide sequestration, chemical fertilization and its associated health concerns, crop production costs, plant productivity, and the mode of land ownership (Rice and Ward, 1996; Rice, 1997). Put simply, traditional shade coffee plantations are typically grown organically by small-scale farmers (the majority of plantations less than 5 ha) and support a high diversity of wildlife in addition to the ecological and biological systems that maintain them. On the contrary, sun coffee often comprises large-scale production systems of monocultural plantations that require extensive agrochemical input and labor with ensuing negative impacts on the environment (Perfecto et al., 1996).

1.3. Managing Habitat on Coffee Plantations

The importance of coffee plantations as secondary habitat for migratory birds has long been recognized (Wilcove et al., 1986; Calvo and Blake, 1998). In a study of the role of coffee plantations for a variety of bird species in the Western Ghats, India, Shahabuddin (1997) has shown that coffee plantations appear to be valuable refuges for many forest-dwelling species. The conversion of a multi-species forest ecosystem to a monocultural coffee plantation resulted in several microhabitat changes and a significant decrease in species composition between forest and plantation habitats. However, species richness was not affected by this conversion in habitat, and plantations were found to be important foraging grounds for avifauna, serving as buffer habitat between fragmented forest (Note: but in some primate species it may act as a barrier; see G. Umapathy in this volume). This study indicates that coffee plantations can play an important role as refuges between natural forest habitat patches, providing a dispersal corridor between patches and serving as marginal habitat for some species and breeding habitat for other species (Shahabuddin, 1997). Coffee plantations can also serve as buffer zones around protected areas; their forest-like habitats reduce edge effects and provide a reservoir area for native fauna. As a result of studies like these, public education programs linking wise consumer purchases of coffee with Neotropical bird conservation on shade coffee plantations in Central America are underway (Anonymous, 1997). Until now, focusing on the conservation of charismatic primates living on coffee plantations was untried.

1.4. The Status of Primates in Nicaragua

Largely due to political and economic instability, very few studies have investigated the conservation status of primates in Nicaragua (Rylands et al., 1995; Rodriguez-Luna et al., 1996a; Crockett et al., 1997). Nicaragua's primate fauna is reported to include three species: the mantled howler monkey (*Alouatta palliata*), the black-handed spider monkey (*Ateles geoffroyi*), and the white-faced capuchin (*Cebus capucinus*) (Coimbra-Filho and Mittermeier, 1981; Wolfheim, 1983; Konstant et al., 1985; Mittermeier et al., 1988; Rowe, 1996). Additionally, it is suspected that there are at least two subspecies of both capuchins (*C. c. limitaneus* and *C. c. imitator*) and spider monkeys (*A. g. geoffroyi* and *A.*

g. frontatus) (Coimbra-Filho and Mittermeier, 1981; Konstant et al., 1985; Mittermeier et al., 1988; Rylands et al., 1995). According to Mace-Lande classifications (Rylands et al., 1995), *A. g. frontatus* is listed as "Vulnerable," while the others are placed in the category of "Lower Risk" (Rodriguez-Luna et al., 1996).

Despite the fact that Nicaragua is the largest nation in Central America, published information on the status of its wildlife remains scarce (Crockett et al., 1997). In the most recent "Mesoamerican Action Plan," there were no accounts cited of the status of Nicaragua's primate populations (Rodriguez-Luna et al., 1996a); moreover, there were no reports of any long-term field studies on primates in Nicaragua, Honduras, or El Salvador (Rodriguez-Luna et al., 1996b). More recently, primatological research has been initiated in the country. Several short-term studies have been conducted on a population of naturally-occurring howlers on Ometepe Island in Lake Nicaragua (Garber et al., 1999). Additionally, we conducted a 14-month field study of the behavioral ecology of howlers living in one of Mombacho's shade coffee plantations.

In one of the few published accounts of Nicaragua's primates, Crockett et al. (1997) reported a preliminary assessment of the conservation status of Nicaragua's primates. They report both sightings of primates and information collected from local residents for 11 of the protected areas listed for Nicaragua in the "Mesoamerican Action Plan." Out of the 11 protected areas visited, howler monkeys were seen or reported in 10 sites. In the Mombacho Volcano Nature Reserve, howler monkeys were heard, and local residents reported their presence. Based on the 11 areas visited, approximately 40 troops were sighted, and several more heard, with troop size ranging from 4 to 20 individuals (mean = 10.7). During this survey, there was only one sighting of spider monkeys occurring in one of the larger reserves, but local reports suggest that they may be present in two or three other protected areas. Capuchins, on the other hand, were not sighted in several areas where they were reported to be present. However, local reports strongly suggest that *Cebus* definitely exists in one or two of the larger reserves, and they may likely exist west of published distributions, in southwest Nicaragua. Based on this preliminary survey, Crockett et al. (1997) argue that while sufficient numbers of primates and large forested areas still exist in Nicaragua, it is important to determine the status of potential areas for protection before critical areas are lost. Nicaragua plays an important role in the conservation of Central America's flora and fauna, and thus, more systematic censuses of the region's flora and fauna need to be conducted to preserve Nicaragua's wildlife.

1.5. Howler Monkey Behavioral Ecology

Howlers (genus *Alouatta*; Figure 2) are the most widely distributed and well studied of New World monkeys (Crockett and Eisenberg, 1987; Neville et al., 1988; Kinzey, 1997). Several systematic, long-term studies of feeding ecology have focused on the Central American mantled howler (Chapman, 1987; Estrada, 1984; Glander, 1978, 1981; Larose, 1996; Milton, 1980; Stoner, 1993), all of which suggest that they are highly selective folivore-frugivores. Foliage is a vital component of the diet, and all howlers regularly consume leafy material (Milton, 1998). However, most populations consume substantial amounts of fruit, typically between 25% and 55% of annual feeding records. Fruit exploitation strongly reflects the availability of fruit in the habitat, with consumption patterns largely tracking seasonal availability (Glander, 1978, 1981; Milton, 1980; Stoner, 1993). Flowers may also be preferred food sources that are heavily exploited during their brief periods of availability. A high degree of dietary diversity,

Figure 2. Adult female mantled howler monkey (*A. palliata*), in a shade coffee plantation in Mombacho Volcano, Nicaragua. Photo by Colleen McCann.

spatially between sites and temporally within highly seasonal sites, characterizes the species (Marsh, 1999). Such flexibility may explain the ability of howlers to colonize many habitats: they have been found in primary and regenerating rainforest, dry deciduous, riparian, coastal lowland, mangrove, and cloud forests (Wolfheim, 1983). Howlers of all species will occupy marginal areas when necessary (e.g., Baldwin and Baldwin, 1972; Schwarzkoph and Rylands, 1989; Limeira, 1997), including areas of shade coffee cultivation (Estrada and Coates-Estrada, 1996; Garber et al., 1999). Nevertheless, they demonstrate a marked preference for primary and riparian habitats, presumably due to a higher density of food species (Glander, 1978; Stoner, 1993).

Although most species of howlers live in small groups (generally less than 10 individuals) with, on average, one adult male and two adult females, members of *Alouatta palliata* differ significantly in troop size and composition (Crockett and Eisenberg, 1987). At other study sites, mantled howlers generally have group sizes of 10 to 20 individuals, with two to four adult males and three to nine adult females. Densities of mantled howlers range from 15 per km^2 (Stoner, 1993) to 92 per km^2 (Milton, 1980), although densities can reach extremely high levels under conditions of habitat destruction and range contraction (Baldwin and Baldwin, 1976). Despite their wide geographical range and apparent flexibility, *A. palliata* are threatened (Crockett and Eisenberg, 1987), and Landsat image modeling suggests that their habitat could be lost by the year 2025

(Kinzey, 1997). Although several populations of mantled howlers persist in Nicaragua (Crockett et al., 1997; Garber et al., 1999), the current conservation status of these populations is uncertain. Given that howlers in Nicaragua are known to exist in coffee plantations, where they cause no crop damage, and that shade coffee encompasses a large portion of the remaining forest cover on Nicaragua's Pacific coast, conservation in shade coffee plantations is likely key in maintaining populations of howlers and other primates in western Nicaragua.

1.6. Conservation Issue

In this study we investigated the conservation status and ecology of mantled howler monkeys (*Alouatta palliata*) living on coffee plantations surrounding the Mombacho Volcano Nature Reserve, Nicaragua. The Reserve is managed by a local non-government organization (NGO), Fundación Cocibolca, dedicated to the preservation of Nicaraguan wildlife. The results of the investigation are currently being applied to the development of management recommendations to Fundación Cocibolca on ways both to protect the monkeys and to increase primate habitat on these agricultural lands. We suggest that the mantled howler monkey can serve as an umbrella species in the Mombacho area, and by managing for its care, the landowners will simultaneously increase wildlife habitat for many other species living in the Mombacho Volcano region.

2. METHODS

2.1. Study Site

The Mombacho region includes the northern expanse of Costa Rican seasonal moist forests (Dinerstein et al., 1995). Mombacho Volcano is a moderate sized, quiescent volcano with one of the two remaining cloud forests found in southwestern Nicaragua (Atwood, 1984; Figure 3). The Mombacho Volcano Nature Reserve is situated on top of the volcano in a 650-ha area between 850 m in elevation and the summit at 1,360 m (Figure 4). The Reserve consists of tall evergreen forests at its lower elevations and elfin cloud forests at the summit. An agricultural belt of coffee plantations surrounds the Reserve in the 300- to 600-m elevation zone. The majority of the coffee plantations grow shade coffee, utilizing the large trees of the seasonal broadleaf forest to shade the growing coffee bushes from direct sunlight. In a preliminary study we observed mantled howler monkeys living in these coffee plantations, but not in the Reserve proper (McCann and Koontz, 1997). The lower slopes of Mombacho support a highly disturbed tropical dry forest with forest type strongly influenced by elevational gradient. The belt of coffee plantations is surrounded by another agricultural zone, comprised mostly of cattle ranches, in the 100- to 300-m elevation zone. Together, the coffee plantations and cattle ranches form the unofficial buffer zone of the Mombacho Reserve.

2.2. Habitat Reconnaissance of Mombacho Coffee Plantations

The habitat reconnaissance of the coffee plantations was conducted in March through June of 1998 and 1999. The habitat assessment consisted of a systematic survey of the vegetation within each coffee plantation using the existing trails. Each coffee plantation

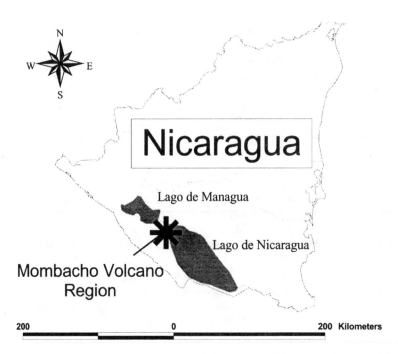

Figure 3. Nicaragua, showing the area of Mombacho with an asterisk.

Figure 4. Mombacho Volcano Nature Reserve and surrounding area. Photo by Colleen McCann.

served as a single sampling unit (N = 25). Approximately 20% of each plantation was surveyed for habitat assessment. This was done to get a rapid assessment of the vegetational make-up of each plantation and to determine if the existing habitat could support primate populations. At every 50 m the vegetation was recorded within a 25-m radius at that sample point. Vegetation was recorded based on eight broad categories (Table 1). The habitat assessment also included a systematic survey of the vegetation within the Reserve proper using the existing trails that extend from the highest altitudes of the Reserve down to the plantations that buffer it. Vegetation categories included scrub, elfin, and cloud forest with tree height ranging from 1 to 25 m.

2.3. Howler Monkey Census

The howler monkey census was conducted concurrently with the habitat reconnaissance. There were two sampling periods during each sample day: the morning sampling periods were conducted between 0630 and 1130 hours; the afternoon sampling periods were conducted between 1400 and 1800 hours. When howlers were sighted, the following data were recorded: the date, the time, the location (in UTM coordinates, determined with a global positioning system [GPS] unit), the elevation (based on available topographic maps), the number of individuals in the group, estimated age and sex of the individuals, the vegetation category they were located in (Table 1), and the tree species they were located in. When howlers were heard howling, the following data were recorded: the time, the date, the GPS location, the compass bearing from the GPS fix to the howling, and the estimated distance from the GPS fix to the howling; these data were used to calculate the estimated longitude, latitude (UTM coordinates), and elevation (derived from topographic maps) of each group's location. Although we did not visit all areas of the plantations surveyed, because howlers' calls travel over long distances, we assumed that we had near-100% coverage of the study area. The same data were recorded if other primate species (*Ateles geoffroyi* and *Cebus capucinus*) were sighted or heard.

Table 1. Vegetation categories used in habitat reconnaissance.

Vegetation Category	Description
Shade Coffee 1	Shade trees less than 15 m in height, open canopy
Shade Coffee 2	Shade trees between 15 and 25 m in height, with mostly or completely closed canopy
Shade Coffee 3	Shade trees more than 25 m in height, closed canopy
Other agriculture	Areas of cultivation such as banana, cacao, or cattle pasture
Regeneration 1	Abandoned coffee plantation
Regeneration 2	Abandoned cattle pasture with tree height less than 15 m
Regeneration 3	Abandoned cattle pasture with tree height greater than 15 m
Forest	Relatively undisturbed area with no evidence of agricultural use, tree height greater than 25 m, and contiguous canopy closure

2.4. Local Interviews

In addition to the survey data, additional information was obtained from local individuals living and working in Mombacho, particularly the coffee plantation laborers. The majority of individuals were migrant workers who typically live in the area and come to the plantation for work. We selected individuals of different ages and levels of employment (supervisor versus laborer) to get a broad view on primate populations in the area past and present. For each plantation two to three worker interviews were conducted (N = 63). A standard set of questions was asked to each person (Table 2).

3. ANALYSIS

The analysis of the spatial distribution of howler monkeys living in the Mombacho region was based on a method adapted from the "centroid cluster analysis method" described in Stoner (1994). This method is especially useful in situations when it is not possible to confirm that each data location represents a distinct group, and thus, overestimates of population numbers is a concern. For example, we might record three bouts of howling within a plantation over several hours (or days) and have no way of knowing if this is the same group or three groups. To improve predictions we used a nearest-neighbor clustering algorithm to construct groups in an unbiased way, based on the spatial distribution of all recorded howler locations. After these statistical groups were established, (1) we mapped the individual locations; (2) we depicted the statistical groups by first calculating the geographic central point ("centroid") for each group and then inscribing a 25-ha home range size for each mapped group (mantled howler home range sizes vary from 5 to 60 ha (Stoner, 1994), and are 15 to 20 ha at La Luz, a shade coffee plantation on Mombacho's southwestern side (Williams-Guillén, unpublished data); we chose a relatively large home range size for this preliminary mapping so that our population estimate for Mombacho howlers would be conservative); and (3) we compared the statistically-derived groups with our field notes to determine if any of the assigned data locations were not possible (e.g., two sighted groups known to be different being assigned to the same statistical group, or two simultaneously heard groups being assigned to the same statistical group). This analysis was used to provide an estimate of the number of groups in the study area and relative abundance among plantations. By

Table 2. Questions asked of interviewees in the Mombacho region.

Topic	Questions
Presence/absence of howlers in plantation	• Are howlers present on this plantation? • If so, how many? • When and where did you last see or hear howlers? • How many monkeys do you usually see in a group?
Attitudes towards primates	• What do people think of monkeys? • Do people hunt the monkeys?
Presence of other primates	• Are there other spider or capuchin monkeys on the plantation? • If not, were they present in the past?
Changes in primate populations	• Are there more or fewer monkeys in the area now than 5 to 10 years ago? • If there have been changes in the numbers of monkeys, why?

multiplying the number of groups by our observed average group size, we calculated an estimate of the howler population size for the Mombacho study area.

4. RESULTS

4.1. Habitat Reconnaissance

There are approximately 25 plantations that surround the Mombacho Reserve. They range in size from 25 to 225 ha, with the average plantation measuring less than 100 ha. We estimate that approximately 20% of each plantation was surveyed for the habitat assessment. The results of the survey showed that the landscape is composed of 64% agricultural lands (56% coffee plantations, 8% other agriculture, e.g., banana, cacao, cattle), 15% regenerating forests (e.g., abandoned cattle and coffee areas), and 21% forests (Table 3). Within the areas surveyed, 61% of the habitat was categorized as low-moderate disturbance ([see Table 3 for description of categories] vegetation categories: C2, C3, R3, F) and 39% was considered highly disturbed (vegetation categories: C1, A, R1, R2) and is poor primate habitat. There were noticeable differences in habitat quality between specific quadrants of the Mombacho region that warranted further analysis. For example, the eastern side of the Volcano had much higher percentage of forested areas within the plantations than did the Pacific side, closer to the capital city, Managua (e.g., 46% and 26% compared to 6% and 8%, respectively).

The Mombacho Reserve is one of the few remaining cloud forests in southwestern Nicaragua. The habitat composition of the Reserve is quite distinct from that of the surrounding area due to the effects of high elevation and wind. Within the Reserve proper, 40% consisted of forest, 10% elfin forest, 34% scrub forest, and 16% agricultural use (15% coffee and 1% other agriculture) despite the fact that the Reserve is a protected area (Roque Espinoza, 1999).

Table 3. Habitat status within the Mombacho Volcano Nature Reserve region.

Quadrant (Number of Plantations)	Vegetation Category (% of Surveyed Habitat)							
	Coffee 1 (C1)	Coffee 2 (C2)	Coffee 3 (C3)	Other Agriculture (A)	Regen. 1 (R1)	Regen. 2 (R2)	Regen. 3 (R3)	Forest (F)
Southwest (N = 7)	14	38	18	6	12	4	2	6
Southeast (N = 4)	11	12	21	3	3	3	1	46
Northwest (N = 10)	19	27	21	10	7	6	2	8
Northeast (N = 4)	16	18	7	11	3	0	19	26
TOTAL AREA (N = 25)	Coffee = 56%			11%	Regeneration = 15%			21%

4.2. Howler Monkey Census

We sighted howlers 62 times and heard them howling 137 times during the census of the 25 plantations. Additionally, seven groups of capuchins (*Cebus capucinus*) were sighted and an additional two heard. No spider monkeys (*Ateles geoffroyi*) were sighted or heard during the census despite that local Nicaraguans report their presence in the Mombacho region.

Using the clustering method described above, a total of 101 howler groups were determined to be present in the survey area. However, after comparing the statistical results with our field notes, the group assignments for 13 of 199 locations needed to be adjusted due to multiple recordings of the same group or due to clustering of observations known to represent different groups. As a consequence, the final analysis yielded 97 groups of howlers living in the survey area (Figure 5). Finally, based on an average group size of 9.9, the total number of individual howlers living in the survey area is therefore estimated to be 960. There are approximately 13 howlers per km^2 in the Mombacho region; densities for *A. palliata* generally range from 15 to 92 individuals per km^2 (Stoner, 1994). Although howler densities in Mombacho are comparatively low, they are within the range of variation observed in other areas of Central America.

The mean group size of the Mombacho howler monkeys was 9.9 $\pm$ 6.2 s.d. and ranged from 1 to 24 (N = 51) (Table 4). The average capuchin group size was 8.8 $\pm$ 6.3 s.d. (range = 2 to 18). The composition of the Mombacho howler population consisted of 16% adult males, 39% adult females, 30% subadults/juveniles, and 15% infants. Out of the 51 groups sighted, 62% had only one adult male present and 75% of the groups sighted had one or more infants present.

4.3. Howler Monkey Distribution

During the initial field season we were informed by local inhabitants that the howler monkeys not only live in the Mombacho Reserve buffer areas, but also range into the Reserve itself, particularly in the dry season. We therefore expanded the survey to include the Reserve proper. Out of the 97 howler groups located in this survey, only 13 were recorded within the Reserve itself, representing only 13% of the population. This emphasizes the importance of the buffer area in sustaining the howler population in the Mombacho region. Although the Reserve is a protected area, the vegetation, primarily

Table 4. Size and composition of howler monkey groups in Mombacho Volcano.

Quadrant	Number of Groups	Average Group Size (±s.d.)	% Adult Males	% Adult Females	% Juveniles	% Infants
Southwest	11	9.0 (±6.3)	15	41	26	15
Southeast	13	12.4 (±6.4)	16	40	29	15
Northwest	18	7.8 (±4.7)	14	33	36	17
Northeast	9	11.6 (±7.7)	18	44	25	13
TOTAL AREA	51	9.9 (±6.2)	16	39	30	15

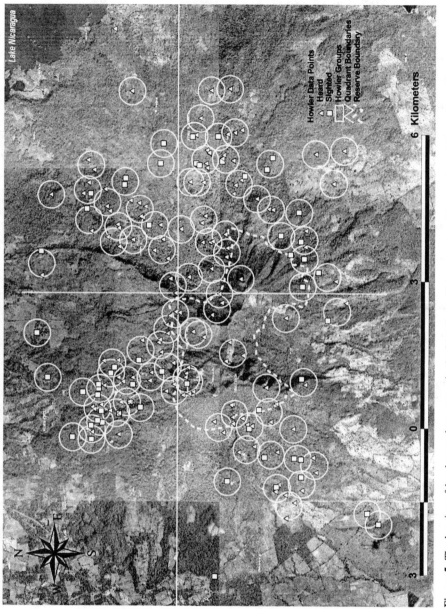

Figure 5. The location of howler monkey groups in the Mombacho Volcano region. Map courtesy of the Instituto Nicaraguense de Estudios Territoriales.

scrub and elfin forest with the majority of trees less than 25 m in height, is quite distinct from the tree community at lower elevations, which comprises semi-deciduous broadleaf forest with an unstratified canopy of 15 to 20 m in height (see tree species list in Appendix 1). Given these characteristics, the Reserve does not appear to be the preferred habitat of the mantled howler monkey and does not provide an abundance of food resources for them compared to the buffering areas surrounding the Reserve (Roque Espinoza, 1999). However, the Reserve may represent alternate habitat during times of food scarcity. Moreover, the ultimate importance of the Reserve may be in its function as a biological corridor for the Mombacho howler monkey population, without which the individual groups would become increasingly isolated and subject to the vagaries of population decline.

Five out of the seven capuchin groups sighted were in the Reserve proper. Additionally, while the remaining two sighted groups were located outside of the Reserve in the buffer area, they were found in forested areas. Thus, the presence of undisturbed forest areas appears to be the preferred habitat of the capuchin monkeys in the Mombacho region.

4.4. Howler Monkey Density and Habitat Quality

The final analysis of habitat quality and howler monkey density seems to indicate that while the number of howler groups appears to be the same in each quadrant of Mombacho, group size is lower in the western region of the volcano where there is a higher degree of human-modified habitat (Table 5). This region is a preferred coffee growing area because it is located on the more accessible Pacific side of the volcano, off of the Pan-American highway less than 30 km from Managua; the western regions therefore experience greater human impact than the eastern side of the volcano. A Spearman rank correlation test demonstrates a significantly positive association between

Table 5. Howler monkey population and its relation to habitat type.

Quadrant	Total Number Groups	Average Group Size (±s.d.)	Estimated Population	% Unimale Groups	% Groups with Infants	% Habitat Highly Perturbed (±s.d.)	% Habitat Mod/Low Perterbed (±s.d.)
Southwest	22	9.0 (±6.3)	198.0	63%	64%	46% (±6.8)	54% (± 7.9)
Southeast	30	12.4 (±6.4)	372.0	54%	85%	25% (±3.1)	75% (± 18.5)
Northwest	24	7.8 (±4.7)	187.2	83%	77%	49% (±3.3)	51% (± 6.6)
Northeast	21	11.6 (±7.7)	243.6	33%	67%	37% (±5.6)	63% (± 9.6)
Western Region	46	8.3 (±5.3)	381.8	76%	70%	48% (±26.2)	52% (± 29.2)
Eastern Region	51	12.0 (±6.8)	612.1	45%	78%	31% (±17.3)	69% (± 28.7)
TOTAL AREA	97	9.9 (±6.2)	960.3	62%	75%	39% (±32.9)	61% (± 51.4)

each quadrant's estimated howler population and the percent of low and moderate disturbance habitat available (r_s = 1.0000, p < 0.001, N = 4).

Additionally, these areas differed not only in the abundance of howlers, but in their group compositions. There was a considerable difference between the western and eastern regions in the number of one-male versus multi-male groups. In the more disturbed western quadrants, 76% of the howler groups only included one adult male, while in the eastern quadrants, 55% of the groups had two or more males present. A Spearman rank correlation demonstrates a significantly negative relationship between the number of males in all observed bisexual groups and the percentage of highly disturbed habitat available within each plantation (r_s = -0.31990, p = 0.0442, N = 40). Given these variables, it is not unexpected that a difference in habitat quality may be resulting in a corresponding difference in howler density, group size, and composition. While howlers have been shown to adapt to disturbed habitats, the level of disturbance can have a significant effect on population dynamics.

4.5. Local Interviews

A total of 63 interviews with local people living in the Mombacho area were conducted during the survey. Regarding the presence of primates in Mombacho, 93% of those interviewed reported their presence on coffee plantations; 89% reported seeing or hearing howlers everyday; 39% reported seeing or hearing capuchins everyday; and only 3% reported seeing or hearing spider monkeys everyday. In response to questions on the size of the howler population living in the Mombacho area, numbers ranging from 20 to greater than 1,000 were reported. When asked about the number of howler groups, 23% reported that greater than three groups currently lived on the plantation where they work while 77% of those interviewed reported one to three groups. When asked about the average group size, 43% reported 5 to 15 howlers per group and 57% reported groups greater than 20.

Questions regarding the current primate population status compared to a decade ago provided insights into past pressures on populations and current trends of protection. In response to the changes in primate populations in Mombacho over time, 61% reported that there are more monkeys now than 10 years ago; 18% reported fewer monkeys; and 21% reported no change in status. Of the 61% that reported more monkeys present, 82% credit the increase to their protected status, while 18% believe it is due to their greater birth rates. Out of the 18% that reported a decrease in population numbers, 60% reported a yellow fever epidemic 20 years ago as the causal factor. The remaining 40% reported that the observed decrease in population size is a result of continued hunting practices.

The overall response to the interview questions revealed that local attitudes towards howler monkeys are relatively positive, howlers do not pose a threat to the agricultural crops in the area. This is an important key factor for creating community involvement in a management plan for preserving wildlife. (Note: see Reynolds et al. for problems with chimpanzees, this volume.)

5. DISCUSSION

Based on our results, at least 960 howlers are living in the Mombacho area, most of them not in the protected area at the volcano's summit, but in the shade coffee plantations that dominate Mombacho's lower elevations. Their abundance in these areas suggests that shade coffee plantations serve as a vital refuge for howlers in Mombacho, and perhaps more widely in Nicaragua's heavily deforested Pacific coast. As a follow up to this work and to gather more information for creating an effective management plan, we conducted a 14-month study of the behavioral ecology of three groups of howlers in Finca La Luz, a fairly typical shade coffee plantation on Mombacho's western flanks (Williams-Guillén and McCann, in preparation). Our results suggest that although patches of regeneration and less disturbed forest are exploited, the Mombacho howlers do indeed rely primarily on trees in areas of active shade coffee cultivation for food, travel, and rest. Shade coffee plantations are, therefore, serving as a vital refuge for howlers in this area.

However, the howlers in these agroforests show important differences from conspecifics in their ecology and demography; variation in these characteristics seems to correlate with the degree of habitat disturbance. Although howlers are abundant in comparable semi-deciduous and semi-evergreen habitat in Costa Rica (Glander, 1978) and Panama (Milton, 1980), they frequently exist at lower densities in high evergreen forests (Estrada, 1984; Stoner, 1994). Densities in Mombacho are more comparable to those of much wetter, less seasonal sites.

Perhaps most intriguing are the unusual group compositions of howlers in Mombacho. In general, *A. palliata* is found in multimale groups with one to three adult males per female (Crockett and Eisenberg, 1987). Howlers in Mombacho, in contrast, show a high frequency of unimale groups, in spite of average group sizes that fall well within the range of variation for the species. The average female to male ratio of all the sighted bisexual groups was 2.94, which is fairly typical of the species (Crockett and Eisenberg, 1987). However, our experience with further study of howlers in Finca La Luz has led us to believe that the more inconspicuous females were routinely undercounted during our census. At the site of our long-term ecological study, the female/male ratio varied from 3.13 to 5.77 females per male, the highest ratios known for this species (Williams-Guillén, unpublished data).

A number of factors could account for the prevalence of unimale groups. Crockett and Eisenberg (1987) suggest that recently-formed howler groups are more likely to have only one male; given that the population of howlers in Mombacho is currently expanding, many of the smaller unimale groups may represent recently established social groups. Ostro et al. (2001) argue that amongst black howlers (*Alouatta pigra*) unimale social groupings are advantageous at low population densities, as there is less need for cooperative defense of breeding opportunities when the frequency of intergroup encounters is reduced. Although this explanation may also account for the prevalence of unimale groups in Mombacho, it does not explain the high female to male ratios seen in Mombacho's one-male groups. Such shifts in social organization may reflect changes in patch size and concomitant shifts in levels of female intragroup feeding competition. Many of the feeding trees used by howlers in shade coffee plantations are both large and widely spaced (Williams-Guillén, unpublished data). Their size may mean that more individuals can feed concurrently in a single patch than at other sites, allowing one or two males to monopolize access to females. However, this hypothesis remains untested.

Despite changes in group composition, with close monitoring and protection Mombacho's howler population will likely persist within these agroforests. On the contrary, the situation for the capuchin and spider monkey populations is less encouraging. While the majority of the interviewees reported the presence of spider monkeys somewhere in the Mombacho region, they were never observed during the census period (however, most recently a few spider monkeys have been sighted by reliable witnesses within the reserve proper; Otterstrom, personal communication). Moreover, capuchins were recorded present on only seven occasions. It is important to note that both spider and capuchin monkeys are commonly observed as pets throughout Nicaragua despite their protected status (Hendrix, 2000). Any degree of hunting of isolated populations in fragmented habitats has serious consequences for the long-term survival of the population (Robinson and Bennett, 2000). Habitat fragmentation coupled with hunting has detrimental effects on small, fragile populations, such as those in Mombacho, which is all too often the cause of the rapid decrease or disappearance of a species from an area. In many cases, the local extinction of a species from an area is irreversible (Rosser and Mainka, 2002) and in the case of Mombacho's capuchin and spider monkeys, with no opportunity of gene flow from nearby populations, it may be too late to halt the process. Thus, critical to the preservation of Nicaragua's biological diversity is the protection of species in addition to the protection of the habitats that support them (Martínez-Sánchez et al., 2001).

6. CONCLUSION

As a result of this study, the following immediate recommendations for maintaining a viable howler population in the Mombacho region and preserving the habitat that supports it can be made. The hunting of wildlife is by far the most serious threat to wildlife populations (Redford, 1992; Oates et al., 2000; Robinson and Bennett, 2000; Bennett et al., 2002; Rosser and Mainka, 2002). To preserve the remaining wildlife of the Mombacho region, all hunting practices must be eliminated immediately. With an ever-increasing fragmentation of forests and the isolation of wildlife populations, small population management becomes a necessary part of wildlife management. With professional wildlife monitoring, the Mombacho howler population can survive as a managed population (Martínez-Sánchez et al., 2001).

The continued practice of shade-grown coffee on the plantations surrounding Mombacho Reserve can have a negligible impact on wildlife if certain practices are followed, such as the preservation of shade tree species that are vital to the survival of the howler population and other wildlife in the region. By encouraging agroforestry practices that have been shown to support wildlife populations, the number of species that can persist in these modified habitats is greatly increased (Oldfield and Alcorn, 1987). In an effort to promote the use of shade grown coffee as an environmentally-sound form of agroforestry, the idea of designating a plantation as certified "monkey-friendly" coffee is in development. In this way the link between the type of coffee plantation and the number of primate populations it supports can be made making direct associations between human agricultural practices and its effect on the persistence of wildlife species (Barborak, 1998).

Ultimately, the survival of the wildlife in Mombacho is dependent on community involvement. If the Mombacho howler population is to survive, it must be closely monitored and supported by the local community. The involvement of the local community is a critical element in the management of this area and can contribute greatly to the development of environmental education in the Mombacho region. (Note: see recommendations by Marsh et al. in final chapter of this volume.) With more than one million Nicaraguans living nearby Mombacho, the importance of promoting conservation action as an educational tool is critical. Efforts to promote the preservation of wildlife resources must be accepted and supported by the local community in order to be effective in influencing the spread of the modification of Nicaragua's remaining forests.

Based on the recommendations above, a management plan is being developed with the ultimate goal of creating a community-based conservation program for the Mombacho region and its wildlife.

7. SUMMARY

Considering Mombacho Reserve's small size, the buffer zone of agricultural lands appears to be of critical importance for the continued existence of many of the Reserve's wildlife species. Establishing and managing key habitat elements in the buffer zone, therefore, has important implications for the long-term protection of the region's biodiversity (Barborak, 1998). While this project focused on surveying howler monkeys living on the coffee plantations, we expect that many other forms of biodiversity found in the plantations will benefit (e.g., Hagen and Johnston, 1992; Rappole, 1995; Greenberg, 1996; Perfecto et al., 1996; Greenberg et al., 1997; Calvo and Blake, 1998). The results of this survey could also generate considerable public interest from the one million Nicaraguans living within 50 km of Mombacho, and it could provide an environmental education platform for Fundacion Cocibolca. In addition, by demonstrating methods for increasing howler habitat on the Mombacho coffee plantations, the development of a management plan for these agricultural lands might serve as a model for increasing primate habitat in other areas of Central America.

8. ACKNOWLEDGMENTS

We thank Archie Carr III, Jim Barborak, Andy Taber, and Felicity Arengo for their guidance and continued support of our efforts. Karen Willet and Gillian Woolmer supervised all GIS analyses while Ari Martinez and David Autry provided assistance in the collection of the census data. Carolyn Crockett provided invaluable advice and feedback throughout the project. We are most grateful for having the opportunity to work with the staff of Fundacion Cocibolca, particularly Jose Manuel Zolotoff Pallais, Roberto Lopez Castillo, Enock Pineda Gonzalez, and Maria Ignacia Galeano Gomez, whose professionalism and dedication to the preservation of Nicaragua's wildlife is inspiring. And most importantly, we thank the Mombacho community for imparting their knowledge of the area and providing us with an opportunity to discuss and develop this important conservation initiative. This initiative would not have been possible without the leadership and support of the following individuals: Raul Lacayo, Gabriel Pasos, Fernando Sequeira and Alejandro Palazio. As board members of Fundacion Cocibolca and as important community leaders their support of conservation efforts in Mombacho specifically, and in Nicaragua in general, is to be commended. Funding for this project was provided by Wildlife Conservation Society's Species Survival Fund #02-198 and the Lincoln Park Zoo's Scott Neotropic Fund #35561.

9. REFERENCES

Anonymous, 1997, One hundred percent shade-grown coffee: By choosing shade coffee you can make a difference, Seattle Audubon Society, Seattle, Washington.

Atwood, J. T., 1984, A florisitic study of Volcan Mombacho department of Granada, Nicaragua, Ann, Missouri, *Bot. Gard.* **71**:191–209.

Baldwin, J. D., and Baldwin, J. I., 1972, Population density and use of space in howling monkeys (*Alouatta villosa*) in southwestern Panama, *Primates* **13**(4):371–379.

Baldwin, J. D., and Baldwin, J. I., 1976, Primate populations in Chiriqui, Panama, in: *Neotropical Primates: Field Studies and Conservation*, R. W. Thorington, Jr. and P. G. Heltne, eds., Proceedings of a symposium on the distribution and abundance of Neotropical primates, National Academy of Sciences, Washington, D.C., pp. 20–31.

Barborak, J., 1998, Buffer zone management: Lessons for the Maya Forest, in: *Timber, Tourists, and Temples*, R. B. Primack, D. B. Bay, H. A. Galletti, and I. Ponciano, eds., Island Press, Covelo, CA, pp. 209–221.

Bennett, E. L., Eves, H., Robinson, J. G., and Wilkie, D., 2002, Why is eating bushmeat a biodiversity crisis? *Conserv. Practice* **3**(2):28–29.

Calvo, L. M., and Blake, J., 1998, Bird diversity and abundance on two different shade coffee plantations in Guatemala, *Bird Cons. Int.* **8**:297–308.

Coates, A., 1997, *Central America: A Natural and Cultural History*, Yale University Press, New Haven.

Coimbra-Filho, A. F., and Mittermeier, R. A., 1981, *Ecology and Behavior of Neotropical Primates, Volume 1.*

Chapman, C. A., 1987, Flexibility in diets of three species of Costa Rican primates, *Folia Primatol.* **49**:90–105.

Crockett, C. M., and Eisenberg, J. F., 1987, Howlers: Variation in group size and demography, in: *Primate Societies*, B. B. Smuts, D. L. Cheney, R. M. Seyfarth, R. W. Wrangham, and T. T. Struhsaker, eds., University of Chicago Press, Chicago, pp. 56–68.

Crockett, C. M., Brooks, R. D., Meacham, R. C., Meacham, S. C., and Mills, M., 1997, Recent observations on Nicaraguan primates and a preliminary conservation assessment, *Neotrop. Primates* **5**(3):71–74.

Dinerstein, E., Olsen, D. M., Graham, D. J., Webster, A. L., Primm, S. A., Bookbinder, M. P., and Ledec, G., 1995, A conservation assessment of the terrestrial ecoregions of Latin America and the Carribean, *The World Bank*, Washington, D.C.

Estrada, A., 1984, Resource use by howler monkeys (*Alouatta palliata*) in the rain forest of Los Tuxtlas, Veracruz, Mexico, *Int. J. Primatol.* **5**:105–131.

Estrada, A., and Coates-Estrada, R., 1996, Tropical rainforest fragmentation and wild populations of primates at Los Tuxtlas, Mexico, *Int. J. Primatol.* **17**:759–783.

Garber, P. A., Pruetz, J. D., Lavallee, A. C., and Lavallee, S. C., 1999, A preliminary study of mantled howling monkey (*Alouatta palliata*) ecology and conservation in Isla de Ometepe, Nicaragua, *Neotrop. Primates* **7**:113–117.

Glander, K. E., 1978, Howling monkey feeding behavior and plant secondary compounds: A study of strategies, in: *The Ecology of Arboreal Folivores*, G. G. Montgomery, ed., Smithsonian Institution Press, Washington, D.C., pp. 561–574.

Glander, K. E., 1981, Feeding patterns in mantled howling monkeys, in: *Foraging Behavior: Ecological, Ethological, and Psychological Approaches*, A. C. Kamil and T. D. Sargent, eds., Garland Press, New York, pp. 231–257.

Greenberg, R., 1996, Birds in the tropics: The coffee connection, in: *Birding*, **XXVII**(6):472–480.

Greenberg, R., Bichier, P., and Sterling, J., 1997, Bird populations in rustic and planted shade coffee plantations of eastern Chiapas, Mexico, *Biotropica* **29**(4):501–514.

Hagen, J. M., and Johnston, D. M., 1992, *Ecology and Conservation of Neotropical Migrant Landbirds*, Smithsonian Institute Press, Washington, D.C.

Heckadon-Moreno, S., 1997, Spanish rule, independence, and the modern colonization frontiers, in: *Central America: A Natural and Cultural History*, A. G. Coates, ed., Yale University Press, New Haven, pp. 177–214.

Hendrix, S., 2000, Where the wild things are, *Int. Wildlife* **30**(5):20–27.

Illueca, J., 1997, The Paseo Pantera agenda for regional conservation, in: *Central America: A Natural and Cultural History*, A. G. Coates, ed., Yale University Press, New Haven, pp. 241–257.

Kinzey, W. G., 1997, *Alouatta*, in: *New World Primates: Ecology, Evolution, and Behavior*, W. G. Kinzey, ed., Aldine de Gruyter, New York, pp. 174–185.

Konstant, W., Mittermeier, R. A., and Nash, S. D., 1985, Spider monkeys in captivity and in the wild, *Primate Conserv.* **5**:82–109.

Larose, F., 1996, Foraging strategies, group size, and food competition in the mantled howler monkey, *Alouatta palliata*, Ph.D. Dissertation, University of Alberta.

Limeira, V. L. A. G., 1997, Behavioral ecology of *Alouatta fusca clamitans* in a degraded Atlantic forest fragment in Rio de Janeiro, *Neotrop. Primates* **5**:116–117.

Marsh, L. K., 1999, Ecological effect of the black howler monkey (*Alouatta pigra*) on fragmented forests in the Community Baboon Sanctuary, Belize, PhD thesis, Washington University, St. Louis, MO.

Martínez-Sánchez, J. C., Maes, J. M., van den Berghe, E., Morales, S., and Casteneda, E., 2001, *Biodiversidad Zoologica en Nicaragua*, Proyecto Estrategia Nacional de Biodiversidad y su Plan de Accion, MARENA/PNUD, Managua, Nicaragua.

McCann, C., and Koontz, F., 1997, Mombacho Volcano Nature Reserve, Nicaragua, 2–9 September 1997, Unpublished report, Wildlife Conservation Society.

Milton, K., 1980, *The Foraging Strategy of Howler Monkeys*, Columbia University Press, New York.

Milton, K., 1998, Physiological ecology of howlers (*Alouatta*): Energetic and digestive considerations and comparison with the Colobinae, *Int. J. Primatol.* **19**:513–548.

Mittermeier, R. A., 1991, Hunting and its effect on wild primate populations in Suriname, in: *Neotropical Wildlife Use and Conservation*, J. G. Robinson and K. H. Redford, eds., The University of Chicago Press, Chicago, pp. 93–110.

Mittermeier, R. A., 1996, Introduction, in: *The Pictorial Guide to the Living Primates*, N. Rowe, ed., Pogonias Press, East Hampton, New York, pp. 1.

Mittermeier, R. A., Ryland, A. B., Coimbra-Filho, A. F., and de Fonseca, G. A. B., 1988, *Ecology and Behavior of Neotropical Primates, Volume 2*, World Wildlife Fund, Washington, D.C.

Moguel, P. and Toledo, V. M., 1999, Biodiversity conservation in traditional coffee systems of Mexico, *Conserv. Biol.* **13(1)**:11–21.

Neville, M. K., Glander, K. E., Braza, F., and Rylands, A. B., 1988, The howling monkeys, genus *Alouatta*, in: *Ecology and Behavior of Neotropical Primates, Vol. 2*, R. A. Mittermeier, A. B. Rylands, A. F. Coimbra-Filho, and G. A. B. de Fonseca, eds., World Wildlife Fund, Washington, D.C., pp. 349–453.

Oates, J. F., Abedi-Lartey, M., McGraw, S. W., Struhsaker, T. T., and Whitesides, G. H., 2000, Extinction of a West African red colobus monkey, *Cons. Bio.* **14(5)**:1,526–1,532.

Oldfield, M. L., and Alcorn, J. B., 1991, Conservation of traditional agroecosystems, *Bioscience* **37(3)**:199–206.

Ostro, L. E. T., Silver, S. C., Koontz, F. W., Horwich, R. H., and Brockett, R., 2001, Shifts in social structure of black howler (*Alouatta pigra*) groups associated with natural and experimental variation in population density, *Int. J. Primatol.* **22(5)**:733–748.

Pendergrast, M., 1999, *Uncommon Grounds: The History of Coffee and How it Transformed the World*, Basic Books, New York.

Perfecto, I., Rice, R., Greenberg, R., and Van der Voort, M. E., 1996, Shade coffee: A disappearing refuge for biodiversity, *Bioscience* **46(8)**:598–608.

Pimental, D., Stachow, U., Takacs, D. A., and Brubaker, H. W., 1992, Conserving biological diversity in agricultural/forestry systems, *Bioscience* **42(5)**:354–362.

Rappole, J. H., 1995, *The Ecology of Migrant Birds: A Neotropical Perspective*, Smithsonian Institute Press, Washington, D.C.

Redford, K. H., 1992, The empty forest, *Bioscience* **42(6)**:412–422.

Rice, R. A., 1997, The coffee environment of northern Latin America: Tradition and change, in: *Proceedings of the 1st Sustainable Coffee Congress*, R. A. Rice, A. M. Harris, and J. McClean, eds., Smithsonian Migratory Bird Center, USA, pp. 5–114.

Rice, R. A., and Ward, J. R., 1996, Coffee, conservation, and commerce in the western hemisphere: How individuals and institutions can promote ecologically sound farming and forest management practices in northern Latin America, Smithsonian Migratory Bird Center and Natural Resources Defense Council, Washington, D.C.

Robinson, J. G. and Bennett, E. L., 2000, *Hunting for Sustainability in Tropical Forests*, Columbia University Press, New York.

Rodriguez-Luna, E., Cortez-Ortiz, L., Mittermeier, R. A., and Rylands, A. B., 1996a, *Plan de Accion para los Primates Mesoamericanos: Borrador de Trabajo*, IUCN/ SSC Primate Specialist Group, Universidad Veracruzana, Xalapa, Mexico.

Rodriguez-Luna, E., Cortez-Ortiz, L., Mittermeier, R. A., Rylands, A. B., Wong-Reyes, G., Carillo, E., Matamoros, Y., Nunez, F., and Motta-Gill, J., 1996b, Hacia un plan de accion para los primates Mesoamericanos, *Neotrop. Primates* **4(suppl.)**:119–133.

Roque Espinoza, A. A., 1999, Report on the relative density and population dynamics of howler monkeys (*Alouatta palliata*) in the Mombacho Volcano Nature Reserve, Unpublished Report, Wildlife Conservation Society.

Rosser, A. M. and Mainka, S. A., 2002, Overexploitation and species extinctions, *Cons. Bio.* **16(3)**:584–586.

Rowe, N., 1996, *The Pictorial Guide to the Living Primates*, Pogonias Press, East Hampton, NY.

Ryan, J. C., 1992, *Life Support: Conserving Biological Diversity*, Worldwatch Institute, Washington, D.C.

Rylands, A. B., Mittermeier, R. A., and Rodriguez-Luna, E., 1995, A species list for the New World primates (Platyrrhini): Distribution by country, endemism, and conservation status according to the Mace-Lande system, *Neotropical Primates* **3(suppl.)**:113–164.

Schwarzkopf, L. and Rylands, A. B., 1989, Primate species richness in relation to habitat structure in Amazonian rainforest fragments, *Biol. Cons.* **48(1)**:1–12.

Shahabuddin, G., 1997, Preliminary observations on the role of coffee plantations as avifaunal refuges in the Palni Hills of the Western Ghats, *Journal of the Bombay Natural History Society* **94**:11–21.

Stoner, K. E., 1993, Habitat preferences, foraging patterns, intestinal parasitic infections, and diseases in mantled howler monkeys, *Alouatta palliata,* Ph.D. Dissertation, University of Kansas.

Stoner, K. E., 1994, Population density of the mantled howler monkey (*Alouatta palliata*) at La Selva Biological Reserve, Costa Rica: A new technique to analyze census data, *Biotropica* **26(3)**:332–340.

Talbot, J., 1995, The regulation of the world coffee market: Tropical commodities and the limits of globalization, in: *Food and Agrarian Orders in the World Economy*, Praeger, Westport, pp. 139–168.

Toledo, V. M., and Moguel, P., 1997, Searching for sustainable coffee in Mexico: The importance of biological and cultural diversity, in: *Proceedings of the 1ˢᵗ Sustainable Coffee Congress*, R. A. Rice, A. M. Harris, and J. McClean, eds., Smithsonian Migratory Bird Center, USA, pp. 165–173.

Toledo, V. M., Ortiz, B., and Medellín, S., 1994, Biodiversity islands in a sea of pastureland: Indigenous resource management in the humid tropics of Mexico, *Etnoecol.* **3**:37–50.

Wallace, D. E., 1997, Central American landscapes, in: *Central America: A Natural and Cultural History*, A. G. Coates, ed., Yale University Press, New Haven, pp. 72–96.

Wilcove, D. S., McLellan, A. P., and Dobson. A. P., 1986, Habitat fragmentation in the temperate zone, in: *Conservation Biology: The Science of Scarcity and Diversity*, M. E. Soule, ed.

Williams-Guillén, K., and McCann, C., 2001, Ranging behavior of howling monkeys (*Alouatta palliata*) in Mombacho Volcano, Nicaragua: A GIS-based approach, *Am. J. Primatol* **54 (suppl. 1)**:70–71.

Williams-Guillén, K., and McCann, C., 2002, Ecology and conservation of mantled howling monkeys (*Alouatta palliata*) in the shade coffee plantations of Mombacho Volcano, Nicaragua, *in preparation.*

Willie, C., 1994, The birds and the beans, *Audubon* **80**:58–64.

Wolfheim, J. H., 1983, *Primates of the World*, University of Washington Press, Seattle.

World Resources Institute, 1998, *Trouble Brewing: The Changing Face of Coffee Production*, Oxford University Press, pp.165–166.

EFFECTS OF HABITAT FRAGMENTATION ON THE CROSS RIVER GORILLA (*GORILLA GORILLA DIEHLI*): RECOMMENDATIONS FOR CONSERVATION

Edem A. Eniang[*]

1. INTRODUCTION

1.1. The Nigerian Tropical Moist Forest and Primate Conservation

Cross River State, Nigeria, is in the tropical moist forest zone (or Guinean forest region) and home to a diverse assemblage of primates (Figure 1). Detailed description of this tropical moist rain forest is given in Onochie (1979), Richards (1981), and Sayer et al. (1992). Increasing human population and the attendant desperate economic situation of the people following years of military rule and failing economy has led to unsustainable forest resource exploitation and destruction (Olajide and Eniang, 2000). Chapman and Lambert (2000) have shown that the tropical forest and the animals they support are being threatened by accelerating rates of forest conversion and degradation. Mittermeier (1988) ranked Nigeria as 8th among the countries of the world with the highest primate diversity and 7th in terms of degree of primate species endemism. Therefore, it becomes extremely necessary to pursue a multifaceted conservation strategy to save the myriads of primates and biodiversity inherent in the tropical moist forest ecosystem. The declining status of the tropical moist forest and wildlife, in general, and decimation of primates, in particular, requires urgent realistic and effective steps, including all people and factors involved. Eniang (1998) listed several factors as being responsible for the decline of non-human primate populations in southeastern Nigeria. The deplorable conservation status of the Cross River gorilla and other primates can be found in Oates (1996). Other identified threats to primates in Nigeria are discussed in Mittermeier (1987) and Starin (1989). Top on this list was habitat destruction. In this chapter, emphasis is placed on the Cross River gorillas (*Gorilla gorilla diehli*) whose survival is threatened not only by habitat destruction but by habitat fragmentation, which

[*] Department of Forestry and Wildlife, University of Uyo, PMB 1017, Uyo, Nigeria, and Biodiversity Preservation Group, 93 Ndidem Usang Iso Road, H. E P. O. Box 990, Calabar, Cross River State, Nigeria, E-mail: edemeniang@yahoo.com

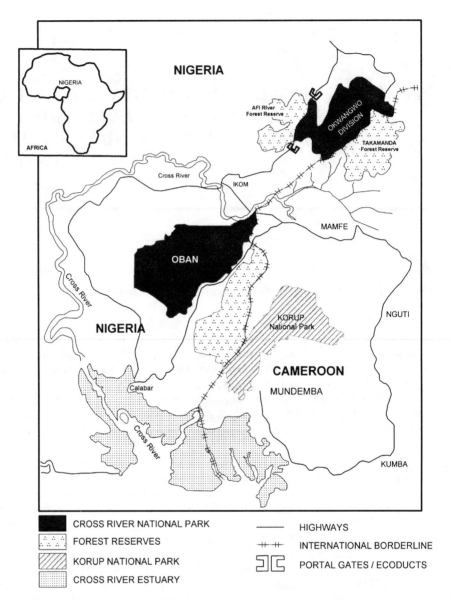

Figure 1. Locations of the Okwangwo and Oban Divisions of Cross River National Park in relation to other conservation areas. Redrawn by Winters Red Star.

has resulted in the isolation of six gorilla subpopulations. These subpopulations exist practically as islands without any possible links in the foreseeable future (Figure 2). The taxonomic and evolutionary history of the Cross River gorilla has been a subject of ongoing research since 1904, while the genetic, demographic status, and origin of the species can be found in Mace (1988), Harcourt et al. (1988, 1989), and Groves (1967, 2001).

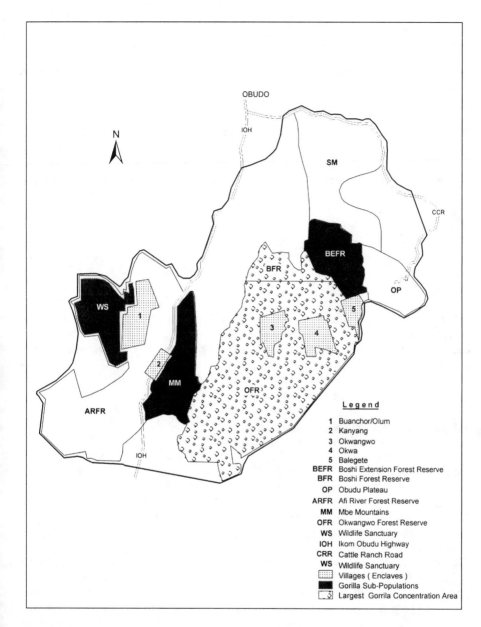

Figure 2. Okwangwo Division showing enclaves, gorilla subpopulations, and the largest gorilla concentration. Redrawn by Winters Red Star.

1.2. Impacts of Dwindling Economic and Sociopolitical Stability

The World Bank in 1980 established that the original wildlife habitat in Africa is about 54 million km^2. Of that, 60% has been lost to agriculture, pastoral activity, and other consumptive and nonconsumptive uses of the forest ecosystems. It is worthy to note here that forest conservation was introduced into Nigeria only in 1887. From then onwards, conservation did not benefit from growing international trade in timber products and an expanding population. Other contributing factors working against conservation include the independence of Nigeria and post-civil war reconstruction.

Currently, Nigeria and Uganda are the African countries that have suffered the most intensive habitat alteration. It has been calculated by the Food and Agriculture Organization (FAO, 1995) that over 90% of the forests have been felled since the beginning of the oil boom in Nigeria. This enormous loss of the forest and its resources was not accompanied with improvement of the economic situation of the people. In fact, Nigeria, which was so economically promising in the early 1970s, with an average annual income of over US$1,000.00 per person, is now rated among the 13th poorest countries in the world with an average annual income of US$365 per person (World Bank, 2002). Among the whole of Nigeria, Cross River State is the most important area environmentally speaking, and it houses the largest tracts of intact tropical moist forest in the country. However, forest fragmentation is an unpleasant reality even in Cross River State. The situation has affected the officially protected Cross River National Park (CRNP) in such a way that the world's most northerly and most endangered gorilla species has suffered from the dramatic effects of habitat fragmentation. For instance, the protected core gorilla habitat, which is in the Boshi-Okwangwo forest of CRNP, is still subjected to persistent and widespread land clearing from enclave[1] communities in addition to other encroachment activities (Figure 2). These communities, three in the core and two in the peripheral zones of the CRNP have been earmarked for resettlement to locations outside the park for as long as 10 years, but the CRNP has neither the financial strength, political will, nor support to accomplish this crucial step to effective protection.

1.3. Cross River Gorillas

Cross River gorillas have been reported since 1904 by Matschie who first described them as a new species (*G. g. diehli*) after a cranial investigation of nine skulls in the Berlin Museum. Since then, the species' taxonomy has been a subject of many scientific studies (Rothschild, 1904, 1908; Coolidge, 1929; Groves, 1967, 1970). The habitat of Cross River gorillas is within the Guinean forest region (White, 1983). Their distribution is shown in Wolfheim (1983). Recent field studies of the species current range have proven their continued existence in the area (Eniang and Nwufoh, 2001). March (1957), Critchley (1968), Harcourt et al. (1989), Oates et al. (1990, 1998), Obot et al. (1997), and Nwufoh (1999) have confirmed their continued existence in the Mbe Mountain Range, Afi Mountains in Afi River Forest Reserve, Boshi and Boshi Extension/Okwangwo Forest (CRNP), and the Takamanda Forest Reserve (Figure 1). Also, Groves and Maisels (1999) have reported their presence in Mone River Forest Reserve to the southeast of the Takamanda Forest Reserve (TFR) Cameroon and around the Mbuli Hills in the east.

In terms of ecology, the Cross River gorillas have not enjoyed adequate ecological studies until recently. The species is currently inhabiting hilly forest areas on altitudes ranging from approximately 150 m to 1,650 m asl along the Obudu-Mbuli Hills axis of the CRNP (personal observation, Mc Farland, 1994; Obot et al., 1997; Nwufoh, 1999). Apart from occasional observation of nests at Debe (salt lick) south of Okwangwo, the

---

[1] Enclave refers to Nigerian villages or local communities located within the Cross River National Park.

fringes of the hills near Bashu, and some areas along the Okorn River Valley, no other records exist of their presence in lowland areas. The whole of northern Cross River State is noted for its seasonal climatic regimes. It has a long dry season between November to March. The rest of the months are taken up by tropical rains with a slight break in August. Vegetation of the area is as described in Obot et al. (1997) and explains why vegetation characteristics have tended to influence gorilla ranging behaviour (McFarland, 1994; Nwufoh, 1999).

Direct personal observations revealed that apart from habitual concentrations of the species within very rugged terrain (hills and valleys) possibly to avoid humans and other human-related perturbations, the gorillas' ranging patterns are influenced by availability of food resources (e.g., fruiting of the most common trees, such as *Santeria* spp., *Dacryodes* spp., and *Irvingia* spp.). This is supported by Obot et al. (1997) and Nwufoh (1999). Herbaceous vegetation, forest cover, and seasonal bush fires have had influence on their ranging behavior. However, the species generally has a stable range. For instance, some groups remain almost permanently for two to four months within a particular hill range. This behavior on its own has given rise to "sure catch" syndrome (i.e., a hunter can precisely tell when and where to locate and kill a gorilla with a reasonable level of certainty). Their migratory route goes from the hills north of Okwa and Okwangwo enclave villages then follows the hills along Anyukwo River in the forest between rivers Bemi and Kanfoh as far as the lower course of the Okorn River. The route goes as far as Bashu Hills across river Anyibiar before crossing into the TFR Cameroon. In the long-run, the migratory route, if confirmed, can be adapted for habituating the gorillas. Their specific movements, routes, and ranges and their potential cross-border migration along these observed routes are a subject of another article by this author and will not be elaborated on here.

Many authors have over the years tried to estimate the gorilla population (March, 1957; Critchley, 1968; Harcourt et al., 1988, 1989; Oates et al., 1990; Obot et al., 1997; Nwufoh, 1999; Groves and Maisels, 1999). Only a thorough and long-term study using advanced technology and strategies will be capable of producing a reliable estimate of the Cross River gorillas. At this point, I estimate between 75 to 100 individuals total within the region, but Bassey and Oates (2001) state there are less than 250 individuals.

It is worthy to note that the following primates in Nigeria are also suffering from the dramatic effects of habitat fragmentation. Preuss's red colobus (*Procolobus badius preussi*) and crowned guenon (*Ceropithecus pogonia pogonias*) in Ikpan Block of the Oban group of forest in Oban Division of CRNP, Niger Delta, red colobus (*Procolobus badius epieni*) in the Niger Delta of Southern Nigeria, Sclatter's guenon (*Cercopithecus sclatteri*) with a more scattered distribution in stubbs Creek forest reserve of Ibeno and Itam, Itu, Uruan Inyang Atakpo, all in Akwa Ibom State, as well as grey checked mangabey (*Lophocebus albigena osmani*) and Preuss's guenon (*Cercopithecus preussi*) in the same habitat with the Cross River gorilla. Other less affected primates are white-throated guenon (*Cercopithecus erythrogaster*) in the nearly established Okomu National Park, red capped mangabey (*Cercocebus torquatus*), putty nosed guenon (*Cercopitheus nictitans*), and mona guenon (*Cercopitheus mona mona*) in Itu Wetlands and Ikpa River Basin, all in Akwa Ibom State Nigeria.

2. STUDY AREA

2.1. Mbe Mountain Range

The Mbe Mountain Range is located in the northern part of Cross River State in the Boki Local Government Area (Figure 3). The area is approximately 100 km^2 and rich in faunal and floral diversity, including the three biggest primates in Nigeria (gorilla, chimpanzee, and mandrill). The primate diversity and other faunal resources of the Mbe Mountains area is given by Eniang and Nwufoh (2001).

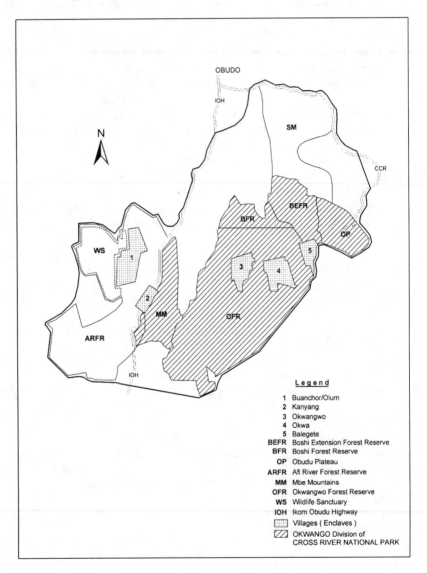

Figure 3. Mbe Mountain Range, Afi River Forest Reserve, and Cross River National Park. Redrawn by Winters Red Star.

The most interesting detrimental aspect of the site is its separation from the adjacent Afi River Forest Reserve. The site was previously connected until the advent of the Ikom-Obudu Federal Highway, which was commissioned in 1978 without an Environmental Impact Assessment. The vegetation of the area is Guineo-Congolian tropical forest with a hilly terrain and a maximum elevation of above 920 m. Urgent biodiversity conservation is necessary for this unique ecosystem. The presence of nearly 30 rural communities, the Ikom Obudu Federal Highway, and numerous village roads, farm trails, tracks, ubiquitous small-holder subsistence farmsteads have rendered the area an isolated habitat. The Mbe Mountain Range is a tropical moist forest with a healthier, less disturbed habitat than Afi Mountain despite the current situation. The area is bound in the east by the Okorn River and in the south by the farmlands of the people of Abo Obisu, Abo Mkpang, and Abo Ogbagante and on the western part by the Ikom-Obudu Federal Highway. The northern axis is straddled by the settlements of Bamba, Wula, and Bokalum. The intact forest areas are still being encroached by the people of these villages who continually clear the forest for banana plantations.

2.2. Afi River Forest Reserve

The Afi River Forest Reserve, which was gazetted into Afi Wildlife Sanctuary (AWS) by the Cross River State Forestry commission in May 2000, is a rocky mountainous area with distinct outcrops, some rising above 1,000 m, and separated by steep-sided, deep valleys (Figure 3). The rugged nature of its terrain has over the years presented a natural barrier against wholesale resource exploitation, land conversion, and occupation, which has augmented continued survival of the Cross River gorillas in spite of the prevailing situation around AWS. AWS, which was originally over 1,000 km^2, has been extensively encroached upon by a number of rural communities whose natives and their activities have impacted negatively on the habitat quality. Human perturbations that threaten and further isolate the gorilla include but are not limited to shifting and subsistence agriculture, annual bush burning, and unguided forest and non-timber resources exploitation.

Escalating levels of hunting and trapping of wildlife, exploitation of non-timber forest species, and logging by an Asian company (West Africa Metal and Plywood Company [WEMPCO]) within adjoining or fringing forest areas, coupled with the area's natural tendency to rock fall through erosion by wind and runoff water, has led to frequent tree falls and bush fires whose impacts are escalated by the steep slopes because of these combined impacts. AWS has become a complex mosaic of vegetation types which creates a diverse vegetational cover throughout the range including numerous gaps, herbaceous vegetation, and grasslands. There are many streams, especially in the rainy season, that drain into the Afi River. There is at present less than 40 km^2 available for the approximately 30 to 40 gorillas thought to be present in the area.

The long isolation of the Afi group from other subpopulations may promote inbreeding depression, which, in the long run, could set a ceiling on the survival prospects of the Afi Mountain gorillas. Since the area also contains a sizeable population of chimps, drills, guenons, spot necked otter, ungulates, rodents, ant eaters, reptiles, and amphibians, insufficient habitat and resources may affect the survival prospects of the gorillas.

2.3. The Cross River National Park

The CRNP came into being through the Federal Military Government Decree No. 36 of 1991, and the Cross River gorilla was selected as the theme animal of the park (the first tropical moist forest national park in Nigeria; Figure 3). This Decree was later repealed by another Federal Military Government Decree No. 46 of 1999. The Decree converted the Nigeria National Parks Services, under whose jurisdiction rests the CRNP and other national parks (Figure 4), into a paramilitary outfit with increased powers and capacity to promote conservation activities within Nigeria (Eniang, 2001).

The CRNP harbors the highest number of the Cross River gorillas with a subpopulation in the Boshi Extension Forest portion (north of Okwangwo Division) and another in the Okwa and Obonyi Hills (in the south of Okwangwo Division) (Figure 2). The two areas are thought to habour between 50 to 60 individuals.

The gorillas generally range in an area of semideciduous, montane, and derived savanna vegetations within approximate latitudes 5°5' to 6°30°N of the equator and longitudes 8°50' to 9°40°E of the Greenwich Meridian. The area borders the international boundary between Nigeria and Cameroon. The gorillas are distributed in isolated subpopulations in and around a complex of hilly escarpments with steep valleys and odd peaks that generally rise higher than surrounding forested areas. Some peaks go as high as 2,000 m and can have semitemperate conditions during the year.

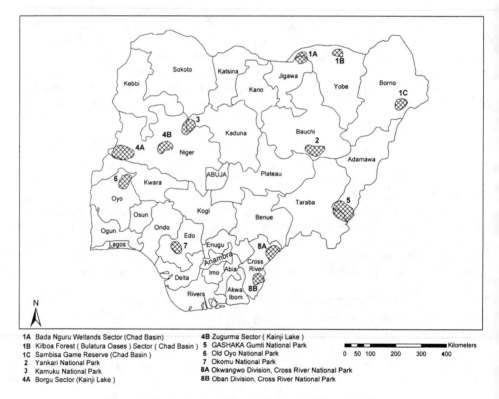

1A Bada Nguru Wetlands Sector (Chad Basin)
1B Kilboa Forest (Bulatura Oases) Sector (Chad Basin)
1C Sambisa Game Reserve (Chad Basin)
2 Yankari National Park
3 Kamuku National Park
4A Borgu Sector (Kainji Lake)
4B Zugurma Sector (Kainji Lake)
5 GASHAKA Gumti National Park
6 Old Oyo National Park
7 Okomu National Park
8A Okwangwo Division, Cross River National Park
8B Oban Division, Cross River National Park

Kilometers
0   50 100   200   300   400

Figure 4. National parks of Nigeria. Redrawn by Winters Red Star.

Unfortunately, the gorilla subpopulations inhabiting CRNP do not have any contact with the Afi Mountain subpopulations and would not have any contact in the very foreseeable future, unless stringent practical conservation actions are embarked upon.

3. CURRENT THREATS

3.1. Hunting and Poaching

Hunting is generally accepted as the principal factor decimating gorilla populations (Harcourt et al., 1989). Hunting of the Cross River gorillas was rampant in the years before the advent of the CRNP. However, hunting was reduced with increasing influence of the CRNP as a result of anti-poaching and increased protection activities (Eniang, 2001). With increasing conservation education by CRNP, Biodiversity Preservation Group, Pandrillus, Nigerian Conservation Foundation, and Cercopan, hunting and poaching of Cross River gorillas was almost forgotten in Cross River State until this author came up with new findings of unreported poaching incidents in enclave communities (Figures 5, 6, and 7). Other primates are also illegally hunted in the park, such as chimpanzee and mandrill (Figures 8 and 9).

Figure 5. A collection of poached gorilla skulls, found in an enclave community in the CRNP. The 1st skull from right shows a bullet hole, while the 2nd from right (the freshest) shows a machete cut on its temple. The humerous bone of the silverback first left is placed on the foreground. Photo by Edem Eniang.

Figure 6. The author examines the skull of a silverback killed in November 2000 in the house of a gorilla poacher. Photo by Edem Eniang.

Figure 7. Skull of a silverback gorilla kept as souvenir by the poacher. Photo by Edem Eniang.

Figure 8. An adult male drill (*Mandrillus leaucophaeus*) is one of the great primates of the CRNP and a potential flagship species for conservation. Photo by Edem Eniang.

Figure 9. A collection of chimpanzee skulls found in an enclave community in the CRNP. The second skull from right and the smallest skull (foreground) both show machete cuts on their temples. Photo by Edem Eniang.

3.2. Habitat Fragmentation

Habitat fragmentation and isolation effect are the greatest threat to the continued survival of gorillas in their natural range in Nigeria. Thus, the Cross River gorillas currently are existing in ecological islands and are prone to declining genetic heterozygosity through population isolation over time. The threat that exists from fragmentation may be greater than from hunting and poaching at their present levels. The long-term impacts of such isolation on the populations will most likely be influenced by such factors as population size in each identified fragment, degree and extent of isolation, time lapse since isolation, and habitat size and quality within each isolated fragment. The effect of habitat destruction and degradation as seen in the present study areas and the potential for accelerating the extinction processes have been discussed by Marsh et al. (1987) and Tilman et al. (1994).

Although fragmentation is severe, bushmeat (or wild animal) hunting is a compounding threat, but not to the degree one might expect. The Boki people, on whose land the gorillas live, have a meat sharing culture that compels a hunter to share selected body parts of gorilla and other larger mammals, such as bush cows, buffaloes, and chimpanzee, with all his relations no matter where they reside. This tradition is still being respected by Boki hunters. Interestingly, they avoid killing larger mammals because they must distribute the meat and it becomes a waste of time, energy, and funds to distribute different preserved body parts (dried carcasses) to more than seven recipients wherever they are located. Instead, I discovered during my most recent field surveys (last quarter 2001 to first quarter 2002) that the larger mammals were the ones whose presence were easily noted, including gorillas, while it was relatively difficult to encounter small- and medium-sized primates and ungulates such as drills, guenons, and duickers. Smaller mammals do not have to be distributed as elaborately as large mammals (Figure 10).

3.3. Land Rights, Tenure, and Ownership

In every Nigerian rural setting, ownership of land rests on the native authority. The people are aware that the Federal government holds absolute rights to all lands and resources and that there is little enforcement of conservation laws. People therefore hold the land and resources as ransom to be compensated for by government through economic development options. The government is always unwilling or unable to meet these needs, thereby making it impossible to adopt drastic or effective conservation actions in those areas. The situation has hampered gorilla conservation efforts in Afi River Forest Reserve, Mbe Mountains, and in CRNP. The land tenure system encourages shifting cultivation and its characteristic slash-and-burn agriculture.

For instance, the most outstanding constraint to effective conservation is the status as a community forest with approximately 14 landlord communities that still depend on this forest for bushmeat as protein. These 14 communities basically surround the Mbe Mountain community, while a few are enclave or villages surrounded by territory of the CRNP. With increasing sociopolitical awareness following the new democratic experiment in Nigeria, the natives are reluctant to hand off this unique habitat without asking unthinkable compensation from the CRNP and whoever is interested. Political issues within the landlord communities have prevented for 10 years all the efforts aimed at including the Mbe Mountain into the CRNP for proper protection. Furthermore, the idea to promote an indigenous Rangers Programme by the Boki Local Government

Figure 10. A collection of eight primates and one duicker head (first from right) smoke dried and kept by a poacher in his bush shed near the CRNP, the first two heads from right are *Cercopithecus erythrotis*, 3rd and 8th heads, *Mandrillus leucophaeus*, 4th and 7th, *Cercopithecus mona*, 5th and 6th, *Cercocebus torquatus*, and 9th, *Cephalohus* spp. Photo by Edem Eniang.

Council in recent times is an idea that leaves a lot of questions, especially as it pertains to the technical know how and sincerity of purpose. The other recognized sites, Afi River Forest Reserve, Okwangwo, Boshi and Boshi Extension Forest reserves are all threatened by similar ecological and sociopolitical situations. The good news is, even given all of the above mentioned pressures, the Mbe Mountain Range is now included in the CRNP.

3.4. Logging/Agricultural Projects

WEMPCO and other small-scale logging firms, as well as illegal operators in the gorilla's range as a whole, have led to more habitat loss, including within buffer zones. The opening of logging routes has led to increased access not only for additional forest exploitation, but also for wildlife poaching. Habitat disturbance has encouraged migration of many species across the border into the TFR Cameroon.

3.5. Trapping and Collection of Non-Timber Forest Products (NTFP)

In the general area, especially around enclave communities, these are always numerous snares set in drift fences for small- and medium-sized mammals, reptiles, and some birds. The trappers report that larger mammals sometimes get caught or injured in such traps. I witnessed this in 2000 when I encountered an adult buffalo with a snare wire embedded in its right hind foot, which was swollen and infected. The trappers open new trails and NTFP collectors use the trails in search of *Gnetum africanum* leaves, *Capolobia lutea* sticks, and *Irvingia gaboonensis* fruits. Such large-scale human presence and exploitative activities continue to push gorillas and other species away from potential habitats into very restricted areas such as hilltops and valleys, thus negatively influencing their ranging behaviors.

3.6. Bush Fires

Annually, particularly in the dry season months (November–March), wild bush fires occur in most areas across the Cross River gorillas' natural range. These fires are human caused, either by arson, by accident, or from carelessness. Other fires are caused by hunters in their bush camps, cattle grazers especially on the Obudu Plateau and the escarpment, and those caused from slash-and-burn farming activities. Annual bush fires in the gorillas' natural range have had far-reaching consequences on gorillas, and all wildlife. Fire has altered the vegetation so much that the transition from forest to savanna has been very rapid in recent years.

3.7. Illiteracy, Poverty, and Lack of Basic Sources of Livelihood

No conservation etiquette can be borne in an illiterate and hungry mind. The human population around the gorillas' natural range in Cross River State have serious problems of illiteracy, poverty, and lack of basic sources of livelihood. Many communities have no functional schools and those that have are overcrowded with little or no facilities for proper teaching and learning. Worse still, the teachers are not paid for months. Since there is little motivation or support for learning, younger people spend their time helping with basic subsistence. Thus, many young men chose to go to the forest for hunting, trapping, and fishing (at times by poisoning the water bodies). The young women collect NTFPs and during the farming season everyone clears more and more forest to plant bananas and other crops. This situation has led to an increasing population of people who are hungry, needy, and illiterate. This group of people cannot accept conservation culture easily. When basic needs are not being met, it is difficult for people to become aware of conservation issues. This awareness is increasing particularly about endangered species, but little is done because obtaining basic subsistence is so difficult.

3.8. Pet Trade

Nigeria has been a trade route for many animal species bound for the international pet trade markets. Animals affected include, but are not limited to, African grey parrots, chimps, gorillas, crocodiles, and snakes. Kano and Lagos states have been recognized as major exit routes while Cross River, Akwa Ibom, Taraba, and Yobe States have had some species repatriated. Most of the traded animals come from less protected forest across the international border with Cameroon and Nigeria.

3.9. Inadequate Scientific Knowledge

The bulk of scientific information already existing on the Cross River gorilla is based more on guess work and hear say, and may not be very reliable for serious planning or management. The time has come to embark upon serious field based scientific research involving experts and advanced scientific techniques. Such efforts should be field based and involve locals. Every effort should be made to avoid inconsistent data collection by training volunteers, students, and local people. To ensure success of research projects, proper consent and support from the government authority is required. There are a number of problems arising from the lack of scientific knowledge having to do with reintroductions. Reintroductions are very difficult, especially if the subspecies is unknown. There have been instances of captive chimps or orphaned chimps escaping into the wild. Reintroductions of surplus species of gorilla, without proper genetic identification, risk contaminating local endangered populations. Thus, all who are involved in captive breeding programs with the intent of reintroduction must caution against jeopardizing the wild populations (Eniang, 1998). Captive facilities should concentrate on conservation education and awareness until the science of conservation is well developed in Nigeria, and knowledge about wild populations are established.

4. STRATEGIES FOR CONSERVATION

4.1. Adapting Traditional Conservation Strategies

In every society, traditional knowledge plays a significant role in all issues. Traditional beliefs and practices of local people in the study area must be recognized and understood with a view to educating locals in a positive manner about conservation or management activities of both the governmental agencies and non-governmental organizations. A complete disregard to these things can bring a great deal of conflict. Opinion and political leaders as well as chiefs and tribal leaders must be included in a participatory manner from the beginning because they have very strong influence on their followers. A conservation project can succeed or fail depending on whether the local leaders support it. The influence of leaders on the local communities is clearly seen in the Afi River Forest Reserve and Mbe Mountain regions.

4.2. Update and Enforce Wildlife Laws

Many wildlife laws are obsolete and are not enforced. Therefore all relevant edicts, legislation, and laws pertaining to forests, wildlife, and resources must be reviewed, updated, and enforced. In many instances culprits have been arrested, detained, and released without appropriate trials and/or punishment for an offence because of contradictory laws. Even if the laws were outstanding, the lack of consistent enforcement renders then useless. Funds must be generated by the government to support enforcement of natural resource and wildlife laws.

4.3. Immediate Demarcation of Boundaries

Both the Afi Wildlife Sanctuary and CRNP require fresh demarcation and maintenance of boundary lines with well established signposts to show their respective boundaries clearly. In some instances, culprits have claimed ignorance of park boundaries and communities demand shifting their boundaries into the park. This is part of establishing and maintaining laws that enforce boundary respect.

4.4. Alternatives to Bushmeat

The "Only Livestock" project developed by the Biodiversity Preservation Group is a project where cattle and other domestic animals are purchased from areas of abundance (producing states) and are sold to natives at reduced costs as alternative sources of protein. In theory, this supply will discourage hunting and poaching within the gorilla habitat. The absence of poaching and hunting pressures will pave the way for a habituation program that will allow for much needed gorilla-based ecotourism. An increase in local income is purported to foster conservation mindedness by protecting a species that is an economic draw. The benefits of gorilla ecotourism are well elaborated by Butynski and Kalina (1998).

Apart from supplying domestic meat from producing states, the "Only Livestock" project also recognizes wise use of biodiversity, and the project recommends the breeding of micro-livestock for communities in critical areas as a kind of backyard farming. Suggested species include rabbits or poultry and adaptable native wildlife species, such as duikers, porcupines, cane rats, giant rats, pythons, crocodiles, and snails to provide the often preferred wild meat. In areas where this project has been successfully established, such as Obudu, Bateriko (Boki), and Ogoja, all in northern Cross River State and in Uyo, Itu, and Etinan (Akwa Ibom State), occupational hunting and poaching are already becoming a "dirty man's job" (out of fashion) and has led to a drastic decrease in the number of full-time hunters in the last decade. The threat to primates arising from bushmeat hunting can be found in Martin (1983), Mittermeier (1987), and Starin (1989).

4.5. Creation of Corridors

There has been no push from the conservation community to establish a corridor connecting gorilla subpopulations in the Afi and Mbe Mountains. One option for conservation is a corridor to link the Mbe subpopulations to the Boshi–Okwangwo subpopulations. While considering another corridor between the Afi River Forest Reserve and Mbe Mountains, the ecological integrity of Mbe Mountain as a gorilla habitat (a multi-community owned land) is still favorable to the gorillas and other wildlife in the area.

A migration corridor must be immediately formulated to allow natural regeneration of forest around the Okwas 1 and 2 and Okwangwo villages. In the long run this will form a connection and migration corridor for gorillas and other species. If well organized and protected, the corridor will form the much needed link between the TFR Cameroon and CRNP. The designs, uses, consequences, and cost of conservation corridors are clearly explained in Simberloff and Cox (1987), Murcia (1995), and Lindenmeyer and Nix (1993).

4.6. Creation of Eco-ducts/Portal Gates

The government, through its Ministry of Environment Works, or the Forestry Commission and other relevant agencies should pursue the creation of an eco-duct or portal gate to form a linkage between Afi Wildlife Sanctuary and Mbe Mountains (Figure 1). Experts in the design of eco-ducts and portal gates can be consulted to tackle this issue in the event that closing the highway or diverting it is not politically or financially feasible in the near future.

4.7. Core Conservation Zone

In the event that it becomes impossible to relocate the enclave communities, then the management of the CRNP should endeavour to establish, maintain and monitor a core conservation zone within the existing management structures. Such a zone should be set aside for strict management which excludes all forms of occupation or exploitation except strictly managed scientific studies. This situation in the long run will pave way for improved conservation through effective monitoring, policing, and protection. The National Park cannot do all this good work without proper funding for project execution, training, and manpower development, operational equipment and facilities as well as patrols, policing and persecution. If all these are put in place, the meager gorilla populations in the wild are capable of natural recovery (Happold, 1991). In the same vein Dorst and Dandelot, 1986; and Tudge, 1992 have all proven that the ultimate success in conservation of endangered species does not lie in captive breeding and reintroduction but in *in-situ* conservation of the primates within their natural ranges. This is true for all the apes of Cross River State since no one has accurate information of their population status. A buffer zone or increased core area could be obtained by the government if land rights could be purchased from communities.

The CRNP on whose shoulder rests the responsibility of gorilla conservation should be tasked with the purchase of land rights from the communities. This land would then be included within the national park for proper protection. Areas that have been over harvested should be replanted or reafforested to enrich its habitat qualities for gorillas.

4.8. Closing Some Portion of the Ikom-Obudu Highway

Closing this highway may prove difficult, but it can be done if we want to ensure the survival of the Afi subpopulation in the next 25 to 50 years and do not want to translocate them. For example, if there are 30 to 40 humans, about 30 to 50 chimpanzees, and over 100 to 150 drills, all using the same 40 km^2 of land, then shortage of habitat and resources will diminish the survival of all the apes and other species over time. Thus, it is recommended that the road be closed either by tunneling for 3 to 5 km or diverted for 5 to 8 km for an alternate route allowing a gap for reafforestation and linkage to Mbe Mountain near Kayang. It would be a costly venture but could be achieved if commitment was on the minds of Nigerians to save our gorillas for the future. The regeneration methods for conservation of bioresources is well documented in (Olajide and Eniang 1999).

4.9. Translocation

In the event that it is impossible to close some portion of the almost abandoned Ikom-Obudu Highway to form a link between Afi Forest Reserve and Mbe Mountains, then the next possible option is to form a core gorilla area by translocating some of the isolated population in Afi into a larger and more protected area, such as the CRNP, Okwangwo Division. Primate translocations have been successful in many species (Konstant and Mittermeier, 1982; Strum and Southwick, 1986; Koontz et al., 1994).

4.10. Exchange Gorillas for Conservation

As a means of securing their gene pool and supplementing *in-situ* conservation, the authority overseeing wildlife resources of Nigeria could embark on an exchange program where some gorillas could be translocated to a safe sanctuary, such as Monkeyland in South Africa or another well-established facility. Monkeyland, or any facility, in return, can provide technical backing and support to the national park in terms of manpower training, protection activities, ecological surveys, and ecotourism. This will go a long way to promote international collaboration and research in the long run, while securing the species' genetic resources for future reintroduction.

4.11. Relocation of Enclave Communities

If it is at all possible, relocating the enclave communities from the CRNP should be considered. These include the three communities of Okwa 1, Okwa 2, and Okwangwo and the 4th community of Bashu from the center of the Cross River gorillas' last natural habitat in Nigeria. Since Voysey (1999) has elaborated the values of gorillas, and the coexistence of humans and gorillas are given by Tutin and Oslisly (1995), it therefore requires urgent and practical conservation strategies to promote the survival prospects of the species in Nigeria. The prospect of relocating entire villages is daunting, however, given the literacy rate and attachment communities may have to the land. If better land can be provided away from other imperiled regions in Nigeria, along with defraying the cost of relocation, can we really begin to hope this is a viable alternative.

4.12. Optimize Conservation Options

Since the three largest primates in Nigeria occur sympatrically in the area, it becomes imperative to properly optimize the conservation efforts and actions on the basis of priorities of the special sites and species (Eniang and Nwufoh, 2001). The possibilities of recovery of a threatened population of primates are explained in Ross (1988, 1992).

Ongoing field research on population abundance, ecology, behavior, and range is being carried out by the Biodiversity Preservation Group. Research and monitoring of gorillas by the CRNP Biological Research Department should be encouraged as well. To date, most of the conservation activities geared towards gorillas in Cross River State have relied mostly on research activities. Hunters and trappers from the area have been hired as research assistants. This is one way to reform people and to encourage conservation within communities through the experiences of these former hunters. These research guides may make good tour guides in the event of developing ecotourism.

5. SUMMARY

The Cross River (*Gorilla gorilla diehli*) is one of the most threatened apes of the African Continent, and it occurs in the Northern Cross River State, Nigeria with another small group in the TFR of South-western Cameroon, which is contiguous to the Okwangwo division of CRNP. Other small populations are isolated by the construction of federal highway that represents a full barrier to inter-populational interactions. The metapopulation so produced, seem to be more threatened by habitat fragmentation thereby causing inbreeding depression and in just a few decades will likely result in the extinction of the Nigeria gorilla. This article therefore attempts to answer the questions What is the degree of threat? What is the population status? Can fragmentation lead to extinction? and What are the strategies for conservation? Considering their current conservation status, it is suggested that translocation of one or more of the remnant populations to form a bigger core population might be of crucial relevance for maintaining a good genetic viability. Other suggestions include forming a protected migratory corridor connecting the Okwangwo Division of the CRNP, Nigeria, and the TFR of Cameroon as well as form another corridor linking the isolated Mbe Mountain subpopulation with Boshi-Okwangwo population. Other practical management strategies are suggested. Four maps are presented to describe and illustrate suggested options for conservation.

6. ACKNOWLEDGMENTS

I must thank Laura K. Marsh for all of her work on this chapter. Thanks also to Hector Hinojosa and Teresa Hiteman for editing and formatting and Winters Red Star for map work. This work was conducted with the support of the Biodiversity Preservation Group.

7. REFERENCES

Bassey, A. E., and J. F. Oates, eds., 2001, *Proceedings of the International Workshop and Conference on the Conservation of Cross River Gorillas*, 6th–9th April, 2001, Calabar, CRS, Nigeria.

Brandon, K. E., and Wells, M., 1992, Planning for people and parks: Design dilemmas, *World Development* **20**:557–570.

Burkey, T. V., 1989, Extinction in nature reserves: The effect of fragmentation and the importance of migration between fragments, *Oikos* **55**:75–81.

Butynski, T. M., and Kalina, J., 1998, Gorilla tourism: A critical look, in: *Conservation of Biological Resources*, E. J. Millner, ed., Gulland and R. Mace, Oxford, Blackwells pp. 294–313.

Chapman, C. A., and Lambert, J. E., 2000, Habitat alteration and the conservation of African primates: A case study of Kibale National Park, Uganda, *Am. J. Primatol.* **50**:169–186.

Critchley, W. R., 1968, Final report on Takamanda gorilla survey, Unpublished report to Winston Churchill Memorial Trust, London.

Coolidge, H. J., Jr., 1929, A revision of the genus *Gorilla, Memoirs of the Museum of Comparative Zoology at Harvard College* **50**:291–381.

Cowlishaw, G., and Dunbar, R., 2000, *Primate Conservation Biology*, The University of Chicago Press, Chicago and London.

Cox, C., 1987, Social behaviour and reproductive status of drills (*Mandrillus leucophaeus*), *Regional proceedings of the American Association of Zoological Parks and Aquaria*, AAZPA, Wheeling WV, pp. 321–328.

Diamond, J. M., 1975, The island dilemma: Lessons of modern biogeographic studies for the design of natural preserves, *Biol. Cons.* **7**:129–146.

Diamond, J. M., 1984, "Normal" extinction of isolated populations, in: *Extinctions*, M. H. Nitecki, ed., University of Chicago Press, Chicago, pp. 191–246.

Dietz, J. M., Dietz, L. A., and Nagagata, E. Y., 1994, The effective use of flagship species for conservation: The example of the lion tamarin, in: *Creative Conservation*, P. J. S. Olney, G. M. Mace, and A. T. C. Feistner, eds., Chapman and Hall, London, pp. 32–49.

Dorst, J., and Dandelot, P., 1986, *Larger Mammals of Africa: A Field Guide*, Collins, London. pp. 51–83.

Eniang, E. A., 1998, *In situ* and *ex situ* conservation of drill (*Mandrillus leucophaeus* R) in the tropical rain forest of South Eastern Nigeria.

Eniang, E. A., 2001, The role of the Cross River National Park in gorilla conservation, *Gorilla J. No. 22*, Berggorilla and Regenwald Direkthilfe, Germany.

Eniang, E. A., and Nwufoh, E. I., 2001, Sites of special scientific interest for the conservation of primates in Nigeria, *Trans. Nig. Soc. Biol. Conserv* **7**:45–54.

FAO, 1995, Food for all, Fifty years of FAO and 15th World Food Day, Rome.

Groves, C., 2001, *Primate Taxonomy*, Smithsonian Institute Press, Washington and London.

Groves, C. P., 1970, Population systematics of the gorilla, *J. Zool.* **161**:287–300.

Groves, C. P., 1967, Ecology and taxonomy of the gorilla, *Nature* **213**:890–893.

Groves, J., and Maisels, F., 1999, Report on the large mammal fauna of the Takamanda Forest Reserve, South West Province, Cameroon, with special emphasis on the gorilla population, Unpublished report to World Wildlife Federation-Cameroon, Yaounde.

Hanski, I., 1998, Metapopulation dynamics, *Nature* **396**:41–49.

Hanski, I., 1999, *Metapopulation Ecology*, Oxford University Press, Oxford.

Happold, D. C. D., 1991, *Large Mammals of West Africa*, *West African Nature Handbooks*, Longman, pp. 73–78.

Harcourt, A. H., 1995, Population viability estimates: Theory and practice for a wild gorilla population, *Cons. Bio.* **9**:134–142.

Harcourt, A. H., and Fossey, D., 1981, The Virunga gorillas: Decline of an "Island" population, *Afr. J. Ecol.* **19**:83–97.

Harcourt, A. H., Stewart, K. J., and Inahoro, I. M., 1988, Nigeria's gorillas: A survey and recommendations, unpublished report.

Harcourt, A. H., Stewart, K. J., and Inahoro, I. M., 1989, Gorilla quest in Nigeria, *Orxy* **23**:7–13.

Koontz, F., Horwich, R., Glander, K., Westrom, W., Koontz, C., Saqui, E. Saqui, H., 1994, Reintroduction of black howler monkeys (*Acouattor pigra*) into the Cockscomb Basin Wildlife Sanctuary, Belize, in: *AZA Annual Conference Proceedings*, Bethesda, Maryland, pp. 104–111.

Konstant, W. R., and Mittermeier, R. A., 1982, Introduction, reintroduction, and translocation of Neotropical primates: Past experiences and future possibilities. *Int. Zoo. Yrbk* **22**:69–77.

Lacy, R. C., 1993, VORTEX: A computer simulation model for population viability analysis, *Wildlife Res.* **20**:45–65.

Lawlor, T. E., 1986, Comparative biogeography of mammals on islands, *Biol. J. Linn. Soc.* **28**:99–125.

Lande, R., Engen. S., and Saether, B. E., 1994, Optimal harvesting, economic discounting, and extinction risk in fluctuating populations, *Nature* **372**:88–90.

Lindenmeyer, D. B., and Nix, H. A., 1993, Ecological principles for the design of wildlife corridors, *Cons. Bio.* **7**:627–630.

Mace, G. M., 1988, The genetic and demographic status of western lowland gorilla (*Gorilla g. gorilla*) in captivity, *J. Zool.* **216**:629–654.

Mace, G. M., and Lande, R., 1991, Assessing extinction threats: Toward a reevaluation of IUCN threatened categories, *Cons. Biol.* **5**:148–157.

March, E. W., 1957, Gorillas of Eastern Nigeria, *Oryx* **4**:30–34.

Marsh, C. W., Johns, A. D., and Ayres, J. M., 1987, Effect of habitat disturbance on rainforest primates, in: *Primate Conservation in the Tropical Rain Forest*, C. W. Marsh and R. A. Mittermeier, eds., Alan R. Liss, New York, pp. 83–107.

McFarland, K., 1994, Update on gorillas in Cross River State, Nigeria, *Gorilla J.* **17**:14–15.

Mittermeier, R. A., 1988, Primate diversity and the tropical forest: Case studies from Brazil and Madagascar and the importance of the Megadiversity countries, in: *Biodiversity*, E. O. Wilson, ed., National Academy Press pp. 145–154.

Mittermeier, R. A., 1987, Effect of hunting on rainforests primates, in: *Primates Conservation in the Tropical Rainforest*, C. W. Marsh and R. A. Mittermeier, eds., Alan R. Liss, publisher, New York, pp. 305–220.

Murcia, C., 1995, Edge effects in fragmented forests: Implications for conservation, *TREE* **10**:58–62.

Nwufoh, E. I., 1999, Survey of gorilla in Boshi Extension portion of the Cross River National Park, Obudu, Nigeria, unpublished report submitted to the World Wildlife Federation-CRNP Project, Nigeria, and to WPTI, Pennsylvania.

Oates, J. F., 1998, The gorilla population in the Nigerian-Cameroon border, *Gorilla Conservation News* **12**:6.

Oates, J. F., White, D., Gadsby, E. L. and Bisong, P. O., 1990, Conservation of gorillas and other species, Appendix 1 to Caldecott, J. O., Oates, J. F., and Ruistenbee, H. J., Cross River National Park (Okwangwo Division): Plan for Developing the Park and Its Support Zone, Godalming, Surrey World Wildlife Federation, UK.

Oates, J. F., 1996, Habitat alteration, hunting, and the conservation of folivorous primates in African forests, *Aust. J. Ecol.* **21**:1–9.

Obot, E., Barker, J., Edet, C., Ogar, G., and Nwufoh, E., 1997, Status of gorilla (*G. gorilla gorilla*) populations in Cross River National Park and Mbe Mountain, Cross River National Park (Okwangwo Division) *Technical Report No. 3*, pp. 14.

Olajide, O., and Eniang, E. A., 1999, Conservation of diverse bioresources of Nigeria: Forest regeneration methods in perspective, in: *Proceedings of the 8th Annual Scientific Conference of Nigeria Society for Biological Conservation (NSBC)*, 13th–17th September, 1999, University of Uyo, Nigeria.

Olajide, O., and Eniang, E. A., 2000, Unguided forest resources exploitation and destruction in Nigeria: Socio-ecological impacts, *Int. J. Env. Devel.* **4(1)**:39–43.

Onochie, C. F. A., 1979, The Nigerian rainforest ecosystem: An overview, in: *The Nigerian Rainforest Ecosystem*, D. U. U. Okali, ed., Proceedings of Mab. UI Conference Centre, 24 – 26 January, 1979.

Pope, T. R., 1996, Socioecology, population fragmentation, and patterns of genetic loss in endangered primates, in: *Conservation Genetics: Case Histories from Nature*, J. C. Arise and J. L. Harwick, eds., Chapman and Hall, London, pp. 119–159.

Richards, P. W., 1981, *The Tropical Rain Forest: An Ecological Study*, Cambridge University Press, London.

Ross, C., 1988, The intrinsic rate of nature increase and reproductive effort in primates, *J. Zool.*, Lond. **214**:199–219.

Ross, C., 1992, Environmental correlates of the intrinsic rate of natural increase in primates, *Oecologia* **90**:383–390.

Rothschild, W., 1904, Notes on anthropoid apes, in: *Proceedings of the Zoological Society of London* **1,904(2)**:413–440.

Rothschild, W., 1908, Note on *Gorilla gorilla diehli* Matschie, *Zoologicae* **15**:391–392.

Sayer, J. A., Harcourt, C. S., and Collins, N. M., 1992, *The Conservation Atlas of Tropical Forests: Africa*, Macmillan, London.

Simberloff, D., and Cox, J., 1987, Consequences and cost of conservation corridors, *Cons. Biol.* **1**:63–71.

Smith, F. D. M., Mary, R. M., Pellew, R., Johnson, T. H., and Walter, K. S., 1993, How much do we know about the current extinction rate? *TREE* **18**:375–378.

Starin, E. D., 1989, Threats to the monkeys of Gambia, *Oryx* **23**:208–214.

Strum, S. C., and Southwick, C. H., 1986, Translocation of primates, in: *Primates: The Road to self–sustaining populations*, K. Benirschke, ed., Springer, New York, pp. 949–958.

Swart, J., Lawes, M. J., and Perrin, M. R., 1993, A mathematical model to investigate the demographic viability of low-density Samango monkey (*Cercopithecus mitis*) populations in Natal, S. Africa, *Ecol. Modelling* **70**:289–303.

Swart J., and Lawes, M. J., 1996, The effect of habitat patch connectivity on Samango monkey (*Cercopithecus mitis*) metapopulation persistence, *Ecol. Modelling* **93**:57–74.

Taylor, B., 1995 The reliability of using population viability analysis for risk classification of species, *Cons. Bio.* **9**:551–558.

Tilman, D., May, R. M, Lehman C. L., and Nowak, M. A., 1994, Habitat destruction and the extinction debt, *Nature* **371**:65–66.

Tudge, C., 1992, Last Animals at the zoo: How mass extinction can be stopped, Oxford University Press, pp. 42–48.

Tutin, C. E. G., 1999, Fragmented living: Behavioural ecology of primates in a forest fragment in the Lope Reserve, Gabon, *Primates* **40**:249–265.

Tutin, C. E. G., and Oslisly, R., 1995, *Homo, Pan, and Gorilla*: coexistence over 60,000 years at Lope in Central Gabon, *J. Hum Evol.* **28**:597–602.

White, F., 1983, The Vegetation of Africa: a Descriptive Memoir to accompany the UNESCO/AETFAT/UNSO Vegetation Map of Africa, Unesco, Paris, pp. 356.

Wolfheim, J. H., 1983, *Primates of the World: Distribution, Abundance, and Conservation*, University of Washington Press, Seattle.

World Bank, 2002, *World Debt Tables*, Washington, DC.

WILD ZOOS: CONSERVATION OF PRIMATES
IN SITU

Laura K. Marsh[*]

1. INTRODUCTION

For the last 20 years, conservation biology has been concerned with populations that live in discontinuous habitat. As deforestation and fragmentation continue at an alarming rate, we must develop new means of protecting remaining isolated populations. For managers, knowing the natural history details of all organisms within a fragment will allow for anticipation of interactions and will better predict potential collapse of the system (Janzen, 1986a). The ecological role of primates in a fragmented forest is critical for many reasons. First, primates have a number of beneficial effects on the forest as seed dispersers, pollinators, and "shapers" of their habitat (Janzen, 1970; Dirzo and Miranda, 1990a; Murcia, 1996; Garber and Lambert, 1998). Second, primates tend to be large arboreal frugivores and account for a high proportion of the biomass in a forest remnant (Chapman, 1995). Primates may be the top of the food chain in many fragments where carnivores have vanished (Marsh, 1999; Terborgh et al., 2001). And finally, primates tend to be important educational and conservation icons for local people (Savage et al., 1997; Weber, 1995; Dietz and Nagagata, 1995).

Primates are important seed dispersers of tropical trees, and thus likely influence the distribution and composition of trees in fragmented forests (e.g., Chapman, 1989, 1995; Coates-Estrada and Estrada, 1986; McKey, 1975; Terborgh, 1986; Gautier-Hion et al., 1985; Leighton and Leighton, 1983; Janson and Emmons, 1990). For example, Foster (1990) hypothesized that if the high frequency of fruits dispersed by mammals and the great densities of primates and other mammals in the Manu floodplain, Peru (Terborgh, 1983; Janson and Emmons, 1990) are related, then "the animals [are] helping to maintain high densities of these plant species through dispersal, and [vice versa] the availability of these fruits contributes to the high densities of primates and other mammals."

The effect of mammals on plant communities can be rapid and dramatic (Dirzo and Miranda, 1990a, b; Estrada and Coates-Estrada, 1986; Redford, 1992). At sites where terrestrial seed predators are no longer present there is a greater abundance of seedlings.

---

[*] Los Alamos National Laboratory Ecology Group (RRES-ECO) Mail Stop M887 Los Alamos, New Mexico 87545, USA. Email: lkmarsh@lanl.gov.

Hence, tree seedling composition is dramatically different in those forests (Dirzo and Miranda, 1990a). Similarly, when a major arboreal seed disperser is absent, as is the case with the loss of *Ateles* in Los Tuxtlas, the spatial distribution of surviving seedlings is altered (Estrada and Coates-Estrada, 1986). Chapman and Chapman (1995) estimated, based on presence or absence of seedlings and saplings under adults, that 60% of the 25 tree species sampled could potentially be lost if all frugivores disappeared or went locally extinct.

Small (4 to 10 ha) isolated islands in Lake Gatun were compared to Barro Colorado Island (160 ha) and the mainland forests of Panamá (Putz et al., 1990; Leigh et al., 1990). Results indicate that the absence of mammalian seed dispersers and seed predators reduced tree diversity. Decreased diversity and increased similarity between small islands is in part due to the success of the remaining large-seeded adult fruiting trees. Without mammalian dispersers, these trees out-compete smaller seeded trees for space, since large seeds tend to germinate with higher success than small seeds and may enhance recruitment probabilities (Janson, 1983; Janzen, 1986b).

Since primates are important to the maintenance of forests, their removal or isolation can have extreme consequences (Marsh and Loiselle, in press). Thus, management efforts to retain primates in the wild are a priority. Debate occurs when primate populations reach critically low numbers. Should these populations be brought in for captive breeding or should they be managed *in situ*? One of the goals of this chapter is to demonstrate the need to better analyze each fragment that houses primate species as potentially important to the population as a whole, even if the fragments themselves may not survive. Central to the discussion of primates in fragments is whether or not all subpopulations are "worth" saving in the context of the overall metapopulation. This chapter argues in support of leaving remaining populations in the wild and explores previous methods of managing primates *in situ* with suggestions for further consideration.

2. CURRENT PRIMATE MANAGEMENT STRATEGIES

Management-in-practice for the conservation of primates *in situ* has taken primarily three forms: preservation (protected area management and establishment, tourism, and community conservation), reintroduction (including captive breeding), and translocation.

2.1. Preservation

Preservation of wild primates has been approached from a large scale with the protection or establishment of parks and large tracks of land (Chapman and Peres, 2001). Strategies to include fragments primarily seek to protect those which are larger than 100 ha (Lovejoy et al., 1986). Primatologists over the last three decades have typically focused on the behavioral ecology of single species within protected areas (Chapman and Peres, 2001). Coupled with metapopulation theory, smaller isolated fragments that house primates, at least for the present, may be overlooked as conservation concerns. Saving the largest, most charismatic species in the fragment is often believed to ensure the protection of all others. This is not always the case, particularly for fragments. Protection is integral to the overall management scheme for primates in fragments, but it must be applied in tandem with other strategies.

Ecotourism, or nature tourism, in many countries has become fashionable as part of the primate preservation package as a means for local people to make conservation dollars while preserving species (Horwich, 1990; Whelan, 1991; Western, 1994). Nature tourism has a presumed benefit to local communities. Many primatologists call for it to aid conservation needs, even without knowledge of whether or not the benefits would be real for their area (e.g., Eniang, this volume). Ecotours promote learning and the wonder of nature, but in some cases it does not translate to the public (Marsh, in press). Like all things in conservation, there have been great successes and great difficulties in preserving a region through nature tourism revenue. Some of the difficulties include tourist safety concerns, damage to the habitat intended for preservation, changing patterns of wildlife through the area, increase in buildings and infrastructure closer to wild areas, and a lack of funds reaching the hands of local people, such that deforestation and internal fragmentation may continue even within the confines of the reserve (Norris et al., 1998, Ceballos-Lascurian, 1996).

Community conservation may have an ecotourism component, but there are many more facets and greater overall potential for long-term success if tourism is left out (Poffenberger, 1994; Bodmer, 1994; Marsh, 2001). Some projects are driven by communities themselves so they can preserve and use the forest for the future. One example of this is the sustainable ecosystems models that many *ejidos* in Mexico use (Marroquin, 1998). Community conservation work may be one of the best ways to preserve wildlife and wild places in the long term (Marsh, 2001).

2.2. Reintroduction

Captive breeding has changed from the idea of being primarily an in-house gene bank system through the frozen zoo concept, to increasing the possibility of population enhancement in the wild by restocking individuals into areas of local extinction with reintroductions (Weise and Hutchins, 1997). The practice of reintroduction can take two primary forms. First, captive bred animals can be reared and trained for release to the wild (e.g., golden lion tamarins, Stoinski et al., 1997) or where species who have reached a critically low number are pulled from the wild, placed in captivity to reproduce, and ultimately returned to suitable habitat (e.g., Rodriguez-Luna, this volume). Both of these methods have been met with mixed results. In particular, captive bred, reintroduced animals may suffer from serious problems. Reintroductions tend to be costly in terms of funding as well as effort, the species are prone to disease outbreaks, there is a chance of domestication during lengthy programs, there is a lack of administrative continuity, and, in general, there has been an overall poor success rate for reintroduction programs (Snyder et al., 1996, but see Hutchins et al., 1997 for rebuttal).

Species including chimps, bonobos, orangutans, golden lion tamarins, black and white ruffed lemurs, lion tailed macaques, olive baboons, mantled howler monkeys, and spider monkeys have all had reintroduction failures and successes (Cowlishaw and Dunbar, 2001). Those that have been successful rely on protected areas with relatively intact habitat. However, less than 5% of tropical forests worldwide are legally protected from human exploitation, and in many countries the amount of protected area is far less (Chapman and Peres, 2001). Numerous primate species have also been reintroduced into varying degrees of "wildness," where some species were released to controlled free-ranging sites, such as St. Catherine's Island (Lindburg et al., 1997). While these species are experiencing a more natural situation than captivity, they serve best as test subjects for understanding reintroductions for fragmented habitats (Rodriguez-Luna, this volume). In all cases, very few individuals are released at a time, making reintroduction difficult if not impossible on a large scale (Koontz, 1997).

2.3. Translocation

The International Union for the Conservation of Nature and Natural Resources (IUCN) defined translocation as the "deliberate and mediated movement of wild individuals or populations from one part of their range to another" (IUCN/SSG, 1995). The concepts behind translocations are positive for fragments. In a metapopulation where a number of individuals are unable to disperse within a landscape, either removal of selected groups or the "rearranging" of their locations would ensure gene flow. Translocations can be expensive, but typically less so than reintroductions (Koontz, 1997). Additionally, translocations are great opportunities for local people to be involved in every aspect, including public outreach (Koontz, 1993).

However, there are risks to translocation, such as loss of individuals through death at the time of transfer, disease or parasite transmission to local resident populations, lack of viability once established in a new location, or endangered or endemic species unable to adapt to their new surroundings (Silver, 1997; Struhsaker and Siex, 1998). There have been relatively few intentional translocations of primates (Silver, 1997). Likewise, there has been little follow-up research on the behavioral responses of translocation, therefore, it is difficult to assess its effectiveness as a conservation tool (Strum and Southwick, 1986; Silver, 1997). Given the potential benefits of translocation, this method may be used increasingly in the future to better preserve wild populations (Silver and Marsh, this volume).

3. VALUE OF SMALL FRAGMENTS (<100 HA)

The challenge of fragments is that we do not have the opportunity to create reserves from the start. If we do have the opportunity, suggestions exist for how to best create them (Laurance and Gascon, 1997). Due to the high costs of creating and managing large preserves, the actual increase in total biodiversity gained in establishing large reserves may not represent the best return on the investment of conservation dollars (Schwartz and van Mantgem, 1997).

It goes without saying in terms of maintaining the most area possible for primates, the largest intact tracts of forest must be preserved (Mittermeier, 1991). If we carry this logic to fragments, the "larger is better" strategy is often at the expense of thousands of small, isolated populations. Spending all of our conservation efforts on only the large fragments works only until that habitat is threatened. In Figure 1, there is a hypothetical situation where the large fragment is selected for preservation and management out of an already fragmented landscape. If this area is near villages (which is typically why the area was fragmented initially), there are no guarantees even in land-owner agreements that the area will always be maintained for conservation (Struhsaker, 1997; Marsh, unpublished data). Over time the matrix has matured (depending on the nature of fragmentation) and the originally selected site has less total area than all other smaller sites combined (Figure 1b). Investing in the landscape with knowledge of the species load residing in even the smallest fragments may prove to be a worthwhile strategy for long-term conservation planning.

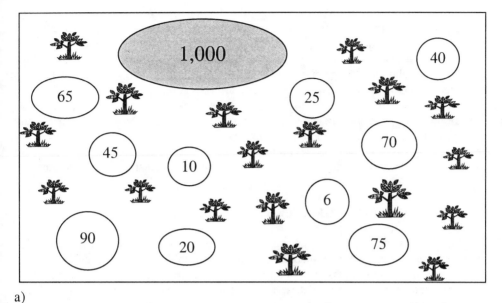

a)

b)

Figure 1. Schematic diagram of a fragmentation scenario. All numbers are hectares. The habitat has been previously fragmented and the largest fragment (1,000 ha) selected for preservation (a). Local villages continue to use the largest fragment, until over time, it contains less suitable habitat than all of the other small, undisturbed fragments combined (b). Illustration by Rhonda Robinson and Julie Hill.

The abundance of small fragments throughout the world provides a unique opportunity to experiment with varying management strategies for differing conservation objectives. Several factors make small fragments valuable: 1) they preserve biodiversity. Endangered primates use and are permanent residents in small fragments (Section II, this volume), 2) subdivided populations may buffer species from extinction (Schelhas and Greenberg, 1996), 3) they provide a refuge for migrating species (Bierregaard et al. 2001), 4) because they are small, they may be more cost effective to manage singly or as part of a landscape (Schwartz and van Mantgem, 1997), and 5) they provide flexible and experimental reserves to better refine conservation management techniques. A key for placing value on a fragment is creating a system of ranking. Laurance et al. (1997) and Marsh (1999) discuss possible ways to accomplish this (c.f., Marsh et al., this volume).

4. MANAGEMENT OF PRIMATES IN FRAGMENTS

In this volume we have established that there are aspects of primates living in fragments that determine their success: primarily their ability to navigate the matrix and the degree of behavioral plasticity in terms of home range size and diet. Although there have been attempts at *a priori* determinations for primate species that will do well in a given fragment, we have not yet established clear management guidelines. Thus, suggestions given for the management of fragments implies some study has been done to determine the conditions within the fragments and their need for active conservation management. Those species that are reluctant to use the matrix or who have few food resources (and are effectively "trapped") are particularly good candidates for active management. It should be remembered that the goal in primate fragment management is to consider the function of the entire ecosystem within and surrounding the fragment. This includes knowledge of other species use or needs (including birds, volant and non-volant mammals, insects, reptiles, amphibians, fish, and plants).

Ultimately, the final product may be very different than what is intended. For instance, a management strategy may be to restore and reconnect the fragment system with surrounding forest (Viana et al., 1997). Another may be to improve the buffer zone and increase connectivity between isolates (Lamb et al., 1997). At the end of the project even if the bigger goals are not met if any improvements are made to a disturbed habitat, it is well worth it. In all cases, without exception, land owner consent and/or aid is a must. Cooperative work with communities will ensure the long-term success of any management plans that are put in to place.

4.1. General

Fragments are important to consider in terms of potentially cascading extinction effects within the surrounding landscape. "Reserve mentality" has been the lull of many primatologists (Brussard et al., 1992). Politicians and the public (and many scientists) believe that once reserves are established, the problem is solved and the lands outside the reserve can be subjected to any sort of use, even if suitable habitat and primates exist outside of the reserve within fragments (Weins, 1996). The difference then in preservation and management of fragments is the active maintenance they potentially need. Not every fragment needs as much maintenance or restoration, so understanding each situation and how the mosaic of matrix and fragments interact is imperative.

There is a growing list of essential components that should be considered before proceeding with the management of a particular species. All of the authors in Marsh et al. of this volume echo many of these growing standards for long-term success in conservation projects. First, the standard package of research and surveys should remain in place, not only to determine management needs but to monitor the populations as time goes on. Second, creating protected areas and working to ensure areas already established are patrolled and wildlife laws are enforced are still critical and missing in most cases (Eniang, this volume). As was stated previously, many of our sites fall outside of the jurisdiction of protection, therefore getting these equally important areas protection status helps when the need for local support comes into play. These components of conservation for primates in fragments are the same for conservation of any species anywhere. The more difficult extension of protected areas is working with local and national governments to improve legislation for the species or habitat in question. Most countries regard scientists as experts, so opinion is not wasted when spoken on behalf of a threatened or endangered primate population. It has become more the job of every researcher to speak up politically and socially within the countries where we work, and this is especially true of nationals working within source countries although there may be consequences for these actions. Finally, probably the most important part of conservation is the support, awareness, and education of the local communities. Although education in and of itself can be a very long and sometimes ineffective prospect (Struhsaker, 1997), there is no doubt that involving communities in projects on their land or in their region is mandatory for long-term success (Marsh, 2001). Conservation has very little to do with species and habitats, and it has everything to do with humans' use and control of fragile areas.

4.2. Wild Zoos

The "wild zoo" concept is one that in no way implies enclosing fragments for the ease of management (Marsh, 2002). It is a metaphor for the intensity of management necessary for fragments harboring primates. There are two main reasons for this concept: 1) it calls attention to the need for active management of fragments as though the animals contained within were part of a captive population, because they may need more direct and active care than establishing them as protected at a local or government level, and 2) it is a way for scientists and conservation organizations like zoological parks interested in supporting *in situ* efforts while defraying costs that are associated with reintroductions and some captive breeding programs. This concept is intended to work along side established conservation protocols and draws heavily from previously discussed techniques (Conway, 1989; Laurance and Bierregaard, 1997; Cowlishaw and Dunbar, 2001). It is the application of a comprehensive action plan within a fragmented landscape that makes it worth summarizing here.

4.2.1. The Primates

1. *Do not manage for only one species.* Different than in a zoo, where each individual species is monitored as a global captive population through studbooks and the Species Survival Plan, a species within a fragment has an ecosystem to depend upon. An example of this is in the Community Baboon Sanctuary (CBS), Belize, the intent for establishing the Sanctuary was to preserve *Alouatta pigra* (Horwich, 1986;

Horwich and Lyon, 1987). Local communities were involved, and they all agreed to protect the monkey on their land. Over 15 years later, all that is left in terms of wildlife is primarily howlers. No restrictions on hunting, clearing, gathering, farming, or logging were made within the Sanctuary boundaries. People continued to use the forest intensively (Marsh, 1999). Although the communities agreed to not cut down large trees that the monkeys needed, the understory was cleared so livestock could have easy access to water and shade. The communities believed they were participating in conservation because tourists came to see the monkeys, regardless that there was disturbed recruitment and replacement of howler food trees throughout the CBS (Marsh, 1999). Without considering the entire fragment as a conservation concern, isolating and managing for one particular species may result in the creation of relic forests (Redford, 1992).

2. *Food Provisioning*. Provisioning primates in an isolated habitat may be particularly important. Provisioning would ensure the health of certain particularly threatened populations, and as provisioning studies have shown, would increase local populations in terms of numbers and viability (Fa 1988, Loy, 1988). Provisioning may be appropriate if populations cannot be translocated or if other methods are being established to increase available food tree species, such as vegetation control or buffer zone, corridor or matrix restoration (see 4.2.2). Some populations may increase disproportionate to the fragment area, in which case, translocation may be the next viable option.

3. *Translocation*. Discussed earlier in this chapter and in Silver and Marsh, this volume, translocation may be necessary for managing the metapopulation for the long term. There may be several possible strategies for translocations and fragments. The first is moving an entire troop from a fragment to a continuous area (Silver, 1997). Second is moving a troop from a small fragment to a larger one. This can be done if the fragment has a similar composition of food tree species or if some of the same primate species already lives within it. Third is moving individuals of a group to another fragment (of any size) for breeding purposes. A studbook or other similar record of breeding should be kept for the metapopulation. The fourth is moving individuals from a continuous forest population into fragments. This again is management for genetic diversity. Finally, the fifth is moving groups from continuous forest into fragments. There may rarely be a need for this, but in the case of a continuous forest suffering human encroachment from the edges or from internal fragmentation (see Introduction, this volume), species who defend territories might have a healthier overall population if the numbers of groups were reduced in the continuous forest. The assumption is the fragments are suitable for the primates in question.

4. *Artificial Insemination (AI)*. In fragmented forests this more intensive step would need to be assessed. There is inherent difficulty in both captive and wild situations to determine ovulation (Gould and Martin, 1986). If this measure was taken, a holding area or field cage might need to be constructed for the period of time when ovulation might be occurring. This enclosure would be very near if not within the fragment where the individuals were captured. AI could be performed *ex situ* (in the field cage) and the individual returned once inseminated. This may also provide an opportunity for zoos to obtain sperm samples for cryopreservation. It is not recommended, however, that sperm from captive animals be used with wild individuals for fear of diluting the wild heterozygotes.

5. *Adopt a Fragment.* The conservation of fragments may be particularly interesting to small zoos who wish to participate in *in situ* primate conservation, but do not have the resources that large zoos do. By adopting projects that focus on primates in fragments, organizations can fund valuable projects and active conservation management on a smaller scale.

4.2.2. The Habitat

In general, there are a few "hands on" management needs for every fragment under conservation consideration, but particularly for those with primates or that support primates (e.g., are a resource for food trees but are not part of the inhabited fragment). This is not an exhaustive list, but merely a call to think about management beyond formal agreements and research. Some may be more expensive than others to do, however, working out the feasibility with the local community may offset some of the expense, particularly if it mutually benefits the humans (e.g., planting fruit trees in the buffer zone). If possible, training local people to manage their fragments for the long term is a priority.

1. *Active vegetation control.* In the case of primate residents, there may need to be active selection and culling of unwanted native species from within the fragment as they invade from the matrix. These species take up space where food tree species, important to the primates, could be growing. Planting fruiting trees and other species that were present in the original forest that are necessary for the primates' diet could be cultivated and reintroduced in place of secondary invaders.
2. *Control exotics and nuisance species.* This includes plants and animals (such as weeds, insects, or avian nest parasites) that may invade after isolation (Luken, 1997). Exotics and some unwanted second-growth species can quickly dominate in a fragment. In terms of primate needs, any imbalance in pollinators, dispersers, or food resources throws their ability to remain in a fragment into question.
3. *Become an expert in edge effects.* Fragments are susceptible to damage from external threats due to edge effects (Janzen, 1986a; Kapos et al., 1997). In small fragments, the entire interior may be subject to edge effect damage (Laurance, 1991). Knowing the conditions for the fragments in question will greatly determine the next steps in maintenance: matrix, buffer zones, and/or corridors.
4. *Manage the matrix.* This is a larger view of 5 and 6 where, if possible, the type of surrounding vegetation can be managed in a way that benefits the primates within the fragments. If it cannot be managed (e.g., selective vegetation removal or planting, removal of cattle or other livestock, relocating agriculture, or creating safe zones for primates around crops), then creating a buffer zone, however small around the fragment, will greatly enhance the site (see 5 below).
5. *Active buffer zone creation and support.* This is different than managing the matrix in that buffer zones can be used by people in a light-use or no-use manner. A well-maintained buffer, even if it is used by humans, benefits the inhabitants. There are various invasions and conditions that change according to the amount of edge exposed, and measurements have been made to determine, in part, the depth of the buffer to prevent invasions within the body of the fragment (Laurance, 1991; Laurance et al., 1997). Edge effects that can be 'absorbed' by a buffer allows for the maturing of core habitat within the primary fragment.

6. *Create corridors*. Although we have evidence that some species will not use corridors even if they are provided (J. Wallis, personal communication), this is an extension of managing the matrix. Connectivity between fragments through forest corridors, however, has shown to be an important factor governing persistence of some mammals in fragmented forest ecosystems (Laurance, 1990; Bierregaard et al., 1992; Arnold et al., 1993; Kozakiewicz, 1993; Hill, 1995; Downes et al., 1997) and for some species of primates (Mamede-Costa and Gobi, 1998). The goal is to create as many connections as possible, such as a) fragment to fragment(s), b) fragment to continuous forest, or c) a combination of the two. The ultimate plan (which may be difficult to achieve in most regions due to topography or sociopolitical barriers) is to establish a pattern of regrowth that may approximate the return of the continuous forest block (Crome, 1997). Many researchers favor this plan as a course of action, although it is a daunting one to approach in most cases (Eniang, this volume). When corridors are difficult or are not cost effective, human-made links, such as rope bridges, can be effective (Horwich, 1998).

7. *Maintain species as a metapopulation*. Although this has been discussed and supported (Section I, this volume), few are taking the action on the ground. One study on the black lion tamarin in Brazil is a very good example of managing the metapopulation (C. Padua, personal communication; Mamede-Costa and Gobi, 1998). The goal is to maintain genetic diversity while leaving the tamarins *in situ*. Some of the primates have been moved to different fragments to increase gene flow (translocation). Those tamarins that have not been moved have received support through development of corridors, human-made links, and in-fragment nest boxes (Mamede-Costa and Gobi, 1998). Fragmented populations of endangered primates should be looked at as a mosaic, but all within the same conservation priority as long as the natural history parameters have been determined for each species in question (Chapman et al., this volume).

Maintaining, monitoring, and managing primates in fragments will take a long-term commitment to the fragments, the metapopulation, and the human communities in the region. Continued work with in-country non-governmental organization, governments, zoos (if they exist), and protected area officials will be essential to the success of a wild zoo program. It is an opportunity to collaborate with many professionals from within source countries as well as with experts in complementary fields, such as nutritionists, educators, horticulturists, zoo professionals, fragment scientists, primatologists, range and landscape managers, and botanical garden staff. If there are many interest groups involved, the costs of starting a landscape-level management program for primates in fragments becomes feasible. Fragment management for primates is not intended to provide species for captive breeding programs. The challenge is to maintain populations in suitable habitat and to practice techniques that may better equip us for the future. McCullough (1996) states: "Knowledge of metapopulation dynamics may allow us to artificially perform the function of dispersal and recolonization of locally extinct [populations]. In this way, we may manage to maintain a metapopulation in the wild and avoid the last resort: establishment of captive breeding colonies." Our commitment to primates means we understand the dire global circumstances they face, and we are willing to step up, roll up our sleeves, and dig in for the long haul.

5. SUMMARY

Primates are important to the maintenance of tropical ecosystems because they are seed dispersers, pollinators, and can have a positive effect on the forests they live in. But forested primate habitat is disappearing at an alarming rate. Because most primates are threatened or endangered and are living at risk in fragments and increasingly disturbed intact habitat, strategies must be in place for their long-term conservation. Traditionally, strategies for the conservation of primates in the wild have been preservation, reintroduction, and translocation. If small fragments are to be valued along with large ones, and metapopulations are the level of conservation concern, then there may be additional methods to consider. In this chapter I proposed the wild zoos methods, where methods used for various situations are brought together specifically for primates in fragments. These are managing for more than one species (other than primates), food provisioning, translocation, artificial insemination, active vegetation control, control of exotics and nuisance species, managing the matrix, edge effect expertise, creation of a buffer zone and/or corridors, and maintaining primate species in fragments as metapopulations. The threats to primates worldwide are serious and continuing. If we can make an active commitment to managing for primates in fragments, we may have a better chance of ensuring future generations of primates throughout the tropics.

6. ACKNOWLEDGMENTS

Thanks again to Hector Hinojosa and Teresa Hiteman for their commitment to this volume and my chapters in particular. Many thanks to the late Al Gentry for sending me down the fragment path in the first place. Jeanne Fair, Scott Silver, and Tim Haarmann provided thoughtful comments to this manuscript. Finally, thanks again to all of the participants at the XVIII IPS Congress in Adelaide. I appreciate your support and thoughts for this work.

7. REFERENCES

Arnold, G. W., Steven, D. E., Weeldenburg, J. R., and Smith, E. A., 1993, Influences of remnant size, spacing pattern, and connectivity on population boundaries and demography in *Euros macropus robustus* living in a fragmented landscape, *Biol. Cons.* **64**:219–230.

Bierregaard, R. O., Jr., Lovejoy, T. E., Kapos, V., Aidos-Santos, A., and Hutchings, R. W., 1992, The biological dynamics of tropical rainforest fragments, *BioScience* **42**:859–866.

Bierregaard, R. O., Jr., Gascon, C., Lovejoy, T., and Mesquita, R., 2001, *Lessons from Amazonia: The Ecology and Conservation of a Fragmented Forest*, Yale University Press, New Haven, pp. 478.

Bodmer, R. E., 1994, Managing wildlife with local communities in the Peruvian Amazon: The case of the Reserva Comunal Tamshiyacu-Tahuayo, in: *Natural Connections, Perspective in Community-based Conservation*, D. Western and R. M. Wright, eds., Island Press, Washington, D.C., USA, pp 113–134.

Brussard, P. F., Murphy, D. D., and Noss, R. F., 1992, Strategy and tactics for censoring biological diversity in the United States, *Cons. Bio.* **6**:157–159.

Ceballos-Lascurian, H., 1996, *Tourism, Eco-Tourism and Protected Areas*, World Conservation Union (IUCN).

Chapman, C. A., 1989, Primate seed dispersal: The fate of dispersed seeds, *Biotropica* **21(2)**:148–154.

Chapman, C. A., 1995, Primate seed dispersal: Coevolution and conservation implications, *Evol. Anthro.* **4(3)**:74–82.

Chapman, C. A., and Chapman, L. J., 1995, Survival without dispersers? Seedling recruitment under parents, *Cons. Bio.* **9**:675–678.

Chapman, C. A., and Peres, C. A., 2001, Primate conservation in the new millennium: The role of scientists, *Evol. Anthro.* **10**:16–33.

Coates-Estrada, R., and Estrada, A., 1986, Fruiting and frugivores at a strangler fig in the tropical rain forest of Los Tuxtlas, Mexico, *J. Trop. Ecol.* **2**:349–357.

Conway, W. G., 1989, The prospects for sustaining species and their evolution, in: *Conservation for the Twenty-First Century*, D. Western and M. Pearl, eds., Oxford University Press, New York, pp. 199–209.

Cowlishaw, G., and Dunbar, R., 2001, *Primate Conservation Biology*, Chicago University Press, Chicago, pp. 498.

Crome, F. H. J., 1997, Researching tropical forest fragmentation: Shall we keep on doing what we are doing? in: *Tropical Forest Remnants. Ecology, Management, and Conservation of Fragmented Communities*, W. Laurance and R. O. Bierregaard, eds., University of Chicago Press, Chicago, IL, USA, pp. 485–501.

Dietz, L. A. H., and Nagagata, E. Y., 1995, Golden Lion Tamarin Conservation Program: A community educational effort for forest conservation in Rio de Janeiro State, Brazil, in: *Conserving Wildlife: International Education and Communication Approaches*, S. Jacobson, ed., University of Columbia Press, New York, pp. 64–86.

Dirzo, R., and Miranda, A., 1990a, Altered patterns of herbivory and diversity in the forest understory: A case of the possible consequences of contemporary defaunation, in: *Herbivory: Tropical and Temperate Perspectives*, P. W. Price, T. W. Lewinson, W. W. Benson, and G. W. Fernandes, eds., Wiley, New York.

Dirzo, R., and Miranda, A., 1990b, Contemporary Neotropical defaunation and forest structure, function, and diversity—A sequel to John Terborgh, *Cons. Bio.* **4(4)**:444–447.

Downes, S. J., Handasyde, K. A., and Elgar, M., 1997, The use of corridors by mammals in fragmented Australian eucalypt forests, *Cons. Bio.* **11**:718–726.

Estrada, A., and Coates-Estrada, R, 1986, Frugivory by howling monkeys (*Alouatta palliata*) at Los Tuxtlas, Mexico: Dispersal and fate of seeds, in: *Frugivores and Seed Dispersal*, A. Estrada and T. H. Fleming, eds., W. Junk, publisher, Dordrecht, pp. 93–105.

Fa, J. E., 1988, Supplemental food as an extranormal stimulus in Barbary macaques (*Macaca sylvanus*) at Gibraltar: Its impact on activity budgets, in: *Ecology and Behavior of Food-Enhanced Primate Groups*, J. E. Fa and C. H. Southwick, eds., Alan R. Liss, publisher, New York, pp. 53–78.

Foster, R. B., 1990, The floristic composition of the Rio Manu flood plain forest, in: *Four Neotropical Rainforests*, A. H. Gentry, ed., Yale University Press, New Haven, pp. 99–111.

Garber, P. A., and Lambert, J. E., eds., 1998, *Primate Seed Dispersal*, Special edition of *Am. J. Primatol.*, **45(1)**:1–141.

Gautier-Hion, A., Duplantier, J. M., Quris, R., Feer, F., Sourd, C., Decoux, J. P., Dubost, G., Emmons, L., Erard, C., Hecketsweiler, P., Moungazi, A., Roussilhon, C., and Thiollay, J. M., 1985, Fruit characters as a basis of fruit choice and seed dispersal in a tropical forest vertebrate community, *Oecologia* (Berlin) **65**:324–337.

Gould, K. G., and Martin, D. E., 1986, Artificial insemination of nonhuman primates, in: *Primates: The Road to Self-Sustaining Populations*, K. Benirschke, ed., Springer-Verlag, New York, pp. 425–444.

Hill, C. J., 1995, Linear strips of rain forest vegetation as potential dispersal corridors for rain forest insects, *Cons. Bio.* **9**:1,559–1,566.

Horwich, R. H., 1986, A Community Baboon Sanctuary in Belize, *Prim. Cons.* **7**:15.

Horwich, R. H., 1990, How to develop a community sanctuary: An experimental approach to the conservation of private lands, *Oryx* **24**:95–102.

Horwich, R. H., 1998, Effective solutions for howler conservation, *Int. J. Primatol.* **19**:579–598.

Horwich, R. H., and Lyon, J., 1987, Development of the Community Baboon Sanctuary in Belize: An experiment in grass roots conservation, *Prim. Cons.* **8**:32–34.

Hutchins, M., Willis, K., and Wiese, R., 1997, Captive breeding and conservation, *Cons. Bio.* **11(1)**:3.

IUCN/SSG, 1995, Draft guidelines for reintroductions, Species Survival Commissions Reintroduction Specialist Group, IUCN-World Conservation Union, Gland, Switzerland.

Janson, C. H., 1983, Adaptation of fruit morphology to dispersal agents in a Neotropical forest, *Science* **219**:187–88.

Janson, C. H., and Emmons, L. H., 1990, Ecological structure of the non-flying mammal community at Cocha Cashu Biological Station, Manu National Park, Peru, in: *Four Neotropical Rainforests*, A. H. Gentry, ed., Yale University Press, New Haven, pp. 314–38.

Janzen, D. H., 1970, Herbivores and the number of tree species in tropical forests, *Amer. Nat.* **104(940)**:501–28.

Janzen, D. H., ed., 1983, *Costa Rican Natural History*, University of Chicago Press, Chicago, IL.

Janzen, D. H., 1986a, The eternal external effect, in: *Conservation Biology: The Science of Scarcity and Diversity*, M. E. Soulé, ed., Sinauer Assoc., Sunderland, MA, pp. 286–303.

Janzen, D. H., 1986b, Mice, big animals, and seeds: It matters who defecates what where, in: *Frugivores and Seed Dispersal*, A. Estrada and T. H. Fleming, eds., Dordrecht, W. Junk, pp. 251–271.

Kapos, V., Wandelli, E., Camargo, J. L., and Ganade, G., 1997, Edge-related changes in environment and plant responses due to forest fragmentation in central Amazonia, in: *Tropical Forest Remnants: Ecology, Management, and Conservation of Fragmented Communities*, W. F. Laurance and R. O. Bierregaard, eds., University of Chicago Press, Chicago, pp. 33–44.

Koonz, F. W., 1993, Trading places, *Wildlife Conservation* **5/6**:53–59.

Koontz, F., 1997, Zoos and in-situ primate conservation, in: *Primate Conservation: The Role of Zoological Parks. Special Topics in Primatology, Vol. 1*, J. Wallis, ed., American Society of Primatologists, pp. 63–82.

Kozakiewicz, M., 1993, Habitat isolation and ecological barriers—The effect on small mammal populations and communities, *Acta Theologica* **38**:1–30.

Lamb, D., Parrotta, J., Keenan, R., and Tucker, N., 1997, Rejoining habitat remnants: Restoring degraded rainforest lands, in: *Tropical Forest Remnants. Ecology, Management, and Conservation of Fragmented Communities*, W. Laurance and R.O. Bierregaard, eds., University of Chicago Press, Chicago, IL, USA, pp 366–385.

Laurance, W. F., 1990, Comparative responses of five arboreal marsupials to tropical forest fragmentation, *J Mammal.* **71**:641–653.

Laurance, W. F., 1991, Edge effects in tropical forest fragments: Application of a model for the design of nature reserves, *Biol. Cons.* **57**:205–215.

Laurance, W. F., and Gascon, C., 1997, How to creatively fragment a landscape, *Cons. Bio.* **11(2)**:577–579.

Laurance, W. F., and Bierregaard, R. O., eds., 1997, *Tropical Forest Remnants: Ecology, Management, and Conservation of Fragmented Communities*, University of Chicago Press, Chicago, pp. 616.

Laurance, W. F., Bierregaard, R. O., Gascon, C., Didham, R. K., Smith, A. P., Lynam, A. J., Viana, V. M., Lovejoy, T. E., Sieving, K. E., Sites, J. W., Anderson, M., Tocher, M. D., Kramer, E. A., Restrepo, C., and Moritz, C., 1997, Tropical forest fragmentation: Synthesis of a diverse and dynamics discipline, in: *Tropical Forest Remnants. Ecology, Management, and Conservation of Fragmented Communities*, W. Laurance and R.O. Bierregaard, eds., University of Chicago Press, Chicago, IL, USA, pp 502–514.

Leigh, E. G., Jr., Wright, S. J., Herre, E. A., and Putz, F. E., 1993, The decline of tree diversity on newly isolated tropical islands: A test of a null hypothesis and some implications, *Evol. Ecol.* **7(3)**:327.

Leighton, M., and Leighton, D. R., 1983, Vertebrate responses to fruiting seasonality within a Bornean rain forest, in: *Tropical Rain Forest: Ecology and Management*, S. L. Sutton, T. C. Whitmore, and A. C. Chadwick, eds., Blackwell Science Publications, Oxford, pp. 181–196.

Lindburg, D. G., Iaderosa, J., and Gledhill, L., 1997, Steady-state propagation of captive lion tailed macaques in North American zoos: A conservation strategy, in: *Primate Conservation: The Role of Zoological Parks. Special Topics in Primatology, Vol. 1.*, J. Wallis, ed., American Society of Primatologists, pp. 131–150.

Lovejoy, T. E., Bierregaard, R. O., Jr., Rylands, A. B., Malcolm, J. R., Quintela, C. E., Harper, L. H., Brown, K. S., Jr., Powell, A. H., Powell, G. V. N., Schubart, H. O. R., and Hays, M. B., 1986, Edge and other effects on isolation on Amazon forest fragments, in: *Conservation Biology: The Science of Scarcity and Diversity*, M. E. Soulé, ed., Sinauer Assoc., Sunderland, MA.

Loy, J., 1988, Effects of supplementary feeding on maturation and fertility in primates, in: *Ecology and Behavior of Food-Enhanced Primate Groups*, J. E. Fa and C. H. Southwick, eds., A. R. Liss, publisher, New York, pp. 153–166.

Luken, J. O., 1997, Conservation in the context of non-indigenous species, in: *Conservation in Highly Fragmented Landscapes*, M. W. Schwartz, ed., Chapman and Hall, New York, NY, USA, pp 107–116.

Mamede-Costa, A. C., and Gobi, M., 1998, The black lion tamarin *Leontopithecus chrysopygus*—Its conservation and management, *Oryx* **32**:295–300.

Marsh, L. K., 1999, Ecological effect of the black howler monkey (*Alouatta pigra*) on fragmented forests in the Community Baboon Sanctuary, Belize, Ph.D dissertation, Washington University, St. Louis, USA.

Marsh, L. K., 2001, Making conservation count: The commitment to long-term community involvement, in: *Proceedings for the American Zoo and Aquarium Association*, St. Louis, MO.

Marsh, L. K., 2002, Wild zoos: Endangered primate survival in fragmented tropical forests, *Am. J. Primatol.* **57(1)**:36.

Marsh, L. K., in press, Wonder lost or wonder lust: Tourists visit monkeys in the wild, *Curator*.

Marsh, L. K., and Loiselle, B. A., in press, Recruitment of howler monkey (*Alouatta pigra*) fruit trees in fragmented forests of northern Belize, *Int. J. Primatol.*

Marroquin, J., 1998, The global environment and Galacia, the new ejidal population center, Marqués de Comillias zone, Ocosingo, Chiapas, in: *Timber, Tourists, and Temples*, R. B. Primack, D. Bray, H. A. Galletti, and I. Ponciano, eds., Island Press, Covelo, CA, pp. 287–292.

McCullough, D. R., 1996, Introduction, in: *Metapopulations and Wildlife Conservation*, D. McCullough, ed., Island Press, Covelo, CA, USA, pp. 1–10.

McKey, D., 1975, The ecology of coevolved seed dispersal systems, in: *Coevolution of Animals and Plants*, L. E. Gilbert and P. H. Raven, eds., University of Texas Press, Austin.

Mittermeier, R. A., 1991, Hunting and its effect on wild primate populations in Suriname, in: *Neotropical Wildlife Use and Conservation*, J. G. Robinson and K. H. Redford, eds., University of Chicago Press, Chicago, IL, pp. 93–110.

Murcia, C., 1996, Forest fragmentation and the pollination of Neotropical plants, in: *Forest Patches in Tropical Landscapes*, J. Schelhas and R. Greenberg, eds., Island Press, CA, pp. 19–36.

Norris, R., Wilbur, J. S., and Marin, L. O. M., 1998, Community based ecotourism in the Maya forest: Problems and potentials, in: *Timber, Tourists, and Temples*, R. B. Primack, D. Bray, H. A. Galletti, and I. Ponciano, eds., Island Press, Covelo, CA, pp. 327–342.

Poffenberger, M., 1994, The resurgence of community forest management in eastern India, in: *Natural Connections, Perspective in Community-based Conservation*, D. Western and R. M. Wright, eds., Island Press, Washington, D.C., USA, pp 53–79.

Putz, F. E., Leigh, E. G., Jr., and Wright, S. J., 1990, Solitary confinement in Panama, *Garden* 2:18–23.

Redford, K. H., 1992, The empty forest, *BioScience* **42(6)**:412–422.

Savage, A., Giraldo, H., and Soto, L., 1997, Developing a conservation action program for the cotton-top tamarin (*Saguinus oedipus*), in: *Primate Conservation: The Role of Zoological Parks, Special Topics in Primatology, Vol. 1*, J. Wallis, ed., American Society of Primatologists, pp. 97–112.

Schelhas, J., and Greenberg, R., 1996, The value of forest patches, in: *Forest Patches in Tropical Landscapes*, J. Schelhas and R. Greenberg, eds., Island Press, CA, pp. xv–xxxvi.

Schwartz, M. W., and van Mantgem, P. J., 1997, The value of small preserves in chronically fragmented landscapes, in: *Conservation in Highly Fragmented Landscapes*, M. W. Schwartz, ed., Chapman and Hall, New York, pp. 379–394.

Silver, S. C., 1997, The feeding ecology of translocated howler monkeys (*Alouatta pigra*) in Belize, Ph.D dissertation, Fordham University, New York, USA.

Snyder, N. F. R., Derrickson, S. R., Beissinger, S. R., Wiley, J. W., Smith, T. B., Toone, W. D., and Miller, B., 1996, Limitations of captive breeding programs in endangered species recovery, *Cons. Bio.* **10(2)**:338–348.

Stoinski, T., Beck, B., Bowman, M., and Lihnhardt, J., 1997, The Gateway Zoo Program: A recent initiative in golden lion tamarin reintroductions, in: *Primate Conservation: The Role of Zoological Parks. Special Topics in Primatology, Vol. 1*, J. Wallis, ed., American Society of Primatologists, pp. 113–130.

Struhsaker, T. T., 1997, *Ecology of an African Rain Forest: Logging in Kibale and the Conflict between Conservation and Exploitation*, University Press Florida, Gainseville, FL, USA.

Struhsaker, T., and Siex, K. S., 1998, Translocation and introduction of the Zanzibar red colobus monkey: Success and failure with an endangered island endemic, *Oryx* **32**:277–284.

Strum, S. C., and Southwick, C. H., 1986, Translocation of primates, in: *Primates: The Road to Self-Sustaining Populations*, K. Benirschke, ed., Springer-Verlag, New York pp. 949–958.

Terborgh, J., 1983, *Five New World Primates*, Princeton U. Press, New Jersey.

Terborgh, J., 1986, Community aspects of frugivory in tropical forests, in: *Frugivores and Seed Dispersal*, A. Estrada and T. H. Fleming, eds., Dordrecht, W. Junk pp. 371–384.

Terborgh, J., Lopez, L., Nuñez, P., Rao, M., Shahabuddin, G., Orihuela, G., Riveros, M., Ascanio, R., Adler, G. H., Lambert, T. D., and Balbas, L., 2001, Ecological meltdown in predator-free forest fragments, *Science* **294**:1,923–1,926.

Viana, V. M., Tabanez, A. A. J., and Batista, J. L. F., 1997, Dynamics and restoration of forest fragments in the Brazilian Atlantic moist forest, in: *Tropical Forest Remnants. Ecology, Management, and Conservation of Fragmented Communities*, W. Laurance and R.O. Bierregaard, eds., Univ. Chicago Press. Chicago, IL, USA pp 351–365.

Weber, W., 1995, Monitoring awareness and attitude in conservation education: The Mountain Gorilla Project in Rwanda, in: *Conserving Wildlife: International Education and Communication Approaches*, S. Jacobson, ed., Columbia U. Press, New York pp.28–48.

Western, D., 1989, Conservation without parks: Wildlife in the rural landscape, in: *Conservation for the Twenty-First Century*, D. Western and M. Pearl, eds., Oxford Univ. Press, New York, NY USA pp 158–165.

Weise, R. J., and Hutchins, M., 1997, The role of North American zoos in primate conservation, in: *Primate Conservation: The Role of Zoological Parks. Special Topics in Primatology, Vol. 1*, J. Wallis, ed., American Society of Primatologists, pp. 29–42.

Western, D., 1994, Ecosystem conservation and rural development: The case of Amboseli, in: *Natural Connections. Perspective in Community-based Conservation*, D. Western and R. M. Wright, eds., Island Press, Washington, D.C. USA pp.15–52.

Whelan, T., 1991, *Nature Tourism: Managing for the Environment*, Island Press, Covelo, CA, pp. 223.

Wiens, J. A., 1996, Wildlife in patchy environments: Metapopulations, mosaics, and management, in *Metapopulations and Wildlife Conservation*, D. McCullough, ed., Island Press, Covelo, CA, USA pp. 53–84.

FRAGMENTATION: SPECTER OF THE FUTURE OR THE SPIRIT OF CONSERVATION?

Laura K. Marsh, Colin A. Chapman, Marilyn A. Norconk, Stephen F. Ferrari, Kellen A. Gilbert, Julio Cesar Bicca-Marques, and Janette Wallis[*]

1. INTRODUCTION

The study of primates in fragments is primarily driven by recognition that many tropical landscapes have already experienced, or soon will experience, deforestation and fragmentation. This recognition brings a desire to conserve the primates that are affected. From this perspective, we want to provide managers, researchers, and students with the information that they will need to maximize species survival and consolidate remnant populations. Fragmentation need not be a daunting prospect. As primatologists and fragmentation scientists, we want to strike a proactive—not a reactive stance. This is difficult given that deforestation is generally out of our hands. The perception is scientists do not have power to change the system of destruction, but as professionals working to prevent the extinction of primates worldwide, we can be very influential. We can do something to change the system. We need a collective voice, a voice that will challenge the need for deforestation in the first place, not one that only deals with the effects of fragmentation. We need a voice that will champion already fragmented areas and elevate their importance in the global scheme of conservation.

To see the world as a few remaining large chunks worth saving at the expense of all other smaller sites may no longer be reasonable. As many authors in this volume have argued, all remaining forested areas currently used by primates are of conservation value, regardless of size. To create solidarity among fragmentation scientists, we need to

[*] L. K. Marsh: Los Alamos National Laboratory Ecology Group (RRES-ECO) Mail Stop M887 Los Alamos, New Mexico 87545, USA. C. A. Chapman: Department of Zoology, University of Florida, Gainesville, Florida, 32611, USA and Wildlife Conservation Society, Bronx, New York, 10460, USA. M. A. Norconk: Department of Anthropology and Biological Anthropology Program, School of Biomedical Sciences, Kent State University, Kent, Ohio, 44242, USA. S. F. Ferrari: Department of Genetics, Universidade Federal do Pará, Caixa Postal 8607, 66.075-900 Belém – PA, Brasil. K. A. Gilbert: Department of Sociology and Criminal Justice, Southeastern Louisiana University, Hammond, Louisiana 70402, USA. J. C. Bicca-Marques: Pontifícia Universidade Católica do Rio Grande do Sul, Faculdade de Biociências, Av. Ipiranga 6681 Pd 12A, Porto Alegre, RS 90619-900, Brazil. Janette Wallis, Department of Psychiatry and Behavioral Sciences, University of Oklahoma, U.S.A. Correspondence to L. K. Marsh (email: lkmarsh@lanl.gov).

Primates in Fragments: Ecology and Conservation
Edited by L. K. Marsh, Kluwer Academic/Plenum Publishers, 2003

establish tools that can be used in conservation. In this sense, we have already conceded there will be fragments, thus making our planning reactive. However, there are more directions we can take to increase the well-being of primates in the wild, both in fragmented landscapes and for those remaining (for now) in contiguous forest. For the future of primates in fragments, it is clear that we need more studies such as those reported in this volume.

1.1. Semantics

For professionals working in fragmented forests, creating a common voice depends heavily on word choice. This is particularly important if we are to successfully encourage governments to enforce wildlife laws, to reroute oil and logging concessions, and to create new legislation governing fragmented landscapes. The message to the local communities, however, may be too strong, and conservation efforts for one species alone may result in the neglect of other species as well as the habitat (Marsh, this volume). Communities need a familiar face they can trust if conservation measures require sacrifices ranging from gathering a different array of food items to changing the way the people make a living (Marsh, 2001). Thus, the top-down approach may not necessarily instill a conservation ethic within local communities.

Species terms such as "flagship" (often a highly endangered "sexy" species used to act as a representative or "ambassador" of a particular region), "keystone" (species which provide resources for many members of a community, generally in low abundance, and without which the community may be driven toward extinction), and "indicator" (species when present in the environment define the wellness of that environment) should be used only with great clarity of definition for fragmented ecosystems (Schelhas and Greenberg, 1996, McCullough, 1996, Laurance and Bierregaard, 1997). Use of these terms to generalize or popularize conservation efforts may ultimately delay or diminish the preservation of other species by calling attention to a single species instead of the entire system. The focus on "sexy mega-vertebrates" argues that ecosystems will be saved "on down" if these larger species are preserved. In reality this approach may ensure ecosystem collapse if active habitat preservation is not part of the conservation plan. An example of this comes from the Community Baboon Sanctuary (CBS), Belize. One of the major landholders in the CBS discussed his conservation plans for *Alouatta pigra* (paraphrased from Creole): "I plan on being able to run my cattle in more area, but I won't bother the monkeys. We'll just clean up the bush under all of the forest, and the monkeys can still have their trees. That way the cattle can get some more shade and access the river, and maybe eat up some of the plants left behind" (Marsh, personal communication).

Education programs in the CBS have stressed howler monkey preservation, not habitat preservation. As a result, the forests are highly disturbed and are not regenerating. In other words, howlers are not having a positive influence through seed dispersal in these sites (Marsh, 1999). Howler monkey populations have increased steadily since the creation of the CBS. Considering that they are endangered in Belize, this would appear to be a positive trend. However, the ever decreasing size of some fragments and the high rate of disturbance within most sites suggest the howler metapopulation may decline once it reaches carrying capacity because of increased competition for space and resources. Similarly, although forests are at present viable for the howlers, their fruit resources are not regenerating in a sustainable manner, so habitat quality for howler monkeys and other large frugivores in the CBS may ultimately decline as well, even if the forest is "saved" (Marsh, 1999). If the goal is to only preserve howler monkeys, the whole fragmented forest ecosystem must still be taken into account. Conservation efforts that leave tracts of forest for a target or "flagship" species may be reducing the success of other species, eliminating the potential for survival. If a forest is left standing with original trees intact, but is essentially "empty" of its principal seed dispersers (Redford, 1992), those trees will not be effectively recruiting into the population. Conservation efforts, in the case of fragmented forests, cannot just target "keystone" species alone, especially since these are extremely difficult to identify. Remnant forest without effective dispersers, or with good dispersers and disrupted ecological systems, must be managed holistically if conservation goals are to be met. To predict the cascading effects of a disturbance that is allowed to continue is difficult, but this is clearly a priority area for research since the majority of fragments are on lands "managed" by individuals or communities.

Perhaps the best way to decide if habitats are viable is to look at the least common denominators, such as the outcome of dispersal (seedling/sapling regeneration), or presence of interior forest species, such as certain butterflies, ants, or amphibians (Gilbert, 1980, Stouffer, 1998, Gascon and Zimmerman, 1998). In the case of sapling regeneration, Chapman and Onderdonk (1998) found that the presence or absence of seedlings and saplings under adult conspecifics does provide an initial assessment of the extent of biodiversity loss that might result from a significant reduction in seed-disperser populations. That is, if a great number of seedlings or saplings grow under conspecifics, but are rare in the habitat, there is a reduction in diversity. Likewise, an absence of seedlings or saplings under conspecifics leads to higher overall plant diversity. As fragmentation primatologists we need to have an understanding of all impacts on our target species.

2. ISSUES OF MANAGEMENT: AN INTEGRATION

What are the most critical pieces of information that we need to provide managers? The most basic questions that must be answered are what are the range of sizes and areas used by inhabitants of the fragments, and what are the mechanisms that permit survival? If we can reliably understand the mechanisms that allow some species to be able to prosper in fragments, then we can start to make generalizations. The ability to generalize is needed, because we are often asked to construct management plans in situations were little background information is available. With this information, we can start to set priorities and understand the balance between investing scarce conservation funding in managing fragmented landscape versus working on preserving larger undisturbed sections of land.

2.1. Primate Responses

Throughout this volume authors have discussed species that will and will not use the matrix surrounding fragments. There are species that simply will not use fragments in some regions, while in others, the same species use them and thrive in them. Making predictions about a primate's survival capabilities without studying them in a fragment can prove quite erroneous and potentially costly for conservation. For instance, *Macaca fascicularis* is considered to have a preference for secondary forest (Crockett and Wilson,

1980, Cowlishaw and Dunbar, 2000). However, one cannot extrapolate this to other *Macaca* species living in fragmented forest, as they may not disperse at all through secondary growth (Umapathy and Kumar, this volume).

While howler monkeys (*Alouatta*, regardless of species) seem to manage in even the smallest fragments (Bicca-Marques, this volume), other genera have more varied responses. *Ateles paniscus* in the Biological Dynamics of Fragmented Forests Project (BDFFP), Brazil (Gilbert, this volume), *Ateles geoffroyi* in Punta Laguna, Mexico (Ramos-Fernandez and Ayala-Orozco, this volume), and *Ateles marginatus* in southern Amazonia (Ravetta, 2001, Ferrari et al., this volume) had very different responses to fragmentation. In the BDFFP, Gilbert reported that *Ateles* groups were absent in all of the fragments (1, 10, and 100 ha) except for one female spider monkey who lived in a 10-ha fragment alone. *Ateles* were only found in the continuous forest (>1,000 ha) in the BDFFP. Ramos-Fernandez and Ayala-Orozco, on the other hand, demonstrate that *Ateles* can thrive in fragments 60 to 200 ha in size. In southern Amazonia, *A. marginatus* occupies some, but not all fragments of large size (300 to 500 ha). Does this demonstrate a species-specific difference or individual flexibility? Are we seeing true species differences? Likely, the difference once again lies in part in the matrix. In the BDFFP, the forest had less than 20 years to regenerate, and for many of the fragments, the secondary growth has been removed periodically to keep them in isolation (Gilbert, this volume). Also, the diversity of tree species, particularly the species whose fruit is preferred by *Ateles*, is low in the forest and in the secondary growth (Williamson and Mequita, 2001). In Mexico, the secondary forest has regrown for 30 to 50 years and is utilized by the spiders regularly even though they maintain their core home ranges within the original fragments. In southern Amazonia, however, evidence was found of significant local variation in the distribution of *A. marginatus*, indicating that its presence in a fragment may be more closely related to habitat quality than fragment size or matrix composition.

These examples bring up subsidiary questions: how do we define a fragment? Are we still talking about a fragment if the secondary forest has grown up to provide natural resources for the primates? When does a fragment no longer have negative consequences for the residents? To create a very basic definition, fragmentation scientists should consider degree of isolation, connectivity, and characteristics of the surrounding matrix. This is especially important for comparative discussions, such as *Ateles* example above because interpretation of the area available may depend on our definitions of 'fragment' (c.f., Section 3.2.3).

Another example of the complexity of primate response to fragmentation is illustrated by the reaction of blue monkeys (*Cercopithecus mitis*) and redtail monkeys (*C. ascanius*) to fragmentation in Western Uganda. Redtail monkeys frequent forest fragments and move between fragments using available forest corridors and cross unforested areas, whereas blue monkeys, which are similar in body size, diet, and social organization, are never found in fragments and even appear to avoid the edges of the main forest of Kibale National Park (Onderdonk and Chapman, 2000, Chapman et al., this volume). One might predict that both of these species would be well suited for fragment living since they have a preference for secondary forest (Thomas, 1991). Yet there are many more factors involved with fragment living than simple compatibility with secondary forest. Here the ability to utilize the matrix is key once again.

Perhaps only species that are known to prefer secondary forest, such as marmosets and tamarins, can potentially be predicted to do well in human-disturbed habitats. Chiarello (this volume) suggests secondary forest is key for *Callithrix geoffroyi*, *C.*

flaviceps, and *Leontopithecus chrysopygus* that are found throughout the sites in the Atlantic Forest region, Brazil. *Mico argentatus* appears to be well adapted to fragmented habitats, but was unexpectedly absent from the smallest fragment in Ferrari et al. (this volume). This example further reinforces how unwise it is to generalize. In general, species like marmosets, tamarins, or howlers and capuchins will be more likely to be present and more abundant in fragments than species such as spider monkeys or bearded sakis. However, one must remain cautious about assumptions because even with the "reliable" fragmentation species, there are exceptions (c.f., Rodriguez-Toledo et al., this volume).

2.2. Primate Persistence

Why are some species able to readily move among fragments, while other species appear to be lifetime residents of a single fragment? For those species that appear to be permanent residents, do animals occasionally disperse among fragments at the time when a maturing individual would typically leave its natal group? How successful is this dispersal, both in terms of mortality risk and permitting genetic exchange? Is an understanding of metapopulation dynamics useful in such situations? The matrix in which the forest fragments are embedded is highly variable across regions. The ability of primates to travel among fragments is strongly determined by the nature of this landscape. The risk of predation is likely much lower for an arboreal primate that has to travel across cattle fields than one that has to travel through small agricultural fields where people and hunting dogs are common. This "matrix effect" may explain the differences found between studies of fragments in Kibale National Park in Uganda (Onderdonk and Chapman, 2000) and in Lopé Forest Reserve in Gabon (Tutin et al., 1997). At Lopé, mangabeys were found at similar densities in forest fragments and in continuous forest, while they were absent from fragments around Kibale. Furthermore, all primate species from Lopé were found in forest fragments to some degree, while two Kibale species were absent from the neighboring fragments (mangabeys and blue monkeys). At Kibale, the matrix surrounding forest fragments is actively used by people, while at Lopé, humans are absent from the surrounding matrix. The importance of the matrix is increasingly being realized.

It is not a surprise that there are proportionally more fragment studies that include howler monkeys since they are more tolerant of disturbed habitats than any other platyrrhines (Chiarello, Ferrari et al., Gilbert, Rodriguez-Toledo et al., Rodriguez-Luna et al., Marsh and Silver, Serio-Silva and Rico-Gray, McCann, Bicca-Marques, this volume; Figure 1). For the Neotropics, the challenge is to develop management plans that support howlers, but do not exclude other platyrrhines. If other species live with howlers, it might be wise to develop plans that support more sensitive species first. Howlers seem to show a tendency for increasing density with decreasing fragment size, and an inverse relationship between density of howlers and species richness, in particular, as related to the presence of *Ateles* who are the howlers' main competitors (Ferrari et al., Rodriguez-Luna et al., Gilbert, this volume). In some instances, however, smaller fragments may have fewer howlers, thus it may be a question of landscape and whether or not howlers feel comfortable moving between fragments (Rodriguez-Toledo et al. this volume). Ultimately, the goal is to maintain diversity, so a fragment packed with howlers does not necessarily indicate the health of a fragmented ecosystem, and in fact may be showing the decline of it (Terborgh et al. 2001).

Figure 1. *Alouatta pigra* traveling on open ground across a villager's yard in the Community Baboon Sanctuary, Belize. Photo by Laura K. Marsh.

2.3. Land Tenure

The question that must be addressed immediately is what is the future of these fragments? To date, the human use of fragments and resulting ecological impacts have largely been often ignored by primatologists. Most fragments have no protected status; they are on private land and are used by local landowners. Thus, fragments change structure and composition as landowners use the forest for grazing and hunting, extract timber or fuelwood, or allow fallow land to regenerate. This fact has not been fully appreciated, probably because a number of previous studies have been conducted in forest fragments that are protected (i.e., they are within a protective reserve; Lovejoy et al. 1986, Tutin et al., 1997, Tutin, 1999). While these studies in protected reserves have provided us with many insights, they are not typical of most fragmented landscapes and they may have biased our perception of the value of forest fragments to primates. The study of fragment use in Western Uganda (Chapman et al., this volume) suggests that privately owned fragments may have a grim future: in only five years 19% of the original fragments supporting primates were cleared.

Similarly, Umapathy and Kumar (this volume) found that fragments owned by the Forest Department in the Indira Gandhi Wildlife Sanctuary, India, were more likely to be larger and less disturbed or would leave corridors or some kind of connectivity between remnants than in privately owned fragments. Fragments that are surrounded by private land owners, whether enclave communities, plantation, or logging concessions, all suffer from a battery of damage, both internal and external because there is no control over primate species or habitats in those area (Reynolds et al., Eniang, McCann, this volume). Depending on the intensity of use for the surrounding matrix and of the fragments themselves, some species may thrive on private lands (McCann, this volume). In Brazilian Amazonia, ongoing colonization offers an opportunity, at least potentially, to contribute to a more systematic planning of fragmentation, as long as colonists can be convinced before cutting begins.

2.4. Conservation Measures

If intense clearing rates are typical of fragmented landscapes, then mechanisms must be worked out that will convince local people not to clear their fragments. To stop the fragments from being cleared or to create conjoining fragments on neighboring plots will require the cooperation of the local people, since this is their land. Alternative sources of income will have to be found (e.g., community involvement where products are grown sustainable for sale, such as bushmeat species), and fuelwood supplies from elsewhere would have to be made available (e.g., a large-scale woodlot project where wood is delivered to villages for free use). In some critical cases, realistic but effective ways to protect forest fragments may involve stimulating local land owners to establish plantations of fast-growing, non-invasive exotic tree species around and/or between fragments. The use of invasive species, on the other hand, must be completely rejected. These plantations may improve the health of a community from a fragmented landscape in several ways. For example, they may provide fuelwood to local people (avoiding the use of native species) and decrease edge effects on native species if established as "green belts" around fragments or serve as corridors if established between fragments. Certain exotic species may even be used as food sources by some primates (Bicca-Marques and Calegaro-Marques 1994; Ganzhorn 1985; Grimes and Paterson 2000; Ratsimbazafy 2002). This alternative, however, must be analyzed on a case-by-case basis and should only be proposed when planting native species is proved far less effective. In addition, a great deal of effort would have to be placed in education and outreach to obtain the willing support of all the communities. This would require a major conservation effort, on a scale and of a nature that is not typically done.

To convince funding agencies to invest at the level that these sorts of efforts will need requires 1) strong evidence of the importance of fragments and the primates, plants, and animals they support, 2) data on feasible management options, and 3) a change in attitude to more strongly value these systems and animals. These funding agencies will likely be development oriented, such as the World Bank, since they are the only agencies with sufficient capital. In fact, Ferrari et al. and Gonçalves et al. (this volume) work in southern Brazilian Amazonia were funded by the World Bank as part of a nationwide habitat fragmentation research program. Smaller organizations, such as zoos or non-governmental organizations, working collectively toward the same goal may also contribute significantly (Marsh, this volume). Academics can provide information on the importance of these systems and figure out how best to manage forest fragments. Also, if they get the information out to a broad audience (i.e., documentaries, popular articles), they can help change attitudes.

3. FUTURE DIRECTIONS

3.1. Ranking Fragments

While discussing the study of fragmentation, Thebaud and Strasberg (1997) stated: "Theoretical metapopulation models have recently emphasized the importance of dispersal as a process that may promote persistence of species at a regional scale. Unfortunately, quantifying dispersal in fragmented landscapes is no easy task, however, knowledge of how species recolonize fragmented landscapes, and why they differ in their colonization ability, can be essential if we are to understand how habitat fragmentation contributes to declines of species." Assessing and predicting fragmentation effects can be time consuming and frustrating for landscape managers. Understanding food webs and interdependence, the effects of edge and isolation, cascading abiotic and biotic trends, all are necessary for conservation decision making for fragmented habitats. A necessary "next step" for assessing the viability of these patches may be the development of a quantitative approach that would provide a reliable "scale" to estimate disturbance, and perhaps the regeneration potential, of the fragment in question.

This kind of analysis is similar to the Quantitative Habitat Analysis (QHA) tool developed for natural resources management and facility construction (Marsh and Haarmann, 2000a, b, c; Marsh et al., 2001). This tool is an application that combines many criteria, ranks them independently with appropriate weights for more important items, and summarizes the scores for a final "grade." Since there are broad categories (landscape, ecosystem health, contaminants, wildlife, wildfire), each must be analyzed separately and a "common currency" calculated to link all of the parameters. A similar system can be developed for fragmented forests.

Laurance et al. (1997) and Marsh (1999) suggest variables that might be included in an "analysis of disturbance" or "disturbance rate" calculation or ranking system (Table 1). Each variable would need to have values assigned to a subset of possible conditions, or at least should have weighing factors. This proposed calculation does not take into account genetic drift or inbreeding although it could be included. However, "inbreeding" has many definitions and it is difficult to generalize (Jacquard, 1975; Templeton and Read, 1994). This new calculation would be in addition to Population and Habitat Viability Analyses, which primarily focus only on a single species' life history parameters (e.g., VORTEX), and general habitat characteristics (if they are included at all) (Lacy, 1993). The most labor-intensive information necessary for the analysis of disturbance is the "recruitment potential" ratio (Marsh, 1999; Table 1), which requires an assessment of both the adult tree and sapling composition of the fragment. Trees in general should be assigned to "disturbance" categories, or trees that are disturbance specialists. There may be non-equilibrium communities that shift in proportion to other disturbance specialists, but are not necessarily shifting due to disturbance of canopy trees. Disturbance specialists may take over a fragment that has lost the regenerative capabilities of the original canopy trees.

Some of the variables are problematic (Table 1). For instance, should the age of a fragment be included? Age can be viewed in two ways. First, the older the fragment the more likely it will be altered through species loss, habitat degradation, and, potentially, system collapse. On the other hand, matrix surrounding an older fragment may provide supplemental resources for primates. Remaining primates are also adapting to changing fragment conditions over time, although this is generally restricted to certain species. By

Table 1. Ranking system to assess relative values of existing forest fragments (after Laurance et al., 1997 and Marsh, 1999). Numbers represent score for each category. Each variable must be weighed depending on its importance in the given situation.

Variable	Conservation Rank		
	High (5)	Medium (3)	Low (1)
% Habitat in reserves	1%	1 to 10%	>10%
Endemic species	>1 present	1 present	none
T and E species[1]	>1 present	1 present	none
Ecosystem quality	pristine	modified	degraded
Disturbance (internal)			
Grazing	pristine	modified	degraded
Hunting	pristine	modified	degraded
Agriculture	pristine	modified	degraded
Timber	pristine	modified	degraded
Foot traffic[2]	pristine	modified	degraded
NTFP[3]	pristine	modified	degraded
Invasive/exotic sp.	none	few (1 to 2)	>2
Matrix type	forest, mixed, secondary growth	agroforest (e.g., coffee plantation)	agriculture, timber, pasture
Proximity to settlements	>1,000 m	100 to 1,000 m	<100 m
Isolation[4]	<100 m	100 to 1,000 m	>1,000 m
Connectivity	linked for most species	linked for many species	poorly linked
Size	>500 ha	100 to 500 ha	<100 ha
Shape	roughly circular	intermediate shape	irregular
Habitat diversity	>2 habitats	2 habitats	1 habitat
Specialists			
Pollinators	expected present	few present	none
Seed dispersers	expected present	few present	none
Predators	expected present	few present	none
Ownership	government owned	government leased	private
Forest structure	strata intact	missing 1 or more	absent
Recruitment potential[5]	>saplings/adult	<saplings/adult	no saplings/adult

1. T and E species = Threatened and Endangered species
2. Foot traffic = Human thoroughfare on path or trail.
3. NTFP = Non-timber forest product, such as fruits, medicine, or thatch.
4. Isolation = Distance from other forested areas.
5. Recruitment potential = [(# saplings of species x)/(# total saplings)] / [((# trees of species x)/(# total trees)]

contrast, recent fragments tend to exhibit a refuge effect with relatively high densities of most species, which can be positive if we are thinking in terms of metapopulation management. Another problem is size. If we were to use Laurance et al.'s (1997) parameters, a high rank would be a fragment >300 ha, medium 3 to 300 ha, and low <3 ha. For primates, a more logical scale might be >500 ha for high, 100 to 500 ha for medium, and <100 ha for low. The caution here is in making the low value too small. As we have seen in this volume, some primates reside (and a few thrive) in fragments smaller than 10 ha (c.f., section 3.2.3). If we reduce the conservation rank to "low" for all of those fragments less than 100 ha we may be eliminating valuable habitat not only for the primates, but for other remaining species. If the quantitative approach is successful, then by weighing factors in order of importance to the particular site, then size may not be as important to the final determination.

If a calculation was created that took into account all of the variables suggested in Table 1, then a comparative scale would need to be developed to determine the disturbance rate. For instance, if the numbers fell between 0 and 1, those scores that are closest to 1 are sites that have a high number of disturbances and a low rate of regeneration. However, this kind of scale may suggest that all sites are equivalent, thus a different metric may need to be proposed (B. Loiselle, personal communication).

3.2. Research

There are many topics that need to be addressed regarding primates in fragments. We discuss some of them in the following section. In general, this is a field where any information can aid in completing the puzzle that fragmentation has made of conservation efforts. For example, more research needs to be done on the role of primates as seed dispersers in fragmented forests, and their ultimate effect on the regeneration of forest systems. Garber and Lambert (1998) concur: "Although information on post-dispersal survivorship is critical in determining the impact, both ecological and evolutionary, that a primate species may have on influencing plant characteristics and forest regeneration, this phase has historically received the least attention in primate studies." Understanding the role of primates as pollinators, seed predators, and dispersers may greatly influence conservation decisions in the future (Marsh and Loiselle, in press).

3.2.1. Disease

In recent years, there has been growing concern for the risk of disease transmission between humans and nonhuman primates (Wallis and Lee, 1999). As our close relatives, nonhuman primates (particularly apes) are susceptible to many human diseases. Whether air-borne, water-borne, or vector-borne, these pathogens can have devastating effects on a wild primate population. The greatest risk comes from "foreign" pathogens introduced into a naïve environment. This fact produces the ironic phenomenon that field researchers are likely to pose health risks to the very primate subjects they work to conserve. Tourism is another source of concern; the wish to gain close proximity to wild primates means that tourists can inadvertently spread disease. Wallis and Lee (1999) provide a summary of several disease outbreaks that occurred in wild primate research subjects, quite probably as a result of close proximity to humans.

Although most attention has focused on Western visitors bringing pathogens to the wild, it may be that primates living in fragments may have more contact with local humans than those living in contiguous forest and this may adversely affect their health (O'Leary and Fa, 1993; Struhsaker and Siex, 1996; Butynski and Kalina, 1998; Graczyk et al., 2001). In addition to contact with humans, stress may compromise an individual's immune system making that individual more susceptible to infection (Fair and Rickleffs,

2002). Forest fragmentation may be thought of as an additional stressor that can impact immunity to parasites and disease.

The diversity and prevalence of endoparasites in red howler monkey (*Alouatta seniculus*) and black-and-gold howler monkey (*Alouatta caraya*) groups in the central Amazon and northern Argentina were related to fragment size (Gilbert, 1994; Cruz et al., 2000). More individuals per group were infected in 10-ha fragments than in 100-ha fragments and had a greater number of parasite species. Prevalence of infection was also significantly related to population density of all primate species per fragment. This may result from crowding in a small area and repeated use of arboreal pathways with fecal contamination. Both red colobus (*Procolobus badius*) and black-and-white colobus (*Colobus guereza*) in forest fragments near Kibale National Park, Uganda, have a higher prevalence and richness of parasite infections compared to individuals in primary forest (T. R. Gillespie and C. A. Chapman, unpublished data). These infections are often zoonoses in degraded fragments (e.g., a type of hookworm that is often particularly pathogenic and can have fitness effects). The infection risk to colobines in degraded fragments is much higher than in pristine forest (measured in terms of density of infective forms). However, infection risk in a similarly sized protected fragment with minimal human use does not differ from infection risk in pristine forest.

Stuart et al. (1993) compared *Brachyteles arachnoides* and *Alouatta guariba* for internal parasite load in four fragment sites ranging in size between 40 and 37,000 ha in the Atlantic forest of Brazil. They found contrary to expectations, muriquis from the largest and least disturbed forest (with the lowest primate population density) had the highest prevalence and diversity of parasitic infection. Similarly, Martins (2002) found no clear relationship between fragment size, population density, and endoparasite load in *Alouatta belzebul* living in island fragments within a reservoir in eastern Amazonia, Brazil. In fact, the highest load (and the largest number of parasite species) was found in a continuous forest site where population density was average, and less than half that of the smallest fragment (180 ha). In addition to ecological factors, it seems likely that the process of fragmentation reinforced original local variation in parasite distribution. However, depending on the vector, some parasites may be obtained from water flowing through or adjacent to the fragment. Water carries pathogens from human communities outside the area regardless of how big the fragment. Small fragments, then, are more likely to have a prevalence of airborne or vector-borne diseases or parasites.

More research clearly must be done, not only on endoparasitic infection but also in ectoparasites to compare primates within fragments to those in continuous forest. For example, are there more botflys in primates that are in fragments or that have highly disturbed matrices? Are there more parasites endo- or ecto- in proximity to human dwellings? Blood-borne pathogens are difficult to collect in the field, but this is another area where we could use more information as well. Just how likely is it that arboreal primates can "catch" something from humans? This is being looked at more closely in the opposite direction (primate to human) in bushmeat to determine the link for diseases such as HIV or Ebola from handling infected primates (Chitnis et al., 2000; Tutin, 2000; Peeters et al., 2002).

3.2.2. Regional Comparisons

One of the lingering problems in determining the path for conservation in fragmented habitats is the issue of scale. For primates, this challenge lies in creating regional (Latin America, Africa) or national (Nicaragua, Nigeria) management plans. Does it make sense to have consensus for projects on that large of a scale? Do we need to include greater flexibility so managers can adapt for any situation? There are dangers inherent in making the wrong generalizations based on data from another region. What works in Africa may not work in Latin America. While this seems intuitive, it is the smaller scale that is even more problematic. For instance, in East Africa, what makes Budongo and Kibale different? What makes a primate do well in one spot and not in another? We may have a more clear idea after we have studied and compared the situations, but in those sites where we have critical landscapes harboring endangered primates, it is likely management decisions will be made without the luxury of research. Situations like these make us cautious, yet determined to develop generalizations for conservation management.

3.2.3. Fragment Size

The issue of scale also involves the question of fragment size. If we are interested in managing a fragment as a habitat, and not a single species, then size recommendations might be quite different than if we were managing for purely primates but c.f. Marsh, this volume. Size definitions may have to be adjusted for particular sites depending on the overall species richness and what it means to the primate metapopulation as well as the landscape in general. Definitions of fragment size have been determined by each author, where size was calibrated based on the largest fragment in the study set. Based on the fragments studied in this volume, authors worked in small (1 to 10 ha), medium (10 to 100 ha), large (100 to 1,000 ha), and extra large (1,000 to 10,000 ha) fragments, with the majority of studies carried out in the medium size category. Cochrane and Laurance (2002) show that most forest fragments in Amazonia are small (<100 ha), but the majority fall between 1 to 10 ha with 20% to 25% that are 10 to 100 ha. Very few are 1,000 to 10,000 ha and even fewer are upwards of 100,000 ha. For genetic purposes, if we need an average of 500 individuals for viability of primates and other species, then only the extra large fragments (20,000 ha) will suffice (Chiarello, 2000). On the other hand, perhaps we need to rethink the importance of small fragments for primates, especially if this is where they are remaining, and work more closely with metapopulation and landscape scales. As more research is conducted on landscape parameters affecting persistence of species, the closer we will be to understanding and thus predicting species response.

3.2.4. Genetic Studies

Genetic variability is a fundamentally important aspect of the biology of primate populations in fragmented landscapes, but remains very poorly understood on the whole. While the loss of variability in fragmented populations seems inevitable, in most cases, little, if anything, is known of the genetic variability or dynamics of the original populations. Without this, it is difficult to fully assess the effects of fragmentation and, in particular, to predict the outcome of long-term processes such as genetic drift or inbreeding. Continuous forest does not necessarily imply random breeding within an extended gene pool. Silvery marmosets (*Mico argentatus*), for example, tend to be patchily distributed in continuous forest, and it seems likely that nonrandom gene flow is a characteristic of their ecology, one which is reflected in patterns of genetic variability found in the fragmented landscape (Gonçalves et al., this volume).

Molecular techniques and advanced modern technology, such as automatic sequencing, have greatly expanded the potential of genetics studies, especially at the population level. Despite this, studies are still few and far between. Brazil is perhaps the best example because of its exceptional primate diversity. To date, there have been little more than a handful of mostly preliminary conservation-oriented studies for primate fauna of around a hundred subgeneric taxa (e.g., Pope, 1998; Perez-Sweeney et al., 2000; Menezes et al., 2002; Gordo et al., 2002; Gonçalves et al., this volume). Such a disproportionately small number of studies appears to be typical of most countries or biomes.

The importance of understanding the genetic characteristics of fragmented populations for the development of management strategies cannot be underestimated, especially because procedures such as translocation are both costly and risky. In addition to the study of specific populations, it will be important to develop a more systematic theoretical approach, for which long-term studies, such as that of Menezes et al. (2002), will be essential. Reliable viability analyses will depend on robust modeling of processes such as inbreeding and genetic drift, and this can only be supplied satisfactorily with the support of empirical data.

3.2.5. Climate Change

The relationship between fragments, primates, and climate change may at first seem vague; however, the relationships are complex. As climate change research advances, there will be more opportunities to further understand how primates may be affected. Global climate change is occurring. The conversion of tropical forest into pasture, timber lots, agriculture, roads, villages, or bare soils disrupts not only forest and local microclimates, but has serious consequences for the global climate. Reducing forested cover increases the Albedo Effect, or the reflection of the earth's surface (Burroughs, 2001). One of the drivers of global warming is trapped heat in the lower atmosphere. Increasing surface reflection increases the amount of heat sent back up into space, as opposed to being trapped in vegetation.

Deforestation in the tropics contributes to a disruption of the global energy balance not only in terms of heat absorption, but also for carbon cycling (Laurance, in press). Identical to the shifts in microclimates in response to forest fragmentation (Turton and Freiberger, 1997; Camargo and Kapos, 1995), the disturbed global climate has changes in wind patterns, rainfall, clouds and humidity, soil moisture, vegetation response, and other effects, such as sea level rise (Lobban and Schefter, 1997). This increases global warming and possibly the timing of dramatic climatic shifts, such as ENSO (El Niño Southern Oscillation; Burroughs, 2001). Tropical forests, when in balance, may act as carbon sinks, but when they are disturbed or are responding to increased drying from rising temperatures and fragmentation, they likely become a non-trivial source of atmospheric carbon emissions (Laurance et al., 1998; STRI 2002).

There have been a number of modeling studies to predict climate change impacts on tropical deforestation. Most assume complete conversion of regions such as Amazonia and Southeast Asia to pasture or savanna (Laurance, in press). The results vary, but collectively they project a notable decrease in region-wide rainfall (up to 20% to 30%), lower evaporation, cloud cover, and soil moisture, and an increase in albedo and surface temperatures (c.f., Laurance, in press). This is a major area for research that is needed on the ground. For instance, in Kibale, East Africa, rainfall data have been collected from 1910 to 2000 (Chapman, unpublished data). Over the last 90 years, rainfall has steadily increased (approximately 1,390 mm to 1,720 mm). Even though there has been an increase in rainfall at Kibale, the global trend for the tropics is in decline (STRI, 2002). Local variations clearly are caused by habitat shifts, deforestation, or protection of a site, but we have little data for comparative studies.

Tropical forests throughout the world are changing significantly in structure, dynamics, and composition. For instance, tree turnover has doubled throughout the tropics in recent decades, from 1% annually in the 1950s to 2% in the 1990s (STRI, 2002). As forests are fragmented and more edge is created, changes deep within standing forest continue (Laurance, in press). Currently, we are becoming aware of the complex interactions between localized climates and forest fragments, but larger scale effects are poorly understood. Primates may be able to "help" climate if their abilities as seed dispersers, leaf pruners, and pollinators can continue unhindered, but as we have seen in this volume, there are many factors impeding primate effectiveness. As a future direction for fragmentation primatologists, understanding the landscape-scale climatic effects will be increasingly important.

4. POLITICS

There are far more examples of political exploitation in the name of conservation than there are of support. That is not to say that excellent laws do not exist or there are no well thought out designs, it is simply that political will ultimately determines the outcome. There are many examples of this, and the first is from Brazil. There is federal legislation restricting clearance of natural habitat on private land to a given percentage of the property's area. Although there is notice of these practices, more often than not there is a considerable contrast between legal theory and actual practice. In Belize, the government promotes the country as one of the world's best for conservation and ecotourism. In practice, they subscribe (as many countries do) to the "unused forest (standing, intact, or in fragments) equals undeveloped land" equation. The idea is the land is being non-productive if it is not under human control. Until recently, Belize imposed a reasonable 1% tax on all leased government land that was left in forest. The tax has been increased to 6.5% to "encourage" development. Obviously, the forest will not remain standing if people have to pay such a high rate for it. Finally, the impact of hunting of primates and other species in or around fragments has not been examined well. In the BDFFP region, Brazil, ever since the opening of the Zona Franca (off of which *fazendas* and reserves are located), the forest north of Manaus has become much more accessible to weekend hunters. *Fazenda* workers hunt for cattle-killing jaguars in and around fragments, disregarding the BDFFP's protected status. In many countries there is clearly a disconnect between saying conservation and doing it.

If one of the goals of our research in fragments is to affect change and to conserve primates and these systems, our task is much more difficult than the typical ones faced by academics or researchers. Our research activities have to be conducted with application to conservation in mind. Too many studies in conservation biology by well-meaning researchers have had minimal impact because the investigator's focus was too narrow and failed to reflect societal needs, economic realities, and the political climate. To be effective, conservation biologists must have the capacity to integrate theories, methods, and tools from a variety of biological disciplines and place their research in an appropriate socioeconomic and political setting.

To effect change, it is essential that scientists make their findings known to local governments and the public. This is a difficult and time-consuming task that detracts from ones abilities to do traditional academic endeavors. In these circumstances, colleagues in our departments, but in slightly different fields, may not see the value of such efforts. The importance of this issue should not be downplayed because it proves a strong selective pressure discouraging such activities (e.g., they rarely count towards tenure and promotion). However, to affect change our results must become widely known.

5. SUMMARY

This chapter summarizes some of the findings by authors in this volume and explores future topics fragmentation primatologists may need to pursue. We discuss the need for creating a common voice for carrying out research projects and in solidifying common conservation goals. We looked at the issues of management of primates in fragments and compared many different species with an understanding that predictions and generalizations are critical for creating management plans, but that there is not enough data to substantiate them across all taxa. We highlight a few generalizations, if used with caution, that may prove to hold for those species. There are many issues of management that include primate responses and persistence in fragments, and conservation measures that we feel must get into place if we will be successful in managing fragmented populations. We discuss a ranking system that will be useful to fragmentation scientists for determining the "value" of a given site. We cover future research topics as they apply to primates in fragments, such as disease, regional comparisons, fragment size, genetic studies, and climate change. We conclude with a discussion on politics and our role in working with regional and national governments to improve the integrity of the process and effectiveness of conservation.

6. REFERENCES

Bicca-Marques, J. C., and Calegaro-Marques, C., 1994, Exotic plant species can serve as staple food sources for wild howler populations, *Folia Primatol.* **63**:209–211.

Burroughs, W. J., 2001, *Climate Change: A Multidisciplinary Approach*, Cambridge University Press, Cambridge.

Butynski, T. M., and Kalina, J., 1998, Gorilla tourism: A critical look, in: *Conservation of Biological Resources*, E. J. Millner and R. Mace, eds., Oxford, Blackwell Scientific, pp. 294–313.

Camargo, J. L. C., and Kapos, V., 1995, Complex edge effects on soil moisture and microclimate in central Amazonian forest, *J. Trop. Ecol.* **11**:205–211.

Chapman, C. A., and Onderdonk, D. A., 1998, Forests without primates: Primate/plant codependency, *American J. Primatol.* **45**:127–141.

Chiarello, A. G., 2000, Density and population size of mammals in remnants of Brazilian Atlantic forest, *Conserv. Biol.* **14**(6):1,649–1,657.

Chitnis, A., Rawls, D., and Moore, J., 2000, Origin of HIV type 1 in colonial French Equatorial Africa? *AIDS Res. Hum. Retrov.* **16**(1):5–8.

Cochrane, M. A., and Laurance, W. F., 2002, Fire as a large-scale edge effect in Amazonian forest, *J. Trop. Ecol.* **18**:311–325.

Cowlishaw, G., and Dunbar, R., 2001, *Primate Conservation Biology*, Chicago University Press, Chicago, pp. 498.

Crockett, C. M., and Wilson, W. L., 1980, The ecological separation of *Macaca nemistrina* and *M. fascicularis* in Sumatra, in: *The Macaques*, D. G. Lindburg, ed., Van Nostrand Reinhold Co., New York, pp. 148–181.

Cruz, A. C. M. S., Borda, J. T., Patiño, E. M., Gómez, L., and Zunino, G. E., 2000, Habitat fragmentation and parasitism in howler monkeys (*Alouatta caraya*), *Neotrop. Primates* 8:146–148.

Fair, J. M., and Ricklefs, R. E., 2002, Physiological growth and immune response of Japanese quail chicks to the multiple stressors of immunological challenge and lead shot, *Arch. Environ. Contam. Toxicol.* 42:77–87.

Ganzhorn, J. U., 1985, Utilization of eucalyptus and pine plantations by brown lemurs in the eastern rainforest of Madagascar, *Primate Conservation* 6:34–35.

Garber, P. A., and Lambert, J. E., 1998, Primates as seed dispersers: Ecological processes and directions for future research, *Am. J. Primatol.* 45(1):3–7.

Gascon, C., and Zimmerman, B., 1998, Of frogs and ponds and peccaries, *Natural History* 107(6):43–45.

Gilbert, K. A., 1994, Parasitic infection in red howling monkeys in forest fragments, *Neotrop. Prim.* 2:10–12.

Gilbert, L. E., 1980, Food web organization and conservation of Neotropical diversity, in: *Conservation Biology: An Evolutionary-Ecological Approach*, M. E. Soulé and B. Wilcox, eds., Sinauer Assoc., Sunderland, MA, pp. 11–34.

Gordo, M., Farias, I. P., Hrbek, T., and Ferrari, S. F., 2002, The pied bare-faced tamarin (*Saguinus bicolor*) in urban Manaus: a challenge for conservation biologists, *Abstracts of the 7th workshop of the European Marmoset Research Group, Paris*: in press.

Grimes, K., and Paterson, J. D., 2000, *Colobus guereza* and exotic plant species in the Entebbe Botanical Gardens, *Am. J. Primatol.* 51:59–60.

Graczyk, T. K., Mudakikwa, A. B., Cranfield, M. R., and Eilenberger, U., 2001, Hyperkeratonic mange caused by *Sarcoptes scabiei* (Acariformes: Sarcoptidae) in juvenile human-habituated mountain gorillas (*Gorilla gorilla beringei*), *Parasitol. Res.* 87:1,024–1,028.

Jacquard, A., 1975, Inbreeding: One word, several meanings, *Theor. Pop. Biol.* 7:338–363.

Lacy, R. C., 1993, VORTEX: A computer simulation model for population viability analysis, *Wildl. Res.* 20:45–65.

Laurance, W. F., in press, Forest-climate interactions in fragmented tropical landscapes, *Phil. Trans. Royal Soc. Lond. B*.

Laurance, W. F., and Bierregaard, R. O., eds., 1997, *Tropical Forest Remnants: Ecology, Management, and Conservation of Fragmented Communities*, University of Chicago Press, Chicago, pp. 616.

Laurance, W. F., Bierregaard, R. O., Gascon, C., Didham, R. K., Smith, A. P., Lunman, A. J., Viana, V. M., Lovejoy, T. E., Sievirg, K. E., Sites, J. W., Jr., Anderson, J., Tocher, M. D., Kramer, E. A., Restrepo, C., and Moritz, C., 1997, Tropical forest fragments: Synthesis of a diverse and dynamic discipline, in: *Tropical Forest Remnants: Ecology, Management, and Conservation of Fragmented Communities*, W. F. Laurance and R. O. Bierregaard, eds., University of Chicago Press, Chicago, pp. 502–514.

Laurance, W. F., Laurance, S. G., and Delamonica, P., 1998, Tropical forest fragmentation and greenhouse gas emissions, *For. Ecol. Manage.* 110:173–180.

Lobban, C. S., and Schefter, M., 1997, *Tropical Pacific Island Environments*, University of Guam Press, Mangilao.

Lovejoy, T. E., Bierregaard, R. O., Jr., Rylands, A. B., Malcolm, J. R., Quintela, C. E., Harper, L. J., Brown, K. S., Powell, A. H., Powell, G. V. N., Schubart, H. O. R, and Hays, M. B., 1986, Edge and other effects of isolation on Amazon forest fragments, in: *Conservation Biology: The Science of Scarcity and Diversity*, M. E. Soule, ed., Sinauer Associates, Sunderland, MA, pp. 257–285.

Marsh, L. K., 1999, Ecological effect of the black howler monkey (*Alouatta pigra*) on fragmented forests in the Community Baboon Sanctuary, Belize, Ph.D. dissertation, Washington University, St. Louis, USA.

Marsh, L. K., 2001, Making conservation count: The commitment to long-term community involvement, in: *Proceedings for the American Zoo and Aquarium Association*, St. Louis, MO.

Marsh, L. K., and Haarmann, T. K., 2000a, Quantitative Habitat Analysis: A progress report on Phase I of the Work Plan, 16 pp., Los Alamos National Laboratory publication LA-UR-00-6018.

Marsh, L. K., and Haarmann, T. K., 2000b, Quantitative Habitat Analysis: A progress report on Phase II of the Work Plan, 83 pp., Los Alamos National Laboratory publication LA-UR-00-6017.

Marsh, L. K., and Haarmann, T. K., 2000c, Quantitative Habitat Analysis: Final report for FY00, 289 pp., Los Alamos National Laboratory publication LA-UR-00-6016.

Marsh, L. K., Haarmann, T. K., and Bennett, K., 2001, Quantitative Habitat Analysis ArcView Application Interface: Final Report FY01, Los Alamos National Laboratory publication LA-UR-01-6694.

Marsh, L. K., and Loiselle, B. A., in press, Recruitment of howler monkey (*Alouatta pigra*) fruit trees in fragmented forests of northern Belize, *Int. J. Primatol.*

Martins, S. S., 2002, Efeitos da fragmentação de hábitat sobre a prevalência de parasitoses intestinais em *Alouatta belzebul* (Primates, Platyrrhini) na Amazônia Oriental, Masters dissertation, Goeldi Museum/UFPa, Belém.

McCullough, D., ed., 1996, *Metapopulations and Wildlife Conservation*, Island Press, Washington DC.

Menezes, E. V., Silva, A., Ferrari, S. F., and Schneider, M. P. C., 2002, Variabilidade genética de *Alouatta belzebul* (Platyrrhini, Atelidae), da área de influência da Usina Hidrelétrica de Tucuruí-PA: resultados preliminares. *Resumos do XXIV°. Congresso Brasileiro de Zoologia*, p. 657.

O'Leary, H., and Fa, J. E., 1993, Effects of tourists on Barbary macaques at Gibraltar, *Folia Primatol.* **61**:77–91.

Onderdonk, D. A., and Chapman, C. A., 2000, Coping with forest fragmentation: The primates of Kibale National Park, Uganda, *Int. J. Primatol.* **21**:587–611.

Peeters, M., Courgnaud, V., Abela, B., Auzel, P., Pourrut, X., Bibollet-Ruche, F., Loul, S., Liegeois, F., Butel, C., Koulagna, D., Mpoudi-Ngole, E., Shaw, G. M., Hahn, B. H., and Delaporte, E., 2002, Risk to human health from a plethora of Simian immunodeficiency viruses in primate bushmeat, *Emerg. Infect. Dis.* **8**(5):451–457.

Perez-Sweeney, B. M., Pádua C. V., and Melnick, D. J., 2000, The microevolutionary force of genetic drift in black lion tamarin (*Leontopithecus chrysopygus*) populations, *Amer. J. Phys. Anthrop.* **111**(S30):249.

Pope, T. R., 1998, Genetic variation in remnant populations of the woolly spider monkey (*Brachyteles arachnoides*), *Int. J. Primatol.* **19**:95–109.

Ratsimbazafy, J. H., 2002, The role of exotic plant species in the diet of ruffed lemurs (*Varecia variegata variegata*) at Manombo Forest in Madagascar, *Am. J. Primatol.* **57**:30.

Ravetta, A. L., 2001, O Coatá-de-testa-branca (Ateles marginatus) do baixo rio Tapajós, Pará: ecologia e status de conservação, Masters dissertation, Goeldi Museum/UFPa, Belém.

Redford, K. H., 1992, The empty forest, *BioScience* **42**(6):412–422.

Schelhas, J., and Greenberg, R., 1996, The value of forest patches, in: *Forest Patches in Tropical Landscapes*, J. Schelhas and R. Greenberg, eds., Island Press, CA, pp. xv–xxxvi.

STRI, 2002, The effects of human-caused atmospheric changes on tropical forests, Association for Tropical Biology Symposium, Smithsonian Tropical Research Institute, Panama, www.si.edu.

Struhsaker, T. T., and Siex, K. S., 1996, The Zanzibar red colobus monkey *Procolobus kirkii*: Conservation status of an endangered island endemic, *Afr. Primates* **2**:54–61.

Stouffer, P. C., 1998, Survival of the ant followers, *Natural History* **107**(6):40–43.

Stuart, M. D., Strier, K. B., and Pierberg, S. M., 1993, A coprological survey of parasites of wild muriquis, *Brachyteles arachnoides*, and brown howling monkeys, *Alouatta fusca*, *J. Helminthol. Soc. Wash.* **60**(1):111–115.

Templeton, A. R., and Read, B., 1994, Inbreeding: One word, several meanings, much confusion, in: *Conservation Genetics*, V. Loeschcke, J. Tomiuk, and S. K. Jain, eds., Birkhauser Verlag: Basel, pp. 91–106.

Terborgh, J., Lopez, L., Nuñez, P., Rao, M., Shahabuddin, G., Orihuela, G., Riveros, M., Ascanio, R., Adler, G. H., Lambert, T. D., and Balbas, L., 2001, Ecological meltdown in predator-free forest fragments, *Science* **294**:1,923–1,926.

Thebaud, C., and Strasberg, D., 1997, Plant dispersal in fragmented landscapes: A field study of woody colonization in rainforest remnants of the Mascarene Archipelago, in: *Tropical Forest Remnants: Ecology, Management, and Conservation of Fragmented Communities*, W. F. Laurance and R. O. Bierregaard, eds., University of Chicago Press, Chicago, pp. 321–332.

Thomas, S. C., 1991, Population densities and patterns of habitat use among anthropoid primates of the Ituri Forest, Zaire, *Biotropica* **23**:68–83.

Turton, S. M., and Freiberger, H. J., 1997, Edge and aspect effects on the microclimate of a small tropical forest remnant on the Atherton Tableland, northeastern Australia, in: *Tropical Forest Remnants: Ecology, Management, and Conservation of Fragmented Communities*, W. F. Laurance and R. O. Bierregaard, eds., University of Chicago Press, Chicago, pp. 45–54.

Tutin, C. E. G., White, L. J. T., and Mackanga-Missandzouo, A., 1997, The use by rain forest mammals of natural forest fragments in an equatorial African savanna, *Cons. Biol.* **11**:1,190–1,203.

Tutin, C. E. G., 1999, Fragmented living: Behavioural ecology of primates in a forest fragment in the Lopé Reserve, Gabon, *Primates* **40**:249–265.

Tutin C. E. G., 2000, Ecology and social organization of African rainforest primates: Relevance for understanding the transmission of retroviruses, *Bull. Soc. Path. Exotic.* **93**(3):157–161.

Wallis, J., and Lee, D. R., 1999, Primate conservation: The prevention of disease transmission, *Int. J. Primatol.* **20**(6):803–826.

Williamson, G. B., and Mesquita, R. C. G., 2001, Effects of fire on rainforest regeneration in the Amazon basin, in: *Lessons from Amazonia: The Ecology and Conservation of a Fragmented Forest*, R. O. Bierregaard, Jr., C. Gascon, T. E. Lovejoy, and R. Mesquita, eds., Yale University Press, New Haven, pp. 325–334.

SUGGESTED READING

Bierregaard, R. O., Jr., Gascon, C., Lovejoy, T. E., and Mesquita, R., eds., 2001, *Lessons from Amazonia: The Ecology and Conservation of a Fragmented Forest*, Yale University Press, New Haven, 478 pgs.

Cowlishaw, G., and Dunbar, R., 2000, *Primate Conservation Biology*, Chicago University Press, Chicago, 498 pgs.

Fimbel, R. A., Grajal, A., and Robinson, J. G., eds., 2001, *The Cutting Edge: Conserving Wildlife in Logged Tropical Forests*, Columbia University Press, New York, 808 pgs.

Gentry, A. H., ed., 1990, *Four Neotropical Rainforests*, Yale University Press, New Haven, 627 pgs.

Harris, L. D., 1984, *The Fragmented Forest*, University Chicago Press, Chicago, 211 pgs.

Laurance, W. F., and Bierregaard, R. O., Jr., eds., 1997, *Tropical Forest Remnants: Ecology, Management, and Conservation of Fragmented Communities*, University of Chicago Press, Chicago, 616 pgs.

McCullough, D. R., ed., 1996, *Metapopulations and Wildlife Conservation*, Island Press: Washington, D.C., 429 pgs.

Sanderson, J., and Harris, L. D., eds., 2000, *Landscape Ecology: A Top-Down Approach*, Lewis Publishers, New York, 246 pgs.

Schelhas, J., and Greenberg, R., eds., 1996, *Forest Patches in Tropical Landscapes*, Island Press, Washington D.C., 426 pgs.

Schwartz, M. W., ed., 1997, *Conservation in Highly Fragmented Landscapes*, Chapman and Hall, New York, 436 pgs.

Young, A. G., and Clarke, G. M., eds., 2000, *Genetics, Demography, and Viability of Fragmented Populations*, Cambridge University Press, Cambridge, 438 pgs.

INDEX